# THE 1926 MINERS' LOCKOUT

OXFORD HISTORICAL MONOGRAPHS

# The 1926 Miners' Lockout

## *Meanings of Community in the Durham Coalfield*

HESTER BARRON

OXFORD
UNIVERSITY PRESS

OXFORD
UNIVERSITY PRESS

Great Clarendon Street, Oxford, OX2 6DP,
United Kingdom

Oxford University Press is a department of the University of Oxford.
It furthers the University's objective of excellence in research, scholarship,
and education by publishing worldwide. Oxford is a registered trade mark of
Oxford University Press in the UK and in certain other countries



First Edition Published in 2010



Published in the United States of America by Oxford University Press
198 Madison Avenue, New York, NY 10016, United States of America

British Library Cataloguing in Publication Data
Data available

Library of Congress Control Number: 2009026997

ISBN 978–0–19–957504–6

For Ewen
1958–2006

# *Acknowledgements*

In the writing of both the doctoral thesis on which this book is based, and then in the preparation of the manuscript, I have been privileged to work with Jose Harris. As supervisor and then as assistant editor she was a source of constant advice and encouragement. The examiners of the thesis, Duncan Tanner and Philip Waller, also provided suggestions and assistance. I am lucky to have benefited from the support and friendship of two other academics. David Howell gave of his time with unstinting generosity though under no obligation to do so, while the support and guidance of Nick Stargardt has been invaluable.

Tom Asch, Rob Lee, Ross McKibbin, Stella Moss, and Chris Prior all read portions—sometimes substantial—of the original thesis, and I am grateful for their comments. Glyn Prysor was meticulous in his reading of the final manuscript, and the end result is stronger for his observations, advice, and encouragement. Since September 2007, I have been fortunate in finding myself within a department that has been hugely supportive, and I owe a tribute to friends and colleagues at Sussex. I also benefited from comments and suggestions made at conferences or seminars at Exeter, Oxford, Sussex, and Teesside. Chapter 3 first appeared in *Historical Studies in Industrial Relations* in 2006; while an earlier version of Chapter 4 appeared in *Twentieth Century British History* in the same year. I am appreciative that reproduction is permitted here. For the use of images, I am grateful to Beamish Museum, Derrik Scott, Durham County Record Office, Gateshead Council, Libraries and Arts, Leeds University Library, the People's History Museum, and the Shields Gazette. Every attempt has been made to secure the permission of copyright holders. In some cases these have proved impossible to trace, for which I offer my sincere apologies.

I would like to record my thanks to those at the various record offices and libraries that I visited, most of whom gave every assistance. Jo Bath at Beamish Museum made me feel particularly welcome. Chapter 4 draws upon unpublished research materials deposited in Durham University Library by Professor Robert Moore, through the Economic and Social Research Council (ESRC) Qualitative Data Archival Research Centre, and it should be noted that Professor Moore bears no responsibility for the further interpretation and analysis of this material. I am also grateful for the financial support provided by the Arts and Humanities Research Council which allowed me to complete the original thesis, and to the University of Sussex which funded the purchase and reproduction costs of images. Perhaps the most sobering 'archive' trip was when I visited the Durham Miners' headquarters at Redhills in Durham. David Guy, the President of the Durham Miners' Association (DMA), was kind enough to show me a room full of treasures—but treasures of which many are sadly still inaccessible to researchers. I hope that his dream to create a permanent public archive of the history of the DMA is not too long in being realized.

Friends in Durham, Oxford, Brighton, and elsewhere provided refreshment and diversion. Many helped to make archive and library trips a financial possibility and generously put me up. In this respect, it is my parents and my sister Cleo to whom I owe the greatest debt. On top of the love and support that they have always given me, their contribution to this book has also been a practical one, for which I am ever grateful.

Ewen Green died in September 2006, two weeks before the submission of the thesis on which this book is based. In his role as tutor and then supervisor I owe him much, but it is for his friendship that I miss him most. It was he who reacted with such enthusiasm when, sitting in his office at Magdalen, I first tentatively mentioned 'doing something about miners' as a possible subject for my undergraduate finals thesis. He would have been proud that, many years later, this book is the final product of that first discussion and I wish he could have seen it. It is dedicated to his memory.

H.B.

# *Contents*


*List of Illustrations* x
*List of Figures and Tables* xi
*Abbreviations* xii
*Political and Union Leaders in 1926* xiv

Introduction 1

1. The Tensions of Class and Region 21
2. The Testing of Political and Union Loyalties 78
3. The Attitudes of Women 138
4. Religious Identities 165
5. The Influence of Education 199
6. Memory and Experience 225

Conclusion 254

*Bibliography* 273
*Index* 295

# *List of Illustrations*


1. Map of the Durham Coalfield, showing concentration of collieries 2
2. A. J. Cook Addressing a Meeting in 1926 (location unspecified) 86
3. Chopwell Lodge Officials with Banner, 1926 112
4. Cartoon from the *Labour Woman*, August 1926 139
5. Cartoon from *The Miner*, 23 July 1926 197
6. 'Committee Women Distributing Boots to Children after Holidays', by Annie Hillary, aged 12, *Labour Woman*, October 1926 224
7. Monkwearmouth Lodge Banner, 1986 248

# *List of Figures and Tables*

## Figures

2.1. Results of Durham Miners' Association ballot, November 1926 120

## Tables

1.1. Proportion of miners in the Durham coalfield 24
1.2. Wages and hours worked by adults in various industries 29
1.3. Distances travelled by miners to work, 1929 53
1.4. Average earnings per shift, January 1922 (figures provided by the Miners' Federation of Great Britain) 68
1.5. Average earnings per shift, May 1925 (figures provided by the Mineowners' Association of Great Britain) 69
2.1. Numbers of men returning to work (Home Office figures) 79
2.2. Mining union membership, January 1926 (major districts) 80
2.3. General Election results, 1924: County constituencies with over 20 per cent of the male population over the age of 12 engaged in the coal industry 81
2.4. Day wages in Durham, selected grades 90
2.5. Nominations and elections to Chester-le-Street Rural District Council, 1925 99
4.1. 'What opinion have you formed as to the moral standards of your parishioners?' Responses of parish priests to the Episcopal Visitation of 1924 178
4.2. 'How far is spiritual life in the parish affected by political agitation?' Responses of parish priests to the Episcopal Visitation of 1928 181
4.3. Religious affiliations of Durham Miners' Association officials and miners' MPs 186
5.1. Workers' Educational Association tutorial classes in Durham, 1926–7 219

# *Abbreviations*

| | |
|---|---|
| *BC* | *Blaydon Courier* |
| BLSA | British Library Sound Archive |
| BMOA | Beamish Museum, Oral History Archive |
| *CC* | *Chester-le-Street Chronicle* |
| CLC | Central Labour College |
| COPEC | Conference on Christian Politics, Economics, and Citizenship |
| CPGB | Communist Party of Great Britain |
| *DC* | *Durham Chronicle* |
| *DCA* | *Durham County Advertiser* |
| DCL | Dean and Chapter Library, Durham Cathedral |
| DCOA | Durham Coalowners' Association |
| DMA | Durham Miners' Association |
| DRO | Durham County Record Office |
| DUL | Durham University Library |
| GCLOT | Gateshead Central Library, Oral History Transcripts |
| *HPD*(C) | *Hansard Parliamentary Debates* (Commons), fifth series |
| ILP | Independent Labour Party |
| JRUL | John Rylands University Library, Manchester |
| LSE | London School of Economics |
| MAGB | Mineowners' Association of Great Britain |
| MFGB | Miners' Federation of Great Britain |
| NUM | National Union of Mineworkers |
| NUR | National Union of Railwaymen |
| OMS | Organisation for the Maintenance of Supplies |
| *SNCC* | *Stanley News and Consett Chronicle* |
| SWMF | South Wales Miners' Federation |

| | |
|---|---|
| *SWN* | *Seaham Weekly News* |
| TNA:PRO | The National Archives: Public Records Office |
| TUC | Trades Union Congress |
| TUCLC | Trades Union Congress Library Collections, London Metropolitan University |
| TWAS | Tyne and Wear Archives Service |
| WEA | Workers' Educational Association |
| *WW* | *Workers' Weekly* |

Place of publication is London unless otherwise stated.

# *Political and Union Leaders in 1926*

## Officials of the Miners' Federation of Great Britain

| | |
|---|---|
| Herbert Smith | President (also President, Yorkshire Miners' Association) |
| A. J. Cook | General Secretary (held no district office in 1926) |
| Tom Richards | Vice-President (also General Secretary, South Wales Miners' Federation) |
| W. P. Richardson | Treasurer (also General Secretary, Durham Miners' Association) |

## Officials of the Durham Miners' Association

| | |
|---|---|
| James Robson | President |
| W. P. Richardson | General Secretary |
| Peter Lee | Executive Committee Secretary (also chairman, Durham County Council) |
| Tom Trotter | Treasurer |
| J. E. Swan | Compensation Secretary |
| James Gilliland | Durham Miners' Association Agent |

## Durham MPs (county seats):

| | |
|---|---|
| Joe Batey | Spennymoor: Labour, sponsored by the Miners' Federation of Great Britain |
| Herbert Dunnico | Consett: Labour |
| Cuthbert Headlam | Barnard Castle: Conservative |
| Jack Lawson | Chester-le-Street: Labour, sponsored by the Miners' Federation of Great Britain |
| Robert Richardson | Houghton-le-Spring: Labour, sponsored by the Miners' Federation of Great Britain |
| Joshua Ritson | Durham: Labour, sponsored by the Miners' Federation of Great Britain |

| | |
|---|---|
| Major Ropner | Sedgefield: Conservative |
| Ben Spoor | Bishop Auckland: Labour |
| Sidney Webb | Seaham: Labour |
| William Whiteley | Blaydon: Labour, sponsored by the Miners' Federation of Great Britain |
| R. J. Wilson | Jarrow: Labour |

# Introduction

On 30 April 1926, lockout notices were posted outside collieries across Britain, demanding that their workforces accept a reduction in wages and a longer working day. One million miners downed their tools; many would not pick them up again until Christmas.[1] This study examines the dispute as it played out in County Durham, exploring the way in which miners and their families experienced, conceptualized, and identified with a 'mining community'. It attempts to understand collective values and behaviour, focusing particularly on the tensions between identities based around class and occupation, and the rival identities that could disrupt the creation of a cohesive community. Questions of solidarity and the construction of a shared consciousness found particular resonance during the strike, when conflicting symbols and ideologies battled for dominance and men and women found their choices politicized. Not only were family and community loyalties and responsibilities intensified (and not necessarily in accommodating directions), but a national strike also forced a heightened awareness of both personal deprivation and the wider socio-economic environment in which an individual lived and worked. The strike therefore provided a focal point around which fundamental issues of identity were challenged, values had to be prioritized, and concrete choices had to be made.

Many coal owners had been threatening wage cuts and an increase in hours since Britain's return to the gold standard in April 1925 and the consequent downturn in an already fragile British export trade. Then, the Baldwin government had attempted to avert trouble by granting a subsidy to the coal industry, but when this expired

[1] I have used the terms 'lockout' and 'strike' interchangeably throughout this book, though their difference was fiercely contested by contemporaries. In May 1926, the miners of Durham (although not of all coalfields) were locked out. However, as the months passed, many pits reopened, either on new or pre-stoppage terms of employment. For those who continued to stay away from work, their position essentially became that of the striker, and it is as a strike that the dispute is remembered and referred to by many who lived through it.

1. Map of the Durham Coalfield, showing concentration of collieries
*Source*: A. Reid, *Reid's Handy Colliery Guide and Directory for the Counties of Northumberland and Durham* (Newcastle-upon-Tyne, 1923).
Reproduced by permission of Durham County Record Office.

at the end of April 1926 it was made clear that no further help would be forthcoming. Lockout notices were the almost inevitable consequence. The sympathetic general strike that followed was called by the Trades Union Congress (TUC) in support of the miners. It began on 3 May 1926, and the story of the subsequent nine days is one of the most significant episodes in the history of the British labour movement. When it was called off on 12 May, however, the miners' fight continued. Attempts at mediation by various public figures found no resolution to the conflict: the Miners' Federation of Great Britain (MFGB) consistently held out for a national settlement which would neither cut wages nor lengthen the working day; the owners continued to insist on the necessity of both, the details of which to be determined by district.

The resolve of the miners' leaders was reflected in the rank and file, and by early July the number of miners who had returned to work remained below 1 per cent. However, as the strike continued, solidarity began to break down, particularly in the Midlands. In October, George Spencer, the General Secretary of the Nottinghamshire Miners' Association, was expelled from the MFGB conference after negotiating a return to work at some of the Nottinghamshire pits. By mid-November, one-third of the national labour force was back at work and at the end of that month, the MFGB leaders finally agreed to the negotiation of district settlements. Work had resumed in all coalfields by the end of the year.[2]

As an exporting coalfield, Durham was one of the hardest hit by the return to the gold standard, by the threatened wage reductions, and ultimately by the outcome of the lockout. It also stood as an example of the solidarity for which miners were famed. Government estimates suggested that by the beginning of November 1926, six months after the strike began, the proportion of Durham miners who had returned to work remained below 5 per cent. Contemporary critics blamed the influence of class consciousness, and on 25 May 1926, the Bishop of Durham, Herbert Hensley Henson, reflected upon the recent general strike in his diary:

> It is clear that in many cases the men disliked the Strike, and that in many cases it hurt their consciences. Nevertheless, they obeyed. They neither resented the behaviour of the TUC in ignoring their wishes, nor refused to break their contracts in violation of their professed principles. The education in 'class consciousness' has been so successful that neither self-respect nor religion count for anything against class.[3]

[2] Surprisingly, the miners' lockout of 1926 has been the subject of little dedicated historical scholarship, although the collection of essays in J. McIlroy, A. Campbell, and K. Gildart (eds), *Industrial Politics and the 1926 Mining Lockout: The Struggle for Dignity* (Cardiff, 2004) has begun to fill the gap. Other accounts are contained within regional histories, histories of the mining unions, or biographies of key actors. The only full-length study, G. Noel's *The Great Lockout of 1926* (1976) is a self-confessed 'informal sketch'. The general strike itself has attracted somewhat more attention (although still far from overwhelming), the most recent academic treatment being K. Laybourn, *The General Strike of 1926* (Manchester, 1993). Anne Perkins' *A Very British Strike: 3 May – 12 May 1926* (2006), is a simplistic narrative account and adds little of significance to current scholarly debates.

[3] Dean and Chapter Library, Durham Cathedral, diary of H. Hensley Henson, 25 May 1926. Herbert Hensley Henson (1863–1947) was Bishop of Durham 1920–39.

In this instance, Henson—a man deeply unsympathetic to such working-class militancy—directed his tirade against the trade unionists across the country who had come out on strike in support of the miners, but as the pits maintained their silence in the months that followed, he would continue to despair at the apparent victory of class conflict that kept the miners from their work and that was doing so much to damage the country's interests.

## I

The use of community as a tool of analysis poses significant conceptual problems, not least of definition. Well over a century ago, Ferdinand Tönnies defined *Gemeinschaft* as based upon interactive, culturally based, and face-to-face relationships, linked by ties of kinship and descent and a similar occupational culture. He placed it in opposition to *Gesellschaft*, which was characterized by relationships based upon the division of labour and contractual relations between isolated individuals, undertaken for their own self-interest.[4] Since his seminal work, academics have continued to argue over how community should be conceptualized. In 1971, Colin Bell and Howard Newby famously counted ninety-four different definitions of 'community' amongst the writings of various sociologists, anthropologists, and historians.[5] The articles and books of almost another forty years must have added countless more.

Some definitions stress social and/or geographic relationships and/or experiences; others dismiss structural factors entirely. Anthony Cohen suggested that a consideration of an emotional attachment was important, for 'the reality of community lies in its members' perception of the vitality of its culture. People construct community symbolically, making it a resource and repository of meaning, a referent of their identity'.[6] More recently, Tony Nicholson agreed that structural characteristics were misleading. Borrowing from E. P. Thompson's conception of class, he argued that, 'Communities started to happen, and continued

[4] F. Tönnies, *Community and Civil Society*, ed. J. Harris; trans. J. Harris and M. Hollis (Cambridge, 2001), 27. Tönnies argued that evidence of both could always be seen in human interactions, but that *Gesellschaft* was increasingly more apparent than *Gemeinschaft* in the modern world.

[5] C. Bell and H. Newby, *Community Studies: An Introduction to the Sociology of the Local Community* (1971), 29.

[6] A. Cohen, *The Symbolic Construction of Community* (1985), 118.

to happen, from the moment people entered that shared local space; sometimes they assumed a stronger sense of cohesion than at other times, but this ebb and flow in size and strength did not mean that they moved in and out of some mythical state of grace called "community".'[7]

In relation to mining settlements in particular, 'community' continues to be a source of debate. In the 1950s and 1960s, commentators put forward a variety of possible models to explain the homogeneity of mining villages, whether the concept of the 'isolated mass', which focused on geographical and social isolation as the basis for homogeneity;[8] the 'separatist group', in which the impact of socio-cultural factors such as the union affected the strike propensity of different communities of miners;[9] the 'occupational community', where the social relations of work and leisure overlapped;[10] or the miner as the 'traditional proletarian', inhabiting a world defined by class conflict.[11] In the mid-1970s, Martin Bulmer discussed the varying theoretical approaches and attempted to formulate an ideal-type mining community. He argued that although the origin of an occupational community of miners lay in technological and economic organization, its maintenance and persistence was due to other sociological factors, not least 'the *gemeinschaftlich* ties of kinship, residence and friendship' first outlined by Tönnies.[12] More recent scholarship has tended to conceptualize community as primarily an intellectual construct, and scholars of mining have made fruitful use of Benedict Anderson's suggestion of the nation as an 'imagined community'.[13]

The problems inherent in the concept of community led some commentators to reject the term altogether. Alan Macfarlane argued

[7] T. Nicholson, 'Community and Class: The Cleveland Ironstone Field, 1850–1914', *Labour History Review*, 60 (1995), 82.

[8] C. Kerr and A. Siegel, 'The Interindustry Propensity to Strike: An International Comparison', in A. Kornhauser, R. Dubin, and A. Ross (eds), *Industrial Conflict* (New York, 1954), 189–212.

[9] G. V. Rimlinger, 'International Differences in the Strike-Propensity of Coal Miners: Experience in Four Countries', *Industrial and Labor Relations Review*, 12 (1959), 389–405.

[10] R. Blauner, 'Work Satisfaction and Industrial Trends in Modern Society', in W. Galenson and S. M. Lipset (eds), *Labor and Trade Unionism: An Interdisciplinary Reader* (New York, 1960), 339–60.

[11] D. Lockwood, 'Sources of Variation in Working-Class Images of Society', *Sociological Review*, 14 (1966), 249–67.

[12] M. Bulmer, 'Sociological Models of the Mining Community', *Sociological Review*, 23 (1975), 84.

[13] B. Anderson, *Imagined Communities: Reflections on the Origin and Spread of Nationalism* (1983); D. Gilbert, 'Imagined Communities and Mining Communities', *Labour History Review*, 60 (1995), 47–55.

that the term had become worthless, justifying examination only as 'an attempt to understand one of the controlling myths of our time'.[14] Such statements were swift to draw criticism,[15] but, almost twenty years later, Joanna Bourke continued to agree that use of the word was dangerous, resonating either with a backward-looking romanticism or laden with potential power as a political weapon.[16] Certainly the contested nature of the concept has made it almost obligatory for any author embarking upon a social study of a mining community to include a survey of the various theoretical positions before beginning.[17]

But, whatever the conceptual problems, the term cannot and should not be dismissed. However ambiguous, 'community' was a word used both by those who lived in mining villages during the 1920s, and by their descendants who sought to place themselves within a historical context. The concept became a rallying cry during the 1984–5 miners' strike, when the 'community' of the earlier period was imagined and consciously emulated. In December 1984, the wife of a Yorkshire miner described how 'now we stop and talk for hours in street. Community's back together like it were years ago'.[18]

Yet, even in the past, there was no consensus on the meaning of community. Competing definitions were considered by David Smith in his important essay on the Tonypandy riots in South Wales in 1910. He argued that the crowd's rage against seemingly neutral institutions such as grocers' shops (a rage which seemed so incomprehensible to men on the outside, including Winston Churchill) constituted a 'deliberate assault on the civil order of a world that had been made for them'. He suggested that when men and women were 'forced, via the strike, to reassess their own status, they ended by commenting on their relationship to a community defined *for* them in a graphic

[14] A. Macfarlane, 'History, Anthropology and the Study of Communities', *Social History*, 2 (1977), 632. See also M. Stacey, 'The Myth of Community Studies', *British Journal of Sociology*, 20 (1969), 134–47.

[15] See, for example, C. J. Calhoun, 'History, Anthropology and the Study of Communities: Some Problems in Macfarlane's Proposal', *Social History*, 3 (1978), 363–73; S. Macintyre, *Little Moscows: Community and Working-Class Militancy in Interwar Britain* (1980), 176.

[16] J. Bourke, *Working-Class Cultures in Britain, 1890–1960: Gender, Class and Ethnicity* (1994), 136–69.

[17] For example, D. Gilbert, *Class, Community, and Collective Action: Social Change in Two British Coalfields, 1850–1926* (Oxford, 1992), 9–53; A. Campbell, *The Scottish Miners, 1874–1939*, i, *Industry, Work and Community* (Aldershot, 2000), 159–60.

[18] M. Brogden, 'Interviews at Armthorpe, December 1984', in R. Samuel, B. Bloomfield, and G. Boanas (eds), *The Enemy Within: Pit Villages and the Miners' Strike of 1984–5* (1986), 187.

coda of selective destruction that was incomprehensible to those whose *idea* of the community was now threatened by this ugly, intrusive reality'.[19]

A similar conceptual conflict took place in 1926 when, during the lockout, both sides sought to claim 'community' as their own. Whereas the MFGB attempted to create a united, cohesive image of the mining community, both to lift the morale of its supporters and to act as propaganda to the outside world, an alternative community was championed by the government and coal owners, particularly during the nine days of the general strike. In a bulletin broadcast on 8 May, for example, the Prime Minister Stanley Baldwin invoked the power of such rhetoric: 'Can there be a more direct attack upon the Community than that a body, not elected by the voters of the country, without consulting the people, without consulting the trade unionists, and in order to impose conditions never yet defined, should dislocate the life of the nation and try to starve us into submission?'[20] His imagery was echoed throughout the country, and one railway company chairman later penned a letter of thanks to those who had volunteered for government service, capturing the sense that the strikers were somehow deviant from the true 'community': 'With characteristic British courage, Government and people sought new ways and means of carrying on the business of the Country . . . We feel that the grave attack on the Community, and the complete defeat, constitute a landmark in the constitutional history of the country.'[21] The damage that this did to the miners' cause was recognized by the general secretary of the MFGB, A. J. Cook. 'Who is this Community?' he asked in frustration, 'Every time we go into a struggle we are told we are striking against the Community . . . has the "Community" protected itself against the mine-owners who locked out the men and caused the "Community" to suffer?'[22]

[19] D. Smith, 'Tonypandy 1910: Definitions of Community', *Past and Present*, 87 (1980), 162, 179. Original emphasis.

[20] The National Archives: Public Records Office (henceforth TNA:PRO), PRO 30/69/1274, BBC bulletin, 8 May 1926.

[21] TNA:PRO, RAIL 1057/2788, letter from the Chairman of the London, Midland, and Scottish Railway Co., 19 May 1926.

[22] TNA:PRO, PRO 30/69/1274, A. J. Cook, 'Is it Peace?', Nov. 1926. A. J. Cook (1883–1931) was born in Somerset but moved to the South Wales coalfield at the age of 16. He held various union positions there before his election to the MFGB Executive in 1921. He was General Secretary of the MFGB, 1924–30. His role in the dispute is discussed further in Chap. 2, Sect. I.

## II

Within the mining village these different communities have traditionally been assumed to converge into one cohesive identity, perhaps explaining why the miner has been such a popular subject of 'community studies'. In part, this assumption stemmed naturally from the perceived solidarity of the miners, grounded historically in their early support of the trade union movement and their later importance to the Labour Party. In 1968, Roy Gregory opened his study of the miners in the early part of the twentieth century with the assertion that 'the characteristic of the miners that has most impressed the outside world is their solidarity. Their outlook on life may be narrow, they may be inarticulate and slow to understand, they can certainly be obstinate and stubborn, but there is no doubt that they know how to "stick together".'[23] In fact, for the period about which he was writing (1906–14), such solidarity was far from assured, and Gregory acknowledged that the reputation (in his opinion, a correct one) of the miners as the 'Praetorian guard of an explicitly socialist Labour Party' from the late 1920s onwards could colour opinions of earlier times.[24] Events of the later years of the twentieth century would further bolster the reputation of the miners as the epitome of trade-union consciousness and solidarity. The year-long struggle of 1984–5 in particular strengthened the popular image of the miner as possessing an extraordinary commitment to his union and fellow workmen. The dispute saw a frequently articulated pride that no other group of workers could have resisted Margaret Thatcher's wrath for so long. Even Sir Anthony Meyer, the (wet) Conservative MP, was moved to comment that 'wrong as they are, maybe they have something to teach us all about solidarity'.[25]

The romantic, heroic community based around union and pit was therefore once received wisdom in conventional accounts, epitomized in the 1950s' study *Coal is Our Life*, based on the Yorkshire mining town of 'Ashton'. Its authors claimed that 'Ashton' was strongly typical of mining settlements, and painted a timeless picture of a community isolated geographically, socially, and culturally from the outside world, and characterized by the social relations of mining which extended

[23] R. Gregory, *The Miners and British Politics, 1906–1914* (Oxford, 1968), 1.
[24] Ibid., 178.
[25] R. Samuel, 'Preface', in Samuel et al., *Enemy Within*, p. xi.

beyond the workplace to affect gender relationships, family life, leisure, and political beliefs.[26] However, more recent scholarship has argued that mining communities should be seen as more heterogeneous. The first substantial attempt at revision was undertaken by Royden Harrison in 1978 (only three years after Bulmer published his ideal type), when he edited a collection of essays reconsidering the miner's image as the archetypal proletarian of legend.[27] In his introduction, he described his thoughts as he stood on a hill overlooking Coinsborough in Yorkshire:

> Below lay Denaby and Cadeby lying cheek by jowl, yet possessed by different legends, traditions and *seemingly* industrial relations systems; an object lesson in humility to all who indulged in those grand—but also cheap and easy—exercises in comparative labour studies. Even supposing that the social distance between one end of the village and the other was a myth, it was a myth which required to be explained. This could hardly be done through the celebrated insight according to which the militancy of miners is a function of their existence as an 'isolated mass'.[28]

Since then, studies of mining and its people have been increasingly sensitive of regional and local variations and rather better attuned to the need to provide a more nuanced analysis. Greater attention has been given to those previously omitted from a union-based account, such as women and (to a lesser extent) groups such as the passive union member, the non-unionist, and the coal owner.[29] Mike Lieven has argued passionately for such a rethinking of mining history, suggesting that 'community' cannot be understood without a new approach. He observed that Liberal, Conservative, and Independent candidates continued to win about 30 per cent of both votes and seats in local elections of the Rhondda throughout the crisis years of the 1930s:

> this is neither to suggest that the Tories, Liberals or Independents were a threat to the Labour Party's growing dominance, nor to deny the proletarian

[26] N. Dennis, F. Henriques, and C. Slaughter, *Coal is Our Life: An Analysis of a Yorkshire Mining Community* (1956).

[27] R. Harrison (ed.), *Independent Collier: The Coal Miner as Archetypal Proletarian Reconsidered* (Hassocks, 1978).

[28] Ibid., 'Introduction', 12.

[29] For example, Alan Burge has attempted to reassess the previously vilified 'scab' in Burge, 'In Search of Harry Blount: Scabbing between the Wars in One South Wales Community', *Llafur*, 6 (1994), 58–69; Quentin Outram has explored the motives of the coalowners in Outram, 'The Stupidest Men in England? The Industrial Relations Strategy of the Coalowners between the Lockouts, 1923–1924', *Historical Studies in Industrial Relations*, 4 (1997), 65–95.

consciousness of a significant section of the community, but nor is it merely to add on some curious appendages to an established coalfield society . . . [Rather], we cannot hope to understand the culture or texture of valley communities if we ignore such evidence.[30]

Nevertheless, some have argued that this revisionist process is not yet complete. In the mid-1990s, David Gilbert suggested that the events of the 1970s had created the mood for a heroic interpretation of mining history, necessitating the qualification advocated by Harrison and the revision of the image of the miner as the archetypal proletarian. Gilbert argued that since then the events of the 1980s and 1990s had created a new temptation: to interpret mining history in terms of tragedy. Because of this, he suggested that other assumptions had not been subjected to the same critical readings, such as that of miners as 'archetypal communitarians'.[31] As if in answer to this lead, Roy Church and Quentin Outram reassessed the strike propensity of British miners, focusing on the more frequent local disputes that occurred, rather than the rarer national struggles. They found that one of the essential characteristics of mining strikes was that they were very brief, localized, and, typically, non-recurrent. And they found that, despite the British miner's reputation for militancy, such strikes were often poorly supported.[32]

In fact, given the multiplicity of loyalties and conflicting relationships within any colliery or its village, Church and Outram's findings should not, perhaps, be so surprising. The tendency of earlier work to exaggerate the unity of a deeply class-conscious, proletarian workforce has rightly been revised but, since then, the implication has often been that the British miner struggled to identify with *any* sense of collective identity, and the role of the union in the construction of solidarity has been dismissed. In his study of the Lancashire working classes, for example, Trevor Griffiths documented the religious, ethnic, and occupational divisions that cut across the social worlds of both miners and cotton workers. He concluded that 'at several points in working-class life, from the workplace to the ballot box, class, as an influence affecting the choices made, appears to have been secondary, at best'.[33]

30 M. Lieven, 'A "New History" of the South Wales Coalfield?', *Llafur*, 8 (2002), 97.

31 Gilbert, 'Imagined Communities', 50.

32 R. Church and Q. Outram, *Strikes and Solidarity: Coalfield Conflict in Britain, 1889–1966* (Cambridge, 1998).

33 T. Griffiths, *The Lancashire Working Classes,* c.*1880–1930* (Oxford, 2001), 331.

In the light of such research it becomes much harder to understand how collective action could happen at all; how—albeit only on occasion and still accompanied by dissent and conflict—sometimes an astonishing degree of solidarity could be achieved. This book attempts to reconcile these apparently conflicting positions: acknowledging and exploring the variety of different identities that might shape the loyalties of those who lived their lives under the shadow of the colliery wheels; but attempting to answer how, despite such divisions, a sense of community *could* still work, and collective action remain effective, as it so clearly did in 1926.

## III

The basis for the study is the Durham coalfield, and, as with any regional survey, the focus on one particular area raises questions of typicality. The various coalfields in Britain could be characterized in vastly different ways in the interwar years, from the political moderation and paternalistic ethos of many of the Nottinghamshire pits, to the collieries of South Wales, whose leaders might flirt with communism and syndicalism. Intra-regional and local variation also needs to be taken into account. In his work on the Scottish miners, Alan Campbell contended that while both geographers and sociologists have vigorously engaged with the concept of 'region', historians have been generally less reflective in their use of the category, often accepting predetermined regional boundaries as unproblematic. While acknowledging that individual local studies are hardly without their own problems, he warned that an emphasis on the 'region' as a unit of comparative analysis risked overemphasizing its homogeneity.[34]

Durham needs to be approached with similar caution. Coal lay beneath every part of the county bar the far south and west, but the coalfield was home to a rich variety of terrains, both over and underground. In the west, the smaller, older, more traditional colliery villages could be semi-rural in character, their pits dangerous and uncomfortable, with seams that could creep downwards to as little as eighteen inches. The big, mechanized, modern pits in the east drove deep below the North Sea, their workers strung out in the ribbon

34 Campbell, *Scottish Miners*, i. 6–10.

of settlements along the coast. In the north, colliery neighbourhoods merged into the urban conurbations of Gateshead and Newcastle; in the south, miners rubbed shoulders with farmers. Such differences inevitably affected the social lives and relationships of colliery workers and their families. At the same time, Durham made up only a part, albeit the larger one, of a wider coalfield region. In both popular and academic literature, the Durham miners are frequently bracketed with their Northumberland neighbours as part of the Great Northern coalfield, or collectively as miners of the North East.

Such issues have repercussions upon the evaluation of sources. Some of the most significant oral history testimonies drawn upon in this study, for example, are the transcripts of dozens of interviews undertaken by Gateshead Central Library in 1976 to commemorate the fiftieth anniversary of the strike. Remarkably, I have come across no instance of their use in the existing historical literature and they provide some invaluable source material. As a project based in Gateshead, however, most of the respondents came from the north of the county. The interviewers were also particularly interested in the colliery village of Chopwell, which became notorious for its militancy. Such a body of evidence—taken by itself—is therefore unable to take account of what might be significant geographical disparities.

Yet, it remains the case that county borders symbolized an important distinction to the Durham miners themselves. The Durham Miners' Association (DMA) was a county-based union (and perhaps was itself guilty of overemphasizing the homogeneity of the collieries it represented). Even for the miner who took no active interest in the union, the organization of his working life, from the amount he paid in union dues to the cavilling rules that determined where he worked underground, was based on his position as a specifically Durham miner and differed even from neighbouring pitmen across the border in Northumberland.[35]

In fact, County Durham has tended to be neglected by academic historians. William Garside published his history of the Durham miners in 1971, before the revisionist approach towards coalfield history really took off.[36] Over twenty years later, Huw Beynon and Terry Austrin's work attempted to correct this by locating the concerns of the union

35 Every three months, lots ('cavils') were drawn to allocate a miner's underground working place. These could vary enormously and affected how much a man earned as well as how comfortably he did so.

36 W. R. Garside, *The Durham Miners, 1919–1960* (1971).

within a wider study of the Durham mining communities.[37] However, the tendency of many works and articles has been to concentrate on other regions: the South Wales miners, for example, have attracted considerably more commentary than those of Durham, though the suffering of both coalfields in the interwar years and their reputation for union loyalty has meant that they are frequently paired in popular imagination. Collections of regional studies invariably forget Durham. The most recent full-length work on the miners' lockout included five regional case studies. None covered Durham or the North East.[38]

Nevertheless, if Durham was by no means 'representative' (if such a coalfield can exist), a study of its people has broader relevance to wider debates about solidarity and fragmentation within working-class community and culture. In the 1920s, coal-mining was the defining occupation of the region, and the county had more miners, both absolutely and proportionately, than any other British coalfield except South Wales.[39] It was an area in which the lack of female employment meant that a woman also found herself bound to the rhythm of the colliery shifts; where children grew up surrounded by the paraphernalia of colliery life; where the shopkeepers, publicans, and landlords of the colliery villages found themselves dependent on the spending power of the miners. In 1926, households across the county were affected as the coal industry was wracked by the worst industrial dispute it had ever seen. As such, Durham provides an important arena in which to explore issues of class, community, and collective action.

## IV

Finally, it is important to note that even the term 'miner' requires attention. This study concentrates on those workers who were represented by the DMA. Unlike in some coalfields, where mechanics, enginemen, deputies, and other grades of colliery employees were incorporated into

[37] H. Beynon and T. Austrin, *Masters and Servants: Class and Patronage in the Making of a Labour Organisation* (1994).

[38] McIlroy et al., *Industrial Politics*. Another collection of regional essays that also omits Durham is A. Campbell, N. Fishman, and D. Howell (eds), *Miners, Unions and Politics, 1910–47* (Aldershot, 1996).

[39] *Census of England and Wales*, 1921, ii, *County of Durham* (1923), p. xxix.

the main regional union, in Durham they maintained distinct organizations (although overlapping did occur). While evidence relating to such workers has occasionally been cited, I have generally considered them a separate group. Yet, even members of the DMA might be employed in a variety of roles within the coal industry. In July, one local newspaper complained to its readers that the word 'miner' had been grossly misused:

> Only about one-third of the people employed in the industry are really miners—coal hewers. The rest include general labourers, masons, fitters, joiners, skilled mechanics at tub mending, sawyers, waggonwrights, blacksmiths, boilersmiths, horse-shoers, plumbers, saddlers, painters, electricians, lamp repairers, platelayers, smiths' strikers, winding enginemen, locomotive engine drivers, hauliers, ostlers, carters, rolleywaymen, screen and washery engineers, stokers, patternmakers, rope splicers, rope splicers' mechanics, rubbish tippers, ashmen, boiler cleaners, shunters, power-house men, topmen in charge of signals, and a certain number of women. All these, according to the Miners' Federation, are 'miners'. Much sympathy for the men who are supposed to spend all their time in cramped quarters underground is thus misapplied.[40]

But, describing the variety of jobs undertaken at a colliery (or conflating them, as the newspaper claimed that the MFGB was doing) constituted more than a propaganda exercise. Occupational divisions, particularly those between day-wage workers and better-paid pieceworkers, had concrete consequences within the union itself. Men moved up a strict occupational hierarchy, and then, as they got older and their strength waned, back down. Hewers were the elite of the workforce: the men employed at the coalface, cutting out the coal with their picks. Before graduating to hewing they had usually been putters: youths employed to put the freshly won coal into tubs, and then push the tubs to the wider tunnels where pit ponies took over. Both jobs demanded immense physical strength and endurance and their position at the top of the underground hierarchy allowed such men to dominate the union. Jack Lawson was the miners' MP for Chester-le-Street in 1926. In his autobiography he remembered his pride when he became a hewer at the age of 23: 'I was now a "man". For a man is not really a man in Durham until he goes to the coal-face.'[41]

[40] *Seaham Weekly News*, 2 July 1926. Note that no women were employed in the Durham coalfield.

[41] J. Lawson, *A Man's Life*, (2nd edn., 1944), 75. Jack Lawson (1881–1965) was born in Cumberland but moved to the Durham coalfield aged 9 and soon followed his father down the pit. He was Labour MP for Chester-le-Street from 1919 until retirement in 1949 and held various junior posts in the interwar MacDonald governments before becoming Secretary of State for War in Attlee's 1945 ministry.

The romantic image of the coal hewer also came to dominate the image of the miner in popular memory. As his biographer noted, Lawson's portrayal came to represent the typical Durham miner, and James Callaghan presented a copy of the book to Jimmy Carter when the American president visited the North East in 1977.[42] In it the miner is suffused with a sense of dignity and nobility:

> Miners are clean, intelligent, orderly, home-loving men. Their depth of thought, expressed in simple language, sometimes backed by amazing reading, will challenge comparison with any class in Great Britain. Knowing as I do the life and conduct below, and the character that goes with conduct on the surface, even the crudest among them humbles me. To think of them as a whole is to have a tightness at the throat, while the heroism of them and their womenfolk in home matters, as well as the action below, increases admiration until it pains.[43]

Yet, as one commentator has pointed out, 'as well as the popular myth of the miner as the prototypical working-class *avant garde*, is another equally widespread image, that of the miner as repressed proletarian'.[44] In the national imagination, coal-mining might summon up romantic images of blackened faces and masculine toil deep beneath the ground but, in Durham, one man entered the pit when he left school aged 14 and hated his work with a passion. Interviewed in his old age, he remembered a joke that had been popular in the coalfield, in which a murderer stood on the scaffold at Durham Gaol. The hangman turned and said to him, 'You can have a reprieve if you start work at the drift, putting.' The condemned man did not hesitate: 'Pull that lever, lad.'[45]

Against Lawson's autobiography, proudly entitled *A Man's Life*, can be set Bert Coombes' account of life in the South Wales coalfield. His title sums up a very different image: *These Poor Hands*.[46] Both representations continue to be seen in romanticized form, including in scholarly work. As Robert Colls noted in a review of Alan Metcalfe's recent study of the Northern coalfield in the hundred years before the

[42] D. Bythell, 'Lawson, John James', *Oxford Dictionary of National Biography* (Oxford, 2004), xxxii, 900.

[43] Lawson, *Man's Life*, 175.

[44] D. F. Crew, 'Rapport/Bericht', in K. Tenfelde (ed.), *Towards a Social History of Mining in the Nineteenth and Twentieth Centuries: Papers Presented to the International Mining History Congress, Bochum, 1989* (Munich, 1992), 57.

[45] Beamish Museum, Oral History Archive, 1991/82.

[46] B. L. Coombes, *These Poor Hands: The Autobiography of a Miner Working in South Wales* (1939).

First World War: 'He wants his coalfield grim. He wants his miners suppressed. He wants his culture driven to the margins. He also wants his miners expressing themselves in a rich associational life with hearts and lungs and strength and skill.'[47]

Even Lawson's image of the miner was not static. In 1941, he published a biography of the MFGB President, Herbert Smith.[48] A bluff Yorkshireman, Smith spent his leisure time on the terraces of Barnsley Football Club and remained suspicious of the intellectual strain of the working-class movement. His character and interests were strikingly different from those of Lawson, the Methodist autodidact. Yet, there is a place for Smith too in the image of the miner conjured up by Lawson's prose:

> Void of finesse, lacking knowledge of all but mines and miners, Herbert Smith was THE MINER in very truth. He never pretended to be anything else . . . Backs were to the wall in the 1926 conflict. Herbert Smith was not only in no mood for compromise: he simply wasn't built that way. He was adamant. More than that, he was brutally frank. He angered many. They were detached. He was not. He was THE MINER.[49]

It was image of the miner that had resonance across the world. In an investigation of the coal-mining settlement of Yallourn in southern Australia, for example, Meredith Fletcher found that even language might be manipulated to avoid the characteristics associated with being a 'miner'. She discovered a model town carefully managed and controlled by the state government from its foundation in the 1920s. In an effort to dissociate the Yallourn workers from the militancy associated with other Australian coalfields, the language of mining was actively rejected by the authorities. Rather than employing miners to mine coal from the coal mine, in Yallourn 'labourers on coal' were engaged in 'winning' coal from 'the open cut'.[50]

Contrasting images of the miner arose in part from his own ambivalence towards his work. Although mining usually supplied adequate

[47] R. Colls, Review of Alan Metcalfe's *Leisure and Recreation in a Victorian Mining Community: The Social Economy of Leisure in North-East England, 1820–1914* (2006), *Social History*, 32 (2007), 89.

[48] Herbert Smith (1862–1938) began working in the Yorkshire mines at the age of 10. He became President of the Yorkshire Miners' Association in 1906 and was President of the MFGB from 1922–9.

[49] J. Lawson, *The Man in the Cap: The Life of Herbert Smith* (1941), 220.

[50] M. Fletcher, ' "Slaves of the Lamp?": Independence and Control in Two State Coal Mining Communities in Victoria, Australia', in S. Berger, A. Croll, and N. LaPorte (eds), *Towards a Comparative History of Coalfield Societies* (Aldershot, 2005), 104.

wages, it was also dangerous and relatively insecure. It demonstrated masculine qualities of strength and power, and the skill required for underground work meant that blackleg labour could not easily be substituted during strikes. However, the absence of an apprenticeship system meant that mining fitted uneasily into a working-class occupational hierarchy. Dirtier than any other occupation, the blackness which covered men as they emerged from the pit could also be interpreted in different ways. Lawson continued to see it as a symbol of the honesty of toil: 'I held that a man might be proud of his dirty work, all the more proud because it was dirty.'[51] To Mary Wade, a little girl growing up in the Northumberland coalfield in the 1920s, it was something that added to the mystique of the miner: 'As a child . . . pits and pitmen had a magic for me that was tinged with awe. I loved to watch the pitmen coming back from work, with their pit dirt on them, lanterns on their caps, thumbs tucked into their string belts.'[52]

It could also be a source of shame, and when in 1926 the Samuel Report emphasized the desirability of pit baths, one reason it gave was the loss of self respect to the miner who had to travel home in dirty clothes (several coalfields had begun to erect colliery baths by the mid-1920s, but in 1926 Durham had yet to follow suit).[53] The social stigma attached to coal could also disadvantage miners in other ways. Many decades later, one old miner still regretted that as a young man in the 1920s he had met a girl in Birtley, '[but] she wouldn't go out with a pitman, her father was at the brickyard'.[54] It was a lament that continued to echo down the years and, in the 1940s, Mark Benney's semi-fictional tale, *Charity Main*, included a description of one Durham miner who remained forever insecure because his future wife, despite proclamations of love, had not wanted to marry a miner and turned him down in favour of a garage hand. She only returned to him after discovering that she was pregnant.[55]

51 Lawson, *Man's Life*, 65.

52 M. Craddock, *North Country Maid* (Maidstone, 1995 edn.; first pub. 1960), 26.

53 *Parl. Papers, 1926*, xiv (1), Report of the Royal Commission on the Coal Industry (1925), i, Report, 207. The commission was appointed in Sept. 1925 under the chairmanship of Sir Herbert Samuel and reported in Mar. 1926. It recommended the reorganization of the industry and recognized the need for wage cuts but rejected any lengthening of hours. Its impact was overtaken by events when the lockout began at the end of the following month, although a few of its suggestions for reorganization were implemented in the Mining Industry Act of Aug. 1926.

54 Gateshead Central Library, Oral History Transcripts, iii, (HM).

55 M. Benney, *Charity Main: A Coalfield Chronicle* (1946), 59–60.

This tension between pride and shame in the work that miners did underground pervades the literature of the period. During the strike itself, it was exploited by Winston Churchill, then Chancellor, when he responded in the House of Commons to a claim that the miners represented 'the million most magnificent men' in the country:

> If they send their sons back into the industry, and if they are the most magnificent million men in the country, it seems to me rather surprising to turn round the next moment and argue that the conditions under which they labour are so tragic and abhorrent and improper and reactionary that they are really amongst the men most deserving of the compassion of the nation . . . it is idle to tell me that if they are conditions below the general standard that prevail in this country, the million most magnificent men in the country would consent to endure them, or would send their sons there.[56]

It is also a tension that forms the background to this study. The experience of shared dangers and hardship underground is frequently suggested by commentators as a factor contributing to the solidarity exhibited by miners. With regard to those of the Ruhr, for example, S. H. F. Hickey argued that it was 'the shared experience of mine-work [that] formed the basis of that collective experience which gave miners their special identity and consciousness'.[57] The nature of the production process below ground could also have important repercussions upon the community living above, as documented in a famous article by M. J. Daunton.[58] In contrast, the focus in this book is on the social relationships above ground, rather than the occupational aspects of a miner's life: during the strike itself, colliery work ceased. However, its influence rumbled on: in the occupational hierarchy that was replicated within the union; in Durham's geological conditions that labelled the coalfield uneconomic; and, not least, in the memory of the unforgiving nature of underground work, whence the miners had come, and where they knew they would, at some point, be going back.

If the popular conception of the miner remains a romantic one, the events of 1926 prompt a similar reaction. Despite recent academic revisionism, in the popular imagination, '1926' conjures up an image of heroic struggle in the face of immense suffering and inevitable tragedy. To one old man, speaking during the 1984–5 strike, such

[56] *Hansard Parliamentary Debates* (Commons), fifth ser., 197, c. 1548.

[57] S. H. F. Hickey, *Workers in Imperial Germany: Miners of the Ruhr* (Oxford, 1985), 109.

[58] M. J. Daunton, 'Down the Pit: Work in the Great Northern and South Wales Coalfields, 1870–1914', *Economic History Review*, 34 (1981), 578–97.

conflict was a time when lads saw 'what it meant to be a miner'.[59] In focusing on one of the longest labour disputes that Britain has ever seen, and a year still resonant with meaning, this study hopes to avoid such romanticization on its own part, while acknowledging the role that such imagery played in shaping the response of participants and other contemporaries. Perhaps it is pertinent therefore to begin with a reminder that, even at the time, not every miner saw either the strike in particular or collective action in general in such epic terms, though he might come out on strike with his fellows and endure with them to the end. In the 1920s, the *Durham Chronicle* ran a column in which it published 'original local anecdotes' sent in by readers. The truth of many of the 'anecdotes' is dubious to say the least, but their telling is revealing, and several are cited in the pages below. One was sent in from Sherburn in June 1926 and recorded the conversation of two strikers:

> Bill: 'Me feyther [father] says this bloomin' coal strike will torn the world upside doon'.
>
> 'Aa wish it wad [would]', said Geordie, 'Then instead o' gannin [going]' doon the pits te get it, the darned stuff wad faal oot'.[60]

59 R. Samuel, 'Introduction', in Samuel et al., *Enemy Within*, 30.

60 *Durham Chronicle*, 19 June 1926.

# 1

# The Tensions of Class and Region

> Of the millions in London, how many have ever spent half an hour in a mining village? . . . How many voters could answer the simplest questions about the hours of work and average earnings of a miner? . . . Who wants to know about coal? . . . The mining communities are remote, hidden away, mysterious. If there had been several working collieries in London itself, modern English history would have been quite different. (For example, we should not have had the General Strike of 1926.)[1]

If J. B. Priestley believed coal-mining communities to be distanced from wider British life, others argued that the mining communities viewed the outside world as distant and irrelevant in their turn. In 1923, G. D. H. Cole stated that 'the miners' intense solidarity and loyalty to their Unions is undoubtedly the result of the conditions under which they work and live. They are isolated from the rest of the world—even the rest of the Trade Union world; but their isolation ministers to their own self-sufficiency and loyalty one to another'.[2] A less sympathetic Durham commentator, Bishop Welldon, agreed that the lack of occupational variety within the pit villages was one reason for the strength of their Labour identity: 'it is thus that the miners are more difficult than other classes to influence by argument.'[3]

A few decades later, Clark Kerr and Abraham Siegel formalized such ideas into a sociological theory to account for the strike propensity of certain industries.[4] Their concept of the 'isolated mass', of which mining

[1] J. B. Priestley, *English Journey* (1935 edn.), 321–2.

[2] G. D. H. Cole, *Labour in the Coal Mining Industry, 1914–21* (Oxford, 1923), 7.

[3] J. E. C. Welldon, *The English Church* (1926), 191. Welldon (1854–1937) was Dean of Durham 1918–33. He had been the Bishop of Calcutta 1898–1902 and so retained the higher title.

[4] C. Kerr and A. Siegel, 'The Interindustry Propensity to Strike: An International Comparison', in A. Kornhauser, R. Dubin, and A. Ross (eds), *Industrial Conflict* (New York, 1954), 191–5.

communities were an example, echoed the observations of Cole and others in suggesting that a sense of collectivism could be fostered within small villages with a high degree of physical and social isolation. Their theory has had tremendous influence on the way in which academics have conceptualized mining communities and has specific relevance to studies of the lockout. David Gilbert has argued that the separateness of local mining communities meant that in the autumn of 1926, just as in the winter of 1984, people trusted the evidence of their own eyes rather than the messages of imminent defeat suggested by the national media, and so could believe that they were winning the strike.[5] Their ideas also tie in with W. G. Runciman's influential study *Relative Deprivation and Social Justice*, which further outlined on a wider scale the importance of geographical and social boundaries in affecting the choices that people make. Runciman argued that attitudes, aspirations, and grievances largely depend on the frame of reference within which they are conceived. He suggested that the limited reference groups of many members of the working class during the interwar years were able to reduce the level of relative deprivation and prevented severe hardship from leading to significant social or political unrest.[6]

However, Kerr and Siegel's theory is not and cannot be applicable to every mining community: while some miners did live in isolated villages, others lived in towns; while some miners did live in pit rows, others lived in streets populated by shipyard workers and shopkeepers. Even for those living in the isolated villages, among the pit rows, migration within and between coalfields affected long-term associations. There are also dangers in generalizing about any occupational community on a national scale, let alone an international one, as Kerr and Siegel sought to do. The more varied social relations and wider geographical horizons of much of the Nottinghamshire coalfield in the interwar years, for example, have been posited as one reason why its miners were less committed to their union than those of elsewhere.[7] Meanwhile, Hywel Francis and Dai Smith's account of the South Wales Miners' Federation

[5] D. Gilbert, *Class, Community, and Collective Action: Social Change in Two British Coalfields, 1850–1926* (Oxford, 1992), 39.

[6] W. G. Runciman, *Relative Deprivation and Social Justice: A Study of Attitudes to Social Inequality in Twentieth-Century England* (1966), esp. 57–77.

[7] Gilbert, *Class, Community, and Collective Action*. Though Gilbert also stressed that in South Wales and Nottinghamshire responses to the strike were conditioned 'not only by the degree of their socio-geographic isolation and occupational homogeneity but also by their social and institutional histories and by changing understandings of the character of those communities' (p. 12).

(SWMF) abandoned the idea of isolation altogether. Their study is noticeable for its emphasis on the international awareness of the South Wales miners and their proletarian solidarity with the working classes of other countries, manifested particularly through their support for Irish nationalism and the Spanish Republic.[8]

In the 1990s, Gilbert observed that Kerr and Siegel's isolated mass hypothesis was still discussed at any gathering of sociologists, 'if only to be ritually dismissed'.[9] It was an observation that Andrew Taylor could still make over ten years later, noting that 'it is now *de rigueur* for any mining scholar to include [such] a critique'.[10] But, Gilbert also suggested that one reason why their work has become such an important starting point is because even if their answers were flawed, they were 'at least asking a good question . . . [regarding] the relationships between geo-social characteristics and collective behaviour'.[11] This chapter will explore the boundaries of region and social class that defined the identity of the Durham miner in the 1920s, and how this affected the positions adopted and the choices made in 1926. Its central theme is the tension between class-based identities and regional ones, arguing that any explanation of solidarity with reference to a purely class consciousness, casting the miner in his traditional role as the archetypal proletarian, needs to be revised. In Durham in the 1920s other forms of close-knit communal identity also existed, whose character was both more and less than merely class-based: these also affected the way the dispute was played out in the coalfield.

## I

In the 1920s, coal-mining was the defining occupation of County Durham: of every 1,000 males over the age of twelve, 291 were classed as coal miners for census purposes in 1921, an average that masked considerably greater concentrations of miners within some parts of the county (see Table 1.1). Only about 80 per cent of these men worked underground, and fewer than one-third were employed in the

[8] H. Francis and D. Smith, *The Fed: A History of the South Wales Miners in the Twentieth Century* (Cardiff, 1980).

[9] Gilbert, *Class, Community, and Collective Action*, 9.

[10] A. Taylor, 'So Many Cases but So Little Comparison: Problems of Comparing Mineworkers', in S. Berger, A. Croll, and N. LaPorte (eds), *Towards a Comparative History of Coalfield Societies* (Aldershot, 2005), 12.

[11] Gilbert, *Class, Community, and Collective Action*, 9.

**Table 1.1** *Proportion of miners in the Durham coalfield*

| | Total population | Proportion of miners per 1,000 males over the age of 12 |
|---|---|---|
| COUNTY DURHAM | 1,479,033 | 291 |
| URBAN DISTRICTS AND MUNICIPAL BOROUGHS (MB) | | |
| Hetton | 17,277 | 648 |
| Stanley | 25,089 | 647 |
| Annfield Plain | 16,531 | 633 |
| Brandon and Byshottles | 18,610 | 625 |
| Tanfield | 10,387 | 617 |
| Tow Law | 4,071 | 585 |
| Willington | 9,202 | 581 |
| Ryton | 14,263 | 580 |
| Houghton-le-Spring | 10,203 | 528 |
| Crook | 12,706 | 516 |
| Seaham Harbour | 16,957 | 514 |
| Leadgate | 5,161 | 493 |
| Chester-le-Street | 15,590 | 489 |
| Blaydon | 33,052 | 485 |
| Spennymoor | 18,238 | 446 |
| Felling | 26,145 | 334 |
| Shildon | 14,165 | 322 |
| Whickham | 19,155 | 288 |
| Bishop Auckland | 14,290 | 225 |
| South Shields | n/a | 208 |
| Southwick on Wear | 14,641 | 203 |
| Durham, City of (MB) | 17,346 | 167 |
| Benfieldside | 8,974 | 160 |
| Hebburn | 24,168 | 156 |
| Gateshead | n/a | 106 |
| Consett | 12,149 | 97 |
| Sunderland | n/a | 57 |
| Jarrow (MB) | 35,576 | 25 |
| RURAL DISTRICTS | | |
| Easington | 75,642 | 655 |
| Houghton-le-Spring | 27,365 | 588 |
| Chester-le-Street | 71,572 | 575 |
| Lanchester | 34,072 | 560 |
| Durham | 31,584 | 559 |
| Auckland | 61,344 | 542 |
| Sedgefield | 37,155 | 480 |

Table 1.1 *Continued*

| | Total population | Proportion of miners per 1,000 males over the age of 12 |
|---|---|---|
| Sunderland | 30,565 | 414 |
| South Shields | 19,104 | 390 |
| Barnard Castle | 11,908 | 252 |

*Notes*: Darlington, Hartlepool, Stockton, and Weardale Rural Districts, Barnard Castle and Stanhope Urban Districts, and Hartlepool and Stockton Municipal Boroughs were not part of the Durham coalfield and have been excluded from the list, as have the County Boroughs of Darlington, Gateshead, South Shields, Sunderland, and West Hartlepool. All have been included in the calculation of the total number for County Durham.
*Source*: *Census of England and Wales*, 1921, ii, *County of Durham* (1923), pp. xxx–xxxi, 2.

popular sense of the word 'miner'—hewers working with their picks at the coalface[12]—but, amongst many commentators, contemporary and otherwise, there was a strong sense that a miner's occupation set him apart from others of the working class. According to Jack Lawson, 'the truth is that coal-getting is among the world's most highly skilled crafts, taking not years but generations to come to any high standard. There are small men whose fathers and grandfathers were miners who would make a Samson, void of pit-craft, look like an infant in arms'.[13]

To some extent, coal miners really were born and not made, and the hereditary nature of the industry was particularly marked in Durham. Figures gathered for the Samuel Commission revealed that of 20,688 new recruits to the Durham colliery workforce in 1924, 29.3 per cent were boys arriving straight from school, 61 per cent came from elsewhere in the coal industry, and only 9.6 per cent came from other industries. The number of 14-year-olds leaving school to sign on at the colliery is particularly striking, and the Durham figures were exceptional in this regard compared to the other coalfield regions of Britain, contrasting an England and Wales average of 12.5 per cent. Even in South Wales and Monmouth, an area similar to Durham in terms of the dominance of the coal industry, only 11.7 per cent of new recruits were boys leaving school.[14] The high Durham figure may have been due to the system of tied housing, more prevalent in Durham than anywhere else in the

[12] Report of the Royal Commission on the Coal Industry (1925), iii, Appendices (1926), 200.

[13] J. Lawson, *Peter Lee* (1936), 216.

[14] Royal Commission on the Coal Industry, iii, 213. The returns from which the Durham figures were calculated covered 98% of the coalfield; those of Great Britain 79%; and those of South Wales and Monmouth 88%.

country, which meant that fathers risked losing their colliery housing if their sons looked for work elsewhere. Lawson may have spoken proudly about an inherited craft, but Ron Rooney remembered:

> If you went into a colliery house and you had sons they automatically had to go in the pit when they left school. If they didn't you were chucked out of your colliery house. A lot of people didn't realise this—they used to say, 'Like father, like son.' They didn't realise that in those days when the management were management and they controlled it all, that you had to go into the pit.[15]

The family nature of the industry was strengthened by high rates of intermarriage, suggesting a further separation of the mining workforce from other local social groups. Seventeen marriage registers for the twenty-one parishes of Easington Deanery survive for the period January 1925 to December 1926 and an analysis is revealing. Church registers have to be taken with a degree of caution, and mistakes and deceptions must have taken place. Easington was also a parish with a particularly high concentration of miners and so the results are unlikely to be representative of the whole coalfield. However, according to the records, of 574 men in Easington Deanery who took marriage vows at an Anglican altar during the period, 330 recorded their occupation as 'miner'. Of these 330 men, 246 (75 per cent) married women whose fathers also gave their occupation as 'miner'; this number rises to 258 (78 per cent) if other colliery occupations such as shaftsman and deputy are included. Even then the figure remains a conservative one: a vicar might not record a father's occupation if he was deceased; there were also various occupations given by the bride's father such as engineman or boilerminder which were likely to have been colliery positions.[16] In a similar, though more wide-ranging, analysis of Scottish mining regions, Alan Campbell discovered mining communities that were increasingly endogamous. With certain regional exceptions, he found that by the 1920s the majority of mineworkers married the daughters of mineworkers. Even then,

[15] R. Rooney, 'Changing Times', in T. Austrin et al. (eds), *But the World Goes on the Same: Changing Times in Durham Pit Villages* (Whitley Bay, 1979), 37. For similar memories from elsewhere in the coalfield, see H. Ashby, 'Send Your Sons into the Mines . . .', in K. Armstrong and H. Beynon (eds), *Hello, Are You Working? Memories of the 'Thirties in the North-East of England* (Whitley Bay, 1977), 35; K. Armstrong (ed.), *Horden Miners: The Lives of Two Horden Miners in their Own Words* (Peterlee, 1984), 4. The system of tied housing is further discussed in Chaps. 1, Sect. IV, and 2, Sect. IV.

[16] Durham County Record Office (henceforth DRO), marriage registers, var.

only in three of the ten districts did the number exceed 70 per cent; the figure was highest at Larkhall (75.1 per cent).[17]

The separation of the miners from the rest of the population was further compounded by linguistic differences. George Hitchin was born just before the First World War and later remembered that upon starting work at the colliery he had had to learn a new language: 'This was "pitmatic". It was a mixture of the broadest dialect of Durham and a number of words . . . used exclusively by pitmen when below ground . . . [making] conversation between pitmen unintelligible to anyone except another pitman.'[18] 'Pitmatic' was also heard by J. B. Priestley during his journey through the coalfield in the early 1930s, and he remarked that the 'dialect within a dialect' was incomprehensible even to the pitmen's wives.[19]

A little over a decade later, another observer of the coalfields commented on the visual distinctions that set miners apart, coming to the conclusion that 'miners formed a distinctive physical type of their own', recognizable by their stature, their gait, and the physical marks of the pit.[20] Such differences were clear even to a child's eyes. The future poet James Kirkup, born in 1918, remembered going as a small boy to the public library in South Shields, where he would stand patiently waiting for his uncle to read the day's news: 'All I could see were the boots and trouser-legs of the men reading the papers on the other side of the rack. I got to know some of those boots and trousers quite well . . . I knew the miners because they had metal-capped boots.'[21]

At least the stigma of travelling home from work black and dirty was temporarily alleviated in 1926 (except for blacklegs, whose social ostracism was presumably even more conspicuous). But, whether on strike or in retirement, an ex-miner could always be detected by the blue scars he wore, caused by the coal dust which coloured grazes and cuts.

## II

If the occupational community of miners was frequently seen as distinct from the rest of the population, any memory of the 1926 lockout remains

[17] A. Campbell, *The Scottish Miners, 1874–1939*, i, *Industry, Work and Community* (Aldershot, 2000), 192–3.

[18] G. Hitchin, *Pit-Yacker* (1962), 70.

[19] Priestley, *English Journey*, 334.

[20] F. Zweig, *Men in the Pits* (1948), 4–7.

[21] J. Kirkup, *The Only Child: An Autobiography of Infancy* (1957), 35.

inseparable from the general strike in May, testifying to some kind of wider solidarity, however short-lived. For the scale of the working-class response was spectacular. Up to two million men, excluding one million miners, had downed their tools by 8 May and, due to the TUC's policy of bringing out the strikers in staggered waves, still more would have been called out had the strike continued.[22] A. J. Cook spoke of the 'nine-days' wonder': 'What a wonderful response! What loyalty!! What solidarity!!! From John O' Groats to Land's End the workers answered the call to arms to defend us, to defend the brave miner in his fight for a living wage'.[23]

Some of those who struck were at least partly responding to self-interest. The miners' propagandists had continually stressed their belief that the miners were simply the first to be affected in what would become a wholesale attack on wages, hours, and working conditions, and such an argument continued to be made even after the general strike had collapsed. In June, the government introduced the Coal Mines Bill, allowing mineowners to extend by one hour the length of the working day that they could demand of their employees.[24] It was fiercely opposed by the Labour benches, and, at the beginning of July, Jack Jones was suspended from the House of Commons for arousing disorder. He was not a miner, but was involved in the National Union of General and Municipal Workers. A few days earlier he had warned the House that 'this will become a general attack on wages. If the miners go down to work eight hours underground, what chance is there for all the other workers above? Their employers will say, "The miners are working eight hours below the surface of the earth; you will have to work nine hours".'[25]

In response, the government attempted to persuade other industrial workers that, in fact, they had very little in common with mineworkers. Bolstered by figures from the Samuel Report, government propaganda

[22] G. Phillips, *The General Strike: The Politics of Industrial Conflict* (1976), 218–19. The government estimated that 1,580,000 workers (excluding miners) were on strike by 8 May; the Strike Organization Committee estimated two million. Phillips suggests a figure of 1,800,000.

[23] A. J. Cook, *The Nine Days: The Story of the General Strike told by the Miners' Secretary* (1926), 16.

[24] The bill suspended the legislation of July 1919, which limited underground work to seven hours, allowing the working day to be extended to eight hours plus one hour winding time. It received Royal Assent on 8 July 1926.

[25] *Hansard Parliamentary Debates* (Commons), fifth ser. (henceforth *HPD*(C)), 197, cc. 942–3. Jack Jones was MP for West Ham, and not the Jack Jones who would become leader of the Transport and General Workers' Union.

**Table 1.2** *Wages and hours worked by adults in various industries*

| Occupation | Class of labourer | Increase 1909–13 to 1925 | Wage for full week, Sept. 1925 | Hours in full week, Sept. 1925 |
|---|---|---|---|---|
| LABOURERS AND SEMI-SKILLED MEN | | | | |
| Agriculture | Labourers | 85%–90% | 29*s*.–37*s* 6*d*. | 48–54 |
| Building | Labourers | 125% | 55*s*. 7*d*. | 44.25 |
| Coal Mining | Subsistence Wages | 65–90% | 41*s*. 6*d*. (Durham)<br>52*s*. 6*d*. (Yorkshire) | 41.25 (Durham)<br>45 (Yorkshire) |
| | Underground Labourers | 75–80% | 52*s*. 9*d*. (approx.) | 41.25 |
| | Pithead and Screen Men | 80% | 45*s*. (approx.) | 46.5 |
| Electricity | Labourers | 120% | 54*s*. 10*d*. | 47 |
| Engineering | Labourers | 80%–85% | 40*s*. 2*d*. | 47 |
| Gas Works | Labourers | 115% | 52*s*. 11*d*. | 47 |
| Iron and Steel | Labourers | 22.5%–70%* | 37*s*. 4*d*.–52*s*. 4*d*. | 45–56 |
| Local Authority | Labourers | 110% | 53*s*. 5*d*. | 46.5 |
| Railway Work | Goods Porters | 150% | 50*s*. | 48 |
| | Engine Cleaners | 150% | 46*s*. 4*d*. | 48 |
| Road Transport | Carters | 125% | 53*s*. 2*d*. | 48 |
| Shipbuilding | Labourers | 75%–80% | 38*s*. 5*d*. | 47 |
| Tramways | Conductors | 112% | 54*s*. 9*d*. | 48 |
| | Drivers | 100% | 59*s*. 2*d*. | 48 |
| SKILLED MEN | | | | |
| Baking | Table Hands | 120% | 64*s*. 9*d*. | 48 |

**Table 1.2** *Continued*

| Occupation | Class of labourer | Increase 1909–13 to 1925 | Wage for full week, Sept. 1925 | Hours in full week, Sept. 1925 |
|---|---|---|---|---|
| Boot and Shoe Making | Skilled Men | 100%–110% | 60*s.* | 48 |
| Building | Painters | 110%–115% | 73*s.* | 44.25 |
| | Other Skilled | 90%–100% | 73*s.* 5*d.* | 44.25 |
| Coal Mining | Hewers | 67% | 76*s.* (approx.) | 41.25 |
| | Timbermen | 68% | 65*s.* (approx.) | 41.25 |
| Engineering | Fitters | 48%–50% | 56*s.* 6*d.* | 47 |
| Printing | Hand Compositors | 110%–130% | 73*s.* 9*d.* | 48 |
| | Bookbinders | 110%–130% | 73*s.* 4*d.* | 48 |
| Railway Work | Drivers | 125% | 87*s.* 5*d.* | 48 |
| | Guards | 125% | 64*s.* 4*d.* | 48 |
| | Signalmen | 125% | 59*s.* 4*d.* | 48 |
| Shipbuilding | Shipwrights | 45%–55% | 55*s.* 7*d.* | 47 |

* Increase from July 1914.

*Notes*: Many Durham mineworkers also received a house rent free and a supply of coal: the Samuel Report suggested this was the equivalent of about 6*s.* 3*d*. a week. See Report, 285. However, miners incurred other costs, such as the purchase of their own tools; some underground workers also had to buy their own explosives. The miners' most vigorous line of attack on the above figures, however, concerned their assumption that a full week was worked. The Samuel Report calculated that in Durham an average of 76.18 working days were lost per man in 1924, for a variety of reasons, including lack of trade and sickness. Across the British coal industry as a whole, it was estimated that an average of 68.49 days had been lost. See Report, 180.

*Source*: *Parl. Papers*, 1926, xiv (1), Royal Commission on the Coal Industry (1925), i, Report, 156–7.

stressed that in wage terms the miners were better off than nearly every other trade (see Table 1.2).[26] The fact of the general strike itself and its ignominious ending was likely to compound any perceived sense of injustice. Walter Citrine, the acting general secretary of the TUC in May 1926, later damned the conduct of the miners in the 1920s, arguing that they had 'had neither the loyalty to the Congress, nor to their colleagues, nor the appreciation of the sacrifices of the movement, to enable them to rise above their restricted vision of their own coalfields'.[27] Given the animosity between the TUC and mining leaderships, it is no surprise that Citrine was less than complimentary. However, his words must have resonated with many amongst the working class who had no direct connection with the coal industry. Despite claims that the general strike had been necessary to protect the whole of the working class, ultimately it had been the attack on the miners that had triggered it. Millions of workers had lost wages and savings, and many consequently suffered reprisals. At the Labour Party conference in October, a defensive Jimmy Thomas argued on behalf of the railwaymen against a financial levy for the miners. His union had 45,000 men who had not returned to work since the beginning of May, and a further 200,000 who were still only working a three-day week: 'Don't talk about sacrifice when that fact stands out,' he told the gathering.[28] Such figures were partly a result of reduced coal traffic due to the continuing lockout, but the railway companies had also practised systematic victimization. In the North East specifically, the transport workers had borne the brunt of such punishment.[29]

The TUC leaders later admitted that one of the main reasons for calling the general strike had been the memory of charges of betrayal on 'Black Friday' when they had declined to come out in support of the miners in April 1921.[30] However, nine days of solidarity in May 1926, culminating in a shambolic return to work, did not allow them to escape

[26] For the government's use of the Samuel figures, see The National Archives: Public Record Office (henceforth TNA:PRO), LAB27/4; *HPD*(C), 196, cc. 2168–9.

[27] W. Citrine, *Men and Work: An Autobiography* (1964), 204. Walter Citrine (1887–1983) was appointed Acting General Secretary of the TUC in Oct. 1925 following the death of Fred Bramley. He was confirmed as General Secretary in Sept. 1926 and remained in post until 1946.

[28] *Labour Party Conference Report* (1926), 198. Jimmy Thomas (1874–1949) was the politically and industrially moderate General Secretary of the National Union of Railwaymen from 1916 until 1931, when he was dismissed following his decision to join the National Government. In 1926 he was Labour MP for Derby.

[29] See A. Mason, *The General Strike in the North East* (Hull, 1970), 91–4.

[30] Such sentiments were expressed by both Ernest Bevin and Walter Citrine. See J. E. Cronin, *Industrial Conflict in Modern Britain* (1979), 129; Citrine, *Men and Work*, 131.

such charges in 1926 either. In the aftermath of the general strike, the miners' leaders castigated the TUC leadership for calling off the strike, and, at the end of May, Cook addressed a rally in South Wales:

> Never have we been bullied by the employers or the Government to the extent that we were bullied by certain trade union leaders to accept a reduction in wages. The Government know that and the coalowners know it. One man on the other side said to me, 'The TUC will help us,' and the Prime Minister on more than one occasion publicly thanked the TUC.[31]

The miners' leaders restricted their vitriol to the TUC leadership, arguing that these men had betrayed not only the miners but also the members of their own unions who had been willing and eager to come to the miners' aid. In contrast, the sense of betrayal felt by the rank-and-file miners themselves was often directed towards the non-mining working class as a whole. Whereas A. J. Cook's rhetoric cast Jimmy Thomas as the Judas of the movement, it is striking that in the memories of many ordinary miners it is the railwaymen as a body who are blamed. One old miner explained in an oral history interview that 'it started off that everybody was on strike, it was the General Strike, every trade, but the miners were let down by the railwaymen'.[32] Another agreed: 'The railwaymen were supposed to have come out, but they went back.'[33] Nor can these beliefs be attributed to the simplification forced by the passage of years: such opinions circulated at the time as well. In September 1926, a resident of Chester-le-Street submitted an entry for publication in the weekly 'anecdotes' column of the *Durham Chronicle*: 'Two railwaymen were watching a miner over the fence the other day picking the weeds out of his garden. "What's the good, Geordie?", said one, "They'll just grow again." "Aye, be just like you and the TUC," came the reply, "—Ne sunner out than you're in again." '[34] This is despite the fact that such perceptions were misplaced. While the caution of their leader is well known, the railway unions' rank and file responded to the strike with enthusiasm. One sympathetic contemporary history described the members of the National Union of Railwaymen (NUR) and the Associated Society of Locomotive Engineers and Firemen as 'a rock-like centre to the strike', while Home Office statistics analysed by Margaret Morris suggest that 91 per cent of Britain's half a million

[31] Cited in C. Farman, *The General Strike: May 1926* (1972), 248.

[32] Beamish Museum, Oral History Archive (henceforth BMOA), 1999/2.

[33] Ibid., 1993/5.

[34] *Durham Chronicle* (henceforth *DC*), 4 Sept. 1926.

railway workers (excluding clerical and supervisory staff) were on strike by 12 May 1926.[35]

Even if many initially sympathized with the miners, however, goodwill may have diminished as the lockout continued. In the North East the perpetuation of the coal dispute continued to have an impact throughout the year because so many of its other industries such as shipping, rail, and engineering relied on the smooth working of the collieries to maintain production. The Consett Iron Company, for example, increasingly found itself having to lay off workers, and by the first week in June, of 7,265 men receiving outdoor poor relief in the Consett area, only 3,718 were miners.[36] At the opposite end of the county, the South Durham Iron and Steel Company were forced to close their shops in West Hartlepool because of lack of fuel: 1,000 men were affected.[37]

The increase in unemployment had inevitable repercussions. By September, the rent arrears of nearly 600 houses on Chester-le-Street's council estates were averaging over £7 per house. The great majority of tenants were miners but those engaged in other trades might also have found themselves in debt.[38] Private landlords who relied on a tenant's income suffered. Easington Poor Law Union recorded the case of one landlord who unsuccessfully applied for poor relief on the grounds that his tenants were miners and were not paying rent, while in Houghton a doctor was jailed for the non-payment of rates, despite his pleas that he could not pay if his patients did not pay him.[39]

For those not directly involved in the industry, the consequences of the stoppage were not simply financial. The dominance of the coal companies over so many aspects of municipal life meant that disruption could lead to multiple discomforts. By early autumn, there was a severe water shortage at Thornley due to the withdrawal of men working on pumping operations at the colliery. Street lighting went out in Hetton due to the stoppage of the gas works, and at Willington, where gas was normally supplied by the colliery company. One local councillor complained: 'Whenever there has been any industrial trouble this has been the first thing we have had to fight . . . I think the day is past when the public should be held up to ransom and put to the expense

[35] R. W. Postgate, E. Wilkinson, and J. F. Horrabin (eds), *A Worker's History of the Great Strike* (1927), 20; M. Morris, *The General Strike* (Harmondsworth, 1976), 30.

[36] *Stanley News and Consett Chronicle* (henceforth *SNCC*), 10 June 1926.

[37] Mason, *General Strike*, 95.

[38] *DC*, 4 Sept. 1926.

[39] DRO, U/Ea17, Easington poor law union minute book, 19 Aug. 1926; *DC*, 7 Aug. 1926.

of having to buy lamps and candles when they have gas fittings in the house.'[40]

The numerical superiority of the mining workforce and the dominance of the DMA within local power structures, particularly on local councils, may also have increased the resentment of those not directly involved in the coal industry: upon the outbreak of the stoppage, Labour and trade union might swung into action to support the miners, leaving other groups undefended. When Jarrow and Hebburn Co-operative Society donated vouchers worth £100 to striking miners at Hebburn Colliery, for example, complaints were raised that the miners were receiving preferential treatment.[41] It was not an isolated incident, and a week later an initialled letter was sent to the Newcastle-based *Evening Chronicle*: 'As a Co-operator, I wish to make a very strong protest against a free distribution of loaves of bread to the miners. I should like to know who will foot the bill . . . The Co-operative Society belongs to the mass of members, not to a miners' section.'[42]

The rush of fundraising events for miners and their families also meant that other charitable organizations suffered. By the end of May, the secretary of the Newcastle Poor Children's Holiday Association and Rescue Agency announced that its funds were considerably depleted due to the strike.[43] Significantly, the union itself was no longer prepared to support wider charitable work. With money tight and a long battle ahead, the immediate concerns of its members were paramount. As early as 5 May, Brancepeth 2 lodge took the decision to end its regular contribution to the National Blind Institution.[44]

Perhaps the most bitter tales of fracture concern the distribution of coal. Few miners found themselves desperately short of fuel during 1926, being well-equipped to cut into the hillsides and start their own mini-drift mines. These were often supervised by the union: shifts were organized and coal shared out amongst needy residents. Selling on coal for profit was strictly prohibited, although it was frequently done. Many colliery companies also continued to provide their employees with free coal as in normal times, and the colliery waste heaps, upon which men, women, and children could scavenge for 'duff' [waste coal] were frequently reserved for miners. In contrast, the consequences for those

[40] DRO, 2 Oct. 1926; *Durham County Advertiser* (henceforth *DCA*), 17 Sept. 1926.

[41] *Evening Chronicle*, 14 July 1926.

[42] Ibid., 22 July 1926.

[43] *SNCC*, 20 May 1926.

[44] DRO, D/DMA 326/4, Brancepeth 2 lodge, minutes of special meeting, 5 May 1926. 'Lodge' was the term used for smallest unit of the union, based around each individual colliery.

not connected to the coal industry (or who lacked family or neighbours who were) is revealed through court records. In November, William Bissenden of Sherburn found himself in court charged with stealing coal from Sherburn Hill Colliery. When asked why he had not gone to the duff heap he replied that because he was not a miner he had not been able to get a coal ticket.[45] Bissenden's feelings are unrecorded, but it is likely that he felt bitterly the perceived injustice of his own inability to keep warm and the miners' ability to do so, though the strike was in their industry. A more serious incident had occurred earlier in the summer between Edgar Myers, a 30-year-old butcher's assistant, and George Carr, a miner. Myers had been attempting to get some coal from the stock available to the miners, and Carr had seen him, stopped him, and informed the lodge secretary. A furious Myers had later assaulted Carr.[46]

Given such potential for resentment, it is astonishing how much support the miners were able to retain. Many of the memories recorded for the Gateshead oral history project in 1976 are striking in this regard. One interviewee, an old man who had worked as a painter and decorator in the 1920s, set out his view of the strike in a jumble of ideas that included no mention of the miners at all:

> We had a class society . . . And the strike was one of the things which tried, I would say to break this, and it actually did. The First World War sort of proved to the masses that all men were equal, although there were some more equal than others, as the Yankee said, wasn't it Lincoln? . . . In 1926 they were trying to get what I would term now, a square deal.[47]

In his old age at least, this old man simply remembered the strike (and perhaps he was remembering only the general strike) as something that sought to improve the position of the working class as a whole.

Even when respondents did remember the centrality of the miners in the dispute, animosity was rarely expressed, and men and women were frequently keen to emphasize the justice of the miners' cause. One man had been a young builder in 1926 and explained, 'There was no ill-feeling towards the miners because then the miners were on very low wages, in fact I think they only had about *6s. 6d.* a day. They worked under appalling conditions, all the seams in that area were you know, soaking wet . . . And then you never knew when they were going to be

[45] *DCA*, 19 Nov. 1926. [46] Ibid., 18 June 1926.
[47] Gateshead Central Library, Oral History Transcripts (henceforth GCLOT), i (P.O'B.).

off'. Another interviewee, whose husband had been a manual worker not connected with the coal industry, also provided justifications for the miners' actions and similarly underestimated the miner's wage. Remembering the children lining up at the soup kitchens, she described how 'it was heart rending . . . to see them. I mean in them days a miner got about thirty shillings a week. That was his wage and mind he was down longer hours than what they are now. It hit them hard.'[48]

There are dangers in relying too heavily on the Gateshead evidence. Most of the respondents for the project came from the north of the county, closer to the urban sprawl of Gateshead and Newcastle where miners were much more likely to live alongside, interact with, and form friendships with other industrial workers. However, their sentiments can be backed up by the recollections of others living elsewhere. One man had been the 6-year-old son of a local government officer when 'the big strike' occurred, living in Horden, a colliery village almost exclusively populated by miners. His recollections are again noticeable for the lack of any castigation of the role played by the miners. To him, the miners were simply the group that suffered most: '[The strike] involved not only the miners but the workforce of the whole country. To me, as a small child and living amongst miners, it seemed that the pitmen suffered the consequences of the strike more than anybody else. It lasted for six long months and when it was finally broken, the workers were no better off.'[49]

Such recollections still need to be analysed with care, as any sharper feelings of resentment or bitterness may have dulled over time. However, an assertion that the Durham miners received support from their non-mining neighbours is also supported by empirical evidence. Throughout the lockout, local newspapers reported on fundraising activities organized by workers of other trades, such as the concert held in Durham City in November, when performances from employees and drivers of the Northern General Omnibus Company contributed towards the total of £375 they raised for the boot fund during the stoppage.[50] During the general strike itself, despite the popular stereotype of Oxbridge students surging enthusiastically from their dreaming spires to man the trams, most volunteer and blackleg labour came from the clerical and manual workforce and the pool of unemployed.[51] The Northern

[48] Gateshead Central Library, i (Mr B.); ii (Mrs B.). For wage levels, see Tables 1.2 and 2.4.

[49] W. Healy, *Between the Wars: Childhood Memories of Horden* (Durham, 1996), 10.

[50] *DCA*, 26 Nov. 1926.

[51] See Phillips, *General Strike*, 155.

Division was one of only two divisions across the country in which the stated number of people volunteering to serve fell below 20,000 (the other, significantly, was South Wales).[52]

## III

Some of the most eager—as well as the most valuable—supporters of the miners were the local shopkeepers. Such people made up the section of the non-mining community for whom the financial consequences of the strike were potentially the most damaging. Some retailers attempted to turn the strike to their advantage: one advert for Ovaltine placed in a local paper in May proclaimed that 'strike-worn nerves need building up'; a few months later the hot water bottle industry also realized the advertising potential with the slogan 'No Coal? . . . get a Taylor's hot water bottle now!'[53] However, in September the secretary of the Incorporated Association of Retail Distributors told a different story:

> although my reports from the large Stores all over the country reveal no appreciable ill-effects over the last seventeen weeks . . . a very different state of affairs [exists] amongst the small retailers who are dependent upon communities entirely composed of miners and their families . . . The local storekeeper has been compelled to give extended credit to his customers . . . Unless the stoppage is brought to a speedy end it seems that the inevitable result will be the ruination of a large number of these small businesses.[54]

Many local stores were already weakened through the widespread credit given in 1921; others survived the 1926 lockout only to go bankrupt in its aftermath as credit remained unpaid.[55] Indeed, contrary to the words of the secretary cited above, larger stores might suffer too. The Co-operative Society responded generously to the strike, providing credit to miners' lodges and often free distributions of food. The Northern Section of the society was particularly supportive, allowing an increase of £101,307 in credit to members during the stoppage, constituting almost a quarter of the increase of £465,697 within the national society in England as a whole. As a large organization, the Co-op was more easily able to absorb losses; through the dividend it also had a built-in

[52] Ibid., 153.
[53] *DCA*, 21 May 1926; *Seaham Weekly News* (henceforth *SWN*), 29 Oct. 1926.
[54] *Blaydon Courier* (henceforth *BC*), 4 Sept. 1926.
[55] See, for example, GCLOT, i (S.M.).

mechanism that could more easily recoup credit owed. But, even the Co-op could not get through the dispute unscathed: by the end of 1926, the Northern branch of the society was owed over £800,000 by its members; £200,000 more than that owed at the end of 1925. Income from sales dropped by nearly £2 million over the same period and local branches were forced to cut back. In Brandon, for example, the employees of the local Co-operative Society were placed on short time.[56]

However, despite hardships forced by the strike, local tradesmen were often prepared to give practical support to the miners. In June, the proprietors of a fish and chip shop in Burnopfield provided a free dinner for 500 local children; in October, a Willington hotel distributed over thirty gallons of soup to those affected by the coal dispute, and continued to do so regularly.[57] In Springwell, the records of a soup kitchen set up for single men in September reveal donations of loaves from two different bakeries, a supply of corned beef from a local grocers, and vegetables from a local farm, as well as financial donations from the local vicar, the postman, and the Springwell Wesleyans.[58] Even more frequently, local tradesmen became involved in charity events. In June, Robert Sewell, a Willington butcher, organized a five-mile race and provided the prizes: a pound of beef and two eggs for each competitor and ten shillings' worth of beef, four stones of potatoes, and a pound of butter for the winner.[59] Indeed, some retailers were so keen to support the miners that they upset opponents of the strike. In August, a complaint was made at a meeting of the Chester-le-Street Ratepayers' Association against tradesmen who refused to join the association on the grounds that they might be ostracized by some of their customers. 'There was something of the jellyfish about such men [who] put their business before their citizenship,' remarked the speaker, 'and they were unworthy of the franchise.'[60]

Rather than embittered relations between miners and tradespeople, tension was often more apparent between the tradespeople themselves, particularly in the struggle between the Co-operative Society and private

[56] Reports of the Fifty-eighth and Fifty-ninth Annual Co-operative Congresses, May 1926 and June 1927; *DC*, 14 Aug. 1926.

[57] *SNCC*, 3 June 1926; *DCA*, 3 Sept. 1926; 15 Oct. 1926.

[58] DRO, D/X411/172, financial report for Springwell and Mount's single men's soup kitchen, 1926.

[59] *DC*, 12 June 1926. The provision of prizes by local businesses is also remembered by oral history respondents. See, for example, BMOA, 1997/19.

[60] *DC*, 21 Aug. 1926.

proprietors over the allocation of relief vouchers. One owner of a small grocery in Birtley was moved to write to the Minister of Health himself: 'I am asking you [to] help me . . . This week all relief vouchers has been given to the Co-op. Society, and we are in a fix how to pay our rates, which you know is high in this Division. Can you say it is legal to leave all grocers who are ratepayers out of vouchers . . . ?'[61] In Coxhoe, it was the Co-operative Society who complained to their local guardians, pointing out that they were large ratepayers and were not receiving their share of vouchers.[62] In both cases, the relevant guardians claimed that the choice of retailer was always left up to the applicant.

The frequency with which the kindness of local shopkeepers is referred to in oral and written reminiscences suggests that the assistance given by local retailers was greatly appreciated by their mining customers. Describing the soup kitchen set up at Whickham, one old man explained:

> The produce was all given by local trades people: coal from the colliery; a farmer would give us a bag of spuds, also a bag of turnips; a market gardener would give leeks, carrots and parsnips; a butcher would often give a supply, often a barrowful of bones, often with a bit of meat on them, and all labour was voluntary.[63]

Indeed, George Hitchin spoke in reverential terms when he described the support given by the local shopkeepers during the earlier lockout of 1921:

> To see that the bairns had enough to eat blossomed into a social concern, and if the community did not always succeed, the fault could not be laid at the door of the tradesmen of the town. These small shopkeepers gave credit not merely till it hurt but till it crippled them. Some closed down; others went bankrupt; all were owed more than they were ever likely to recover.[64]

It is possible that such generosity has been romanticized over time by old people thinking back to a shared camaraderie during the lockout. One (Northumberland) miner admitted that he viewed events very differently several decades later. He believed that his later judgements were the correct ones; the historian is not so sure. One night he and his brother had crept to a nearby farm:

> Well we got there, opened the gate and there was the open-fronted barn, full of the loot, turnips in their thousands. I was in fear and trembling, waiting for the

[61] DRO, U/CS310, George Hunter to Minister of Health, n.d., late 1926? Grammar as in original.

[62] *Co-operative News*, 12 June 1926.

[63] GCLOT, i (C.L.). Elizabeth Stoves expressed similar sentiments about the soup kitchens in Dawdon, while tradesmen also contributed to the soup kitchens at Dipton. Telephone conversation with E. Stoves, 13 June 2005; M. Ridley, 'Making a Contribution', in Austrin et al., *But the World Goes on the Same*, 59

[64] Hitchin, *Pit-Yacker*, 35–6.

dogs to bark and the cocks to crow. But thinking about it now, I'm sure if the farmer had heard us he would have said, 'It's just some of them miners getting a turnip or two.'[65]

Certainly the contemporary references made by the miners' propagandists suggest that any battle to gain the support of the tradesmen had not been decisively won. Herbert Dunnico, the Labour MP for Consett, announced at one meeting that he had never been able to understand the mentality of the lower-middle-class men and shopkeepers in the area around Stanley and Annfield Plain: 'Almost to a man they voted against Labour every time, but who would profit more than they if the miners had good wages? . . . He thought that the shopkeepers, as a class, were about the biggest mugs he had ever met.' His audience responded with laughter and cheers.[66] A few days later, Joshua Ritson, the miners' MP for Durham, addressed another crowd. 'Who were the best friends of the shopkeepers?' he wondered: 'Has Lord Londonderry taken the place of the miners while they have been without money and purchased from the tradesmen of Durham? Not likely! The friends of the shopkeepers are the workers who spend their money on the spot and pay their way as they go along, and it is time the tradespeople realise that fact.'[67]

Even when shopkeepers did help out their regular customers it did not necessarily indicate a relationship purely based on goodwill. It was precisely because shops stood to lose so much by alienating the mining community that many felt they had no choice. The alternative might be disastrous. 'I'll tell you a funny thing,' one old miner told an interviewer, 'There was a chap had a butcher's shop in South Moor. He joined the Special Police that was to fight the miners. Aye the people of South Moor boycotted his shop and he went bankrupt. You see you've got to be very careful how you treat [people] . . . He thought he was all right joining the Special Police but he was wrong.'[68] Demonstrations of

[65] Tyne and Wear Archives Service, DX 201/2, 'Reminiscences of a Walbottle Colliery Miner'.

[66] *SNCC*, 25 Nov. 1926. Herbert Dunnico (1875–1953) was the son of a Lancashire miner. He worked in a variety of jobs, including in the mines, before becoming Labour MP for Consett 1922–31.

[67] *DC*, 27 Nov. 1926. Joshua Ritson (1874–1955) worked underground at collieries in Cumberland, Northumberland, and Durham before joining the Sunderland police force as a young man. He returned to the pits in his early 30s. He was MP for Durham 1922–31 and 1935–45. Charles Vane-Tempest-Stewart, seventh marquess of Londonderry (1878–1949), was one of the leading landowners in County Durham (as well as in Ulster), and a major coalowner. In 1931 he was appointed Secretary of State for Air in the National Government.

[68] BMOA, 1999/2.

support were all the more important owing to the fact that those who supported the miners' cause were often publicly acknowledged. In late May, for example, dozens of tradesmen who had contributed prizes for a walking competition were listed in the local press; presumably those who had declined were noticeable by their absence.[69] Furthermore, any who took an active stand against the miners were subject to the same community sanctions as blacklegs. In August, 'lively scenes' were reported at Ryhope after three tradesmen refused to buy flags in support of the miners' children's fund. A hostile crowd quickly gathered, broken up only upon the arrival of the police.[70]

Nevertheless, some shopkeepers undoubtedly did feel a genuine sympathy for their mining clientele. One bakery informed the DMA that 'we will be pleased to supply free 1,000 loaves of Bread weekly to your stations in the Sunderland, South Shields and Durham districts . . . We are not seeking advertisement, but merely make this offer as our contribution towards suffering humanity.'[71] One can continue to be sceptical, but assistance was also offered from retailers outside the region who stood to make no obvious gain from the building of goodwill. In May, the DMA received a letter from Fry and Sons Ltd of Bristol offering to distribute 7,500 tins of cocoa to miners' families; the following month a fishmonger of Hull promised to supply the best quality cod to the Durham miners at the exceptionally low price of 2*d*. per pound.[72]

Of course, the adoption of too sympathetic an attitude could also have serious consequences, and the real dilemma facing tradesmen is caught by the testimony of William Turnbull of Thornley, whose appearance in Durham bankruptcy court coincided with the end of the lockout. He explained that his grocery and provision business had been in trouble ever since the 1921 strike:

Registrar: You seem to have given credit to a considerable extent?
Turnbull: Yes, one is bound to do so in a colliery village, otherwise they would not get any trade.
Registrar: But there is a limit . . . ?
Turnbull: I struggled on and tried to pull myself together.
Registrar: In some cases you have given credit to a ridiculous extent.
Turnbull: It was a very difficult position.[73]

[69] *SNCC*, 27 May 1926.
[70] *DCA*, 13 Aug. 1926.
[71] DRO, D/DMA (Acc. 2157 (D)), 208 (box)/3, Hunter Bakeries Ltd to DMA, 5 May 1926.
[72] *DC*, 29 May 1926; *DCA*, 4 June 1926.
[73] *DC*, 4 Dec. 1926.

Even at village level, lives overlapped to such an extent that it becomes reasonable to assume that, for many, loyalties must have done the same. Shopkeepers might have been involved in the mining industry themselves, such as Tommy Lawton's father, a colliery worker who also owned a little shop. 'As long as his stock lasted he allowed people to tick on,' his son later recalled, '[He] never did let any of the people he had allowed to continue ticking pay one penny back and consequently [during the strike] we were worse off than them.'[74] Many more had previous links to colliery work, and during the lockout the *Durham Chronicle* noted the death of 59-year-old Henry Cox. He had worked in the pits until the age of 30, before leaving to set up an oil and hardware business at Houghton. During the 1921 stoppage, he had lent a considerable sum without interest to local lodges.[75] Even those who had never been directly connected to the industry might form ties with the village simply through long association. In September 1926, a Mrs Thompson of Blackgate, Coxhoe, celebrated her 96th birthday. She was the widow of Joseph Thompson who had been a grocer there for many years, noted the local paper, 'She's the oldest inhabitant of the district and is well known.'[76]

## IV

The dominance of an occupational culture in the pit villages was compounded by the lack of a conspicuous middle-class presence. In County Durham at the time of the 1921 census, only thirteen men in every thousand over the age of 12 were engaged in professional occupations (excluding clerical staff). This compared to an average of twenty-two per thousand in England and Wales and in fact no other county had a lower proportion than Durham, though Staffordshire could equal it.[77] The DMA leader Peter Lee believed that this contributed to the miners' sense of solidarity, arguing that 'those who are inclined to blame us and say we stand for class distinction should remember that it has been bred in us, not because our fathers had a desire for it but because the educated and the rich left us very largely alone in our village life'.[78]

74 GCLOT, ii (T.L.). 75 *DC*, 25 Sept. 1926. 76 Ibid., 18 Sept. 1926.

77 *Census of England and Wales, 1921: General Report with Appendices* (1927), 95.

78 *Sunderland Daily Echo*, 14 May 1925, cited in H. Beynon and T. Austrin, *Masters and Servants: Class and Patronage in the Making of a Labour Organisation* (1994), 274. Peter

Even when miners lived in larger settlements where the presence of a middle class was more visible, their paths were unlikely to overlap. As an old man, Tommy Turnbull described South Shields, which—in his memory at least—was characterized by strict class and occupational boundaries:

> Shields was divided into villages, and where you lived depended on what you did for a living. The likes of doctors, lawyers and businessmen lived in Westoe. Shipping people lived on the Lawe Top, seamen near the market, and dockers and railwaymen near Tyne Dock. Pitmen lived in Bolden Lane and Whiteleas.[79]

Any wider awareness of the higher social classes tended to be limited to girls in service. Fewer than one in five Durham women were classed as occupied in the 1921 census, but domestic servants accounted for over a quarter of those who were.[80] Memories of such employment could remain with women for years. Edie Bestford worked at Windlestone Hall in Northumberland and with bitterness remembered being told one day that 'Her Ladyship says that you all have to be in at seven. Because these awful miners are about and you never know what might happen to you.'[81] For some, such experiences proved formative. Another girl was in service with a titled family in the early 1920s and visited her family during the 1921 lockout. She accompanied her father when he went to get coal off the heap, and saw the soup kitchens: 'And I went back to my situation. I thought, "Well, why should some people have all and no worries." And from that day I became a Socialist.'[82] However, these girls, sometimes far from home and in irregular contact with their family, tended to remain isolated. Despite her scorn for her employers, and despite the fact that her father and her brothers were all miners, Edie Bestford felt cut off from their situation: 'I wasn't involved in the worst of the Depression like they were in the villages. Living in Jesmond

Lee (1864–1935) began work in a Lancashire cotton factory before moving to Durham's Sherburn Hill Colliery aged 10. He later worked in a variety of pits in Britain and abroad. In 1924 he was appointed executive and joint committee secretary of the DMA, and then general secretary in 1930. He became president of the MFGB in 1932. He played an active part in local government for most of his life, including as the first Labour Chairman of Durham County Council in 1919, a position held in 1926.

79 J. Robinson, *Tommy Turnbull: A Miner's Life* (Newcastle-upon-Tyne, 1996), 13.

80 *Census of England and Wales, 1921*, ii, *County of Durham* (1923), p. xxxvii.

81 E. Bestford, 'My Father and Brothers are Miners!', in Armstrong and Beynon, *Hello, Are You Working?* 87.

82 GCLOT, i (Mrs C.). Alan Campbell has recorded similar anecdotes with regard to the daughters of Scottish miners. Campbell, *Scottish Miners*, i. 233.

I was out of the way of it.'[83] She spoke of her time in service in the 1930s, but presumably other girls felt similarly in 1926.

Ignorance was frequently reciprocal and many middle-class observers displayed a staggering lack of imagination when speaking of miners and their work. During the sitting of the Samuel Commission, Sir Richard Redmayne was asked whether he thought that a return to eight-hour shifts would lower the miners' standard of living. By 1925, Redmayne had become a national figure within the industry and had been the government's Chief Inspector of Mines 1908–19; he also had specific knowledge of his native Durham, where he had trained and worked in the early part of his career. 'No, I do not think so,' he replied, 'I have worked increased hours of late, and I find my standard of living has not gone down.'[84] Two years earlier, Beatrice Webb had told the women of Seaham about Sidney's first few days in Parliament: 'A fortnight ago he had his first experience of an all-night sitting. He left home at two o'clock in the afternoon, but he did not return until seven o' clock the next morning, except for a short interval for dinner. That is a longer shift than any miner works!'[85] During the strike itself, local newspaper editors faced a tirade of letters penned by middle-class authors furious with the spiralling rates caused by the increased provision of outdoor relief. In July, the *Chester-le-Street Chronicle* published a particularly vociferous one from a 'Birtley Ratepayer':

> Week after week in certain papers one reads of the 'splendid fight of the miners,' etc. But . . . the miners are resting, and compelling the ratepayers, who have no say in the matter, to fight the battle for them . . . seriously, where is the destitution? . . . One hears of a harassed ratepayer who saw no alternative but to sell a costly piano to pay for his rates, etc., and his neighbour on the relief offering him instantly £45 for it.[86]

Yet, when miners and members of the middle class did come into contact with each other, the contrast in living standards could be clear.

83 Bestford, 'My Father', 86.

84 Royal Commission on the Coal Industry (1925), ii, Minutes of Evidence (1926), qq. 4081–2.

85 London School of Economics (henceforth LSE), Stoatley rough school 8/4, B. Webb to Seaham women, 15 June 1923. Beatrice Webb (1858–1943) and Sidney Webb (1859–1947) began their association with Durham following Sidney's nomination and subsequent election as Labour MP for Seaham (1922–9). They never lived permanently in the area, but Beatrice kept up a steady stream of correspondence and newsletters with the women's section of the constituency Labour Party.

86 *Chester-le-Street Chronicle* (henceforth *CC*), 30 July 1926. See also the letter from 'Fairplay' published *BC*, 21 Aug. 1926.

For George Terrans, growing up in Trimdon during the First World War, 'the highlight of the day' was glimpsing the local doctor in his car, the only private vehicle the village regularly saw; Isabella Embleton, a small girl at Kelloe at the same time, was similarly impressed by her local doctor's Ford Chevrolet.[87] In the heated atmosphere of the lockout, such visible disparities of wealth might engender animosity. One female activist from Shotton wrote to Beatrice Webb: 'I'll tell you Mrs. Webb my candid opinion it [the strike] is making people and even children Hate the Rich people, it will grow in the children's minds, how there [*sic*] fathers, and Brothers are kept down, and it will be God Help the Boss class through time, I Hope I don't live to see it.'[88]

However, such sentiments are rare. More frequently, when class distinctions are made in oral or written memoirs, their authors make no further comment. The historian is left wondering how widely an awareness of such differences translated into political sentiments, just as Hensley Henson did when, after chatting to a miner in his grounds at Auckland Park, he offered to show him around Auckland Castle:

> He was greatly impressed and I think appreciated my concern for him. But what can he make of it all? On the one hand, the idle and embittered miners full of Cook's rhetoric: on the other, the bizarre and unintelligible splendours of the Bishop of Durham's house! I would give much to know the old fellow's mind on the contrast.[89]

A few years later, and coming from a different political standpoint, J. B. Priestley visited a women's sewing circle in East Durham: women 'sewing on the razor-edge of life'. He, too, noted the contrast between his position and theirs, but also implied that such a recognition was on his part only: 'They were glad to see me and were neither resentful nor whining, but nevertheless they made me feel like a fat rich man.'[90]

During the strike, even the miners' leaders made relatively few references to the middle classes (as distinct from the coal owners). One of the few who did, or at least whose words were reported by the local press, was Jack Lawson. Although he was naturally a cautious orator, Lawson's words betrayed not the slightest hint of animosity but simply a

[87] Clayport Library, Durham, B/L Ter, 'An Interview with Councillor George William Terrans of Trimdon'; DRO, A14/11, I. A. Embleton, 'When I was Growing Up in a Pit Village'.

[88] LSE, Passfield 2/4/H, S. A. Seymour to B. Webb, 23 Dec. 1926. Grammar as in original.

[89] Dean and Chapter Library, Durham Cathedral, diary of H. Hensley Henson, 21 July 1926.

[90] Priestley, *English Journey*, 333.

deep-rooted pride in his own class. After visiting various feeding centres, he commented that he had seen meals served and eaten in several of the great public schools, but had not the slightest hesitation in saying that the conduct of the Durham miners' children compared very favourably indeed.[91]

Rather than an antipathy towards the higher social classes in general, much of the 'class' resentment fostered amongst the mining community was focused upon that section of the middle and upper classes who were specifically connected with the coal industry. The colliery hierarchy was not an undifferentiated group. Quentin Outram has revealed divisions not only within the Mineowners' Association of Great Britain (MAGB), which was subject to inter-district rivalry, but also between the colliery owners and the royalty owners: for many of the former, the payment of royalties to the landowners was a grievance they shared with their workforce.[92] However, there is little evidence that the miners themselves made much distinction. Hostility was frequently keen, whether directed towards royalty owners, colliery owners, or colliery management. The latter were the group most visible within the villages themselves, and one old miner expressed again and again his (inherited) hatred of the vindictive manager at Sacriston Colliery where his father and grandfather had worked in the 1920s.[93] Those interviewed by Robert Moore expressed similarly bitter sentiments regarding long-dead colliery officials.[94]

Even supposed acts of paternalistic generosity such as the provision of free housing were seen less as acts of kindness than as a means of control.[95] The dependence of the Durham miner on the coal owner for the provision of housing was unparalleled: in 1924, of nearly 68,000 houses provided free of rent to mineworkers in England and Wales, 48,942 were in Durham, and most of the rest were in Northumberland.[96] The allocation of such houses was made according to a man's status within the industry: 'If you were a collier [hewer] and you had three sons . . . There was a colliery house for you. But if you were a putter or

[91] *CC*, 4 June 1926.

[92] Q. Outram, 'Class Warriors: The Coalowners', in J. McIlroy, A. Campbell, and K. Gildart (eds), *Industrial Politics and the 1926 Mining Lockout: The Struggle for Dignity* (Cardiff, 2004), 110–11.

[93] Interview with R. M., Horden, 17 Sept. 2005.

[94] Durham University Library, misc. accession 1996/7:4, qualidata reference QD35/Moore, data gathered by Prof. Robert Moore.

[95] The issue of paternalism is discussed in Chap. 2, Sect. IV.

[96] Royal Commission on the Coal Industry, iii. 248–9.

datal worker with three daughters, well you were put at the bottom of the list.'[97] The condition of much colliery housing was a further grievance, and when the 1901 and 1911 censuses revealed horrible overcrowding in the county Sidney Webb suggested that it was no coincidence that Durham also possessed the greatest proportion of tied housing.[98] The 1921 census painted a similarly bleak picture. It found that almost one-third (29.5 per cent) of the population of Durham and its associated county boroughs were living more than two persons to a room. The situation was worse only in Northumberland, where 30.8 per cent of the population lived in such conditions. After the two North-Eastern counties, London took third place with 16.1 per cent, and the England and Wales average was only 9.6 per cent.[99] By the 1920s, even Henson was concerned enough to write to Londonderry, warning of the resentment caused by the 'terrible' housing prevalent in the mining parishes.[100]

For the inhabitants of such houses, the ostentatious wealth of the upper-class residences of royalty owners must have made a dramatic impact on any living under their shadow. In neighbouring Northumberland during the general strike, one Cramlington miner was amongst several sent to jail for the derailing of the *Flying Scotsman*. Over fifty years later he attempted to defend his actions in his memoirs. One of the pictures he included was a photograph of Alnwick Castle, home to the Duke of Northumberland. The picture is gratuitous to the text, and no comment is made on it.[101] In the year of the Sankey Commission, the Duke of Northumberland received over £82,000 in royalty payments for the coal taken from under his land (although taxes reduced the gross amount substantially); six years later the Samuel Report estimated that across the country the payment of royalties was equal to an average charge of two shillings on the wage slip of every mineworker.[102] Bishop

97 G. Alsop, 'A Kind of Socialism', in Austrin et al., *But the World Goes on the Same*, 25.

98 S. Webb, *The Story of the Durham Miners (1661–1921)* (1921), 130–1.

99 *Census, 1921, General Report*, 51.

100 H. H. Henson, *More Letters of Herbert Hensley Henson*, ed. E. F. Braley (1954), 29.

101 W. Muckle, *No Regrets* (Newcastle-upon-Tyne, 1981), 27.

102 *Parl. Papers, 1919*, xii (1), Coal Industry Commission (1919), ii, Reports and Minutes, Apr. 1919, q. 15033; Royal Commission on the Coal Industry, i, Report, 110. The Coal Industry Commission was appointed in Feb. 1919. It was chaired by Sir John Sankey and consisted of equal numbers of representatives of both miners and owners. Some of its recommendations, such as an underground day of seven hours, found their way into legislation later that year, but proceedings were dominated by the question of industrial reorganization and the chairman ultimately swung his support behind nationalization. However, three interim and four

Henson's Auckland Castle residence was also built with wealth at least partly amassed by coal royalties. When he showed around the striking miner, perhaps it was slightly disingenuous to contemplate what was on 'the old fellow's mind': the contrast between the lifestyle it represented and that of his guest hardly needed further comment.

## V

As well as affecting the miners' conception of occupational and class loyalties, a national strike also forced an awareness of wider geographical identities, calling into question the miners' sense of regional belonging. Historically, Durham had been a dynamic coalfield, and David Byrne, writing about the nineteenth century, has depicted an atmosphere akin to the frontier towns of the Wild West, as the massive extension of industrialization led to an influx of young, single males.[103] Lawson described its feel in 1860: 'Durham was then the great coalfield of the country. It was expanding, and pioneers from all parts of the Kingdom, pushing their way from Wales, Scotland, Ireland and all England, made their human contribution. All the dialects of England and accents of the Kingdom could be heard.'[104] Lawson's words come from his biography of Peter Lee, who was born at Trimdon Grange in 1864. As young parents, Lawson explains, Lee's father and mother sold their furniture twenty-one times in twenty-two years when they moved between Durham and Lancashire alone. One year they moved five times.[105] Lawson's childhood was more settled and his family moved only once, but the 'social melting-pot' of the late-nineteenth-century Durham coalfield remained a theme in his own autobiography.[106]

By 1926, the coalfield had settled down. The 1921 census reveals a population that was overwhelmingly English: of 1,474,143 Durham

final reports were produced. Citing lack of unanimity, the Lloyd George government therefore rejected the creation of a nationalized industry, despite earlier promises to abide by the recommendations of the report. The decision led to bitter charges of 'betrayal' by the miners.

[103] D. Byrne, 'Immigrants and the Formation of the North Eastern Industrial Working Class', *North East Labour History Bulletin*, 30 (1996), 29–36. See also Robert Lee's description of ' "communities" so fractured and fissile that census returns frequently reveal some of them to have contained not a single native-born head of household'. R. Lee, *The Church of England and the Durham Coalfield, 1810–1926: Clergymen, Capitalists and Colliers* (Woodbridge, 2007), 203.

[104] Lawson, *Peter Lee*, 14.

[105] Ibid., 19.

[106] J. Lawson, *A Man's Life* (rev. 2nd edn., 1944), 36.

inhabitants who recorded their country of birth, 96.7 per cent had been born in England, 1.5 per cent in Scotland, 0.8 per cent in Ireland, 0.3 per cent in Wales, and 0.6 per cent elsewhere.[107] This fails to take into account second- or third-generation immigrants, and the handing down of a sense of cultural difference may have been significant amongst Irish families in particular. Trevor Griffiths has noted the longevity of such identities in Lancashire.[108] However, the Irish are rarely, if ever, referred to in oral or written memoirs, newspaper accounts, or union records, and it seems that religious and ethnic differences failed to generate the social and political divisions in Durham that they did elsewhere. Indeed, the lack of such records despite the presence of a small but significant Catholic presence in the coalfield arguably demonstrates the marginality of a distinctive Irish (or at least Irish-Catholic) identity.[109]

Census evidence suggests that immigration even from elsewhere in England was also increasingly limited. The 1921 census dropped the question regarding county of birth from its enquiries, but, ten years earlier, of all those enumerated in County Durham and its associated county boroughs, 78 per cent had also been born there, compared to 75 per cent in 1901, 71 per cent in 1891, and 68 per cent in 1881, revealing a gradual but consistent stabilization of the population.[110] Further evidence supports this impression for the 1920s. Figures submitted to the Samuel Commission counted the number of vacancies (in any industry) registered by employment exchanges for the eighteen months ending 7 September 1925. Of 676 vacancies in Durham filled in such a way, only 2.8 per cent of those engaged were applicants from other districts. Across Britain, the comparative figure was 18.8 per cent.[111] Commission evidence also revealed a remarkably stable mining workforce. Not only did few men arrive from outside the region to enter the pits, but turnover was small. In 1924, new recruits to Durham collieries made up 12.3 per cent of the total number of men working underground in the industry. This was the lowest figure for any coalfield except Bristol (where fewer than 1,000 workers were covered by the returns), and compared to a national percentage of 27.9 per cent. In South Wales and Monmouth, also a depressed coalfield,

107 *Census, 1921*, ii. 97.

108 T. Griffiths, *The Lancashire Working Classes* c.*1880–1930* (Oxford, 2001), 275–6.

109 For estimates of the Catholic population in Durham see the introduction to Chap. 4.

110 *Parl. Papers, 1913*, lxxviii (1), 1911 Census, 15–19; *Parl. Papers, 1902*, cxviii (673), 1901 Census, 78; *Parl. Papers, 1893–4*, cvi (1), 1891 Census, 480; *Parl. Papers, 1883*, lxxx (1), 1881 Census, 454.

111 LSE, Beveridge Coal Commission, xii, MF archives 276.

new recruits constituted 29.6 per cent of the mining workforce in 1924.[112]

The miners' leaders were also strongly rooted in their region. Of the six highest officials of the DMA and the five Durham miners' MPs in 1926, all but three had been born in Durham.[113] At the level of lodge leadership, many men boasted a similarly long association with the area. From September to December 1926, the *Durham Chronicle* ran a series of columns featuring the chairmen, secretaries, and treasurers of local lodges. Thirty mentioned their place of birth, of whom twenty-one were Durham-born compared to nine born elsewhere. At least three of those nine had arrived in Durham as very small children with their parents.[114] At the end of May 1926, the DMA President, James Robson, addressed the MFGB conference and admitted that there was a strong case to be made in support of simply closing down all uneconomic collieries. But, he added:

> if that is so it means a very large number of men will migrate from other coalfields. We are as old as anyone. We have a large number of old collieries, and [their closure] would mean a large number of men have to seek work. We have these men in Durham and Northumberland, reared for generations within three miles, whose great grandfathers lie buried in the local churchyard; families have come down hundreds of years who have sacred associations who would be transferred to Scotland or anywhere else. Just think of such a state of affairs.[115]

Such feelings were not confined to political rhetoric. Ten years later, John Newsom asked a schoolteacher whether there was much objection to young people leaving the village. 'Quite a bit,' came the reply, 'You see the family spirit is very strong in Durham . . . If a boy or girl goes south, to London or the Midlands, it's a big wrench and the parents don't know when they'll see them again.'[116]

Occasionally a wider awareness might be fostered by the visits of relatives from overseas. Peter Lee was again unusual amongst the DMA

[112] Royal Commission on the Coal Industry, iii. 213. The returns from which the Durham figures were calculated covered 98% of the coalfield; those of Great Britain 79%, and those of South Wales and Monmouth 88%.

[113] Joe Batey had been born in Northumberland; Jack Lawson and Joshua Ritson in Cumbria.

[114] *DC*, 10 Apr. – 25 Dec. 1926.

[115] MFGB, *Annual Volume of Proceedings for 1926* (1927), 453. James Robson (1860–1934) had worked in the Durham coalfield since the age of 10. He was President of the DMA 1917–34 and Treasurer of the MFGB 1918–21.

[116] J. Newsom, *Out of the Pit: A Challenge to the Comfortable* (Oxford, 1936), 39.

hierarchy in having spent time working in coalfields in South Africa and the United States as a young man. But, if this was rare, it was not unknown. Lee had a wife and child to bring him back to Durham, but others made their new country their home. Mary Wade remembered that during her childhood in neighbouring Northumberland her family would occasionally have visitors 'who, prior to the outbreak of war had emigrated to America or the British Empire countries. This was their chance to show off, talk of how well they had done "out there" or maybe thank God that they had got away in time. Few, if any, wanted to return home.'[117] In May 1926, the *Chester-le-Street Chronicle* published an interview with a man who had emigrated to Canada twenty-one years earlier and had briefly returned home. He described a country where miners could earn £1 a day. 'So many miners go to work in their automobiles that the Company has had to build a garage to accommodate them,' he boasted, 'It is the recognized thing for a household to be on the telephone, and practically every home is equipped with electricity and central heating.'[118]

The return of the newly prosperous to their old homes, forcing comparisons with friends and relatives, is a common phenomenon across time and across cultures: the brother who took a risk, moved away, and made good became the personification of the what-might-have-been that haunted Willy Loman in Arthur Miller's *Death of a Salesman*.[119] But, such stories had the potential to be powerful, particularly during a seven-month lockout. It may be that there was a generational gap in attitudes, with parents reluctant to see children leave the area, as documented above, but whose children themselves were less concerned about staying. In fact, during the strike the local press consistently reported the movement of young miners from Durham. 'From various parts of the Durham coalfield during recent months there has been an exodus of the cream of young miners to the Colonies,' declared the *Durham County Advertiser* in July. It continued by citing the father of a young miner about to leave for Australia. He told the newspaper that every man in the village would leave tomorrow if only he had had the means.[120] The period of depression heralded by the lockout may also have affected attitudes. Between the 1921 census and that of 1931, nearly 150,000 people migrated from the county, representing

[117] M. Wade, *To the Miner Born* (Stocksfield, 1984), 2. [118] *CC*, 28 May 1926.
[119] A. Miller, *Death of a Salesman* (first performed 1949) (New York, 1949).
[120] *DCA*, 30 July 1926.

10 per cent of the population; one government study estimated that over 129,000 of these had moved after 1926.[121]

## VI

If the degree of movement over county boundaries was limited before the lockout, there had always been a much higher degree of geographical movement within the county itself. Sid Chaplin was born in 1918 and remembered his childhood as a time of constant movement around Durham: before he was 12 he had lived in seven villages and eleven pit rows.[122] For such children, the movement of their fathers was reflected in their attendance at a succession of schools, and very few remained at one school for the whole of their educational life. The admissions' register of the girls' department of Seaham Harbour's Church of England school reveals that of the ninety-seven children in attendance born between 1 April 1912 and 31 March 1913 (therefore reaching their 14th birthday during the 1926–7 school year), most arrived from Seaham Harbour's infants' school at the age of 7. However, of those who did, only ten then remained at the same school until the age of 14. The rest changed schools to attend others in the immediate area, several moved elsewhere in the North East, and a couple left the area altogether.[123]

Miners who worked at the same pit together did not necessarily live in the same village. The lodges of the DMA recruited their members according to the colliery at which they worked, not their place of residence, and membership could therefore be scattered across a wide geographical area. When the North Hetton welfare committee decided to fund a new institute in the 1920s for the miners of Hazard Colliery, for example, it was agreed that instead of one large building it would be better to erect two smaller ones in each of the nearby villages of Moorsley and Rainton, owing to its workforce being 'widely separated'.[124] The miners at Hazard were not exceptional and a clue to the distances that some miners travelled between home and place of work is provided by the answers to lodge questionnaires returned to the DMA in 1929. The union was considering approaching the Durham Coal Owners' Association (DCOA)

[121] *Parl. Papers, 1933–4*, xiii (313), Reports of Investigations into the Industrial Conditions in Certain Depressed Areas, Part II: Durham and Tyneside (1934), 74–5.

[122] S. Chaplin, *A Tree with Rosy Apples* (Newcastle-upon-Tyne, 1972), 114.

[123] DRO, E/NE/F6, admissions register, Seaham Harbour Church of England School (girls' dept).

[124] *DC*, 2 Oct. 1926.

**Table 1.3** *Distances travelled by miners to work, 1929*

| Colliery | Total lodge membership | Number travelling two or more miles to pit | Percentage of workforce travelling two or more miles to pit |
|---|---|---|---|
| Blackhall | 1,520 | 330 | 21.7% |
| Bearpark | 1,241 | 250 | 20.1% |
| Brandon | 836 | 590 | 70.6% |
| Eden | 869 | 33 | 3.8% |
| Hamsterley | 442 | 32 | 7.2% |
| Lambton | 1,447 | 170 | 11.7% |
| Sacriston | 830 | 10 | 1.2% |
| Murton | 2,700 | 110 | 4.1% |
| Usworth | 1,547 | 358 | 23.1% |

*Source*: DRO, D/DMA (Acc: 2157 (D) ): box 294.

to request the provision of transport for men who had to travel far to their work, but first had to establish for itself the distances involved. The few surviving returns reveal a picture which hardly matches the stereotype of each pit having its own colliery village within which all members shared lives and leisure (see Table 1.3). Generally, the bigger pits employed a significant proportion of men who did not live in the immediate area, and at Usworth, for example, the lodge secretary gave the particulars of men who travelled in from Felling, Pelaw, Heworth, Wrekenton, Jarrow, Hebburn, Birtley, Castletown, Sunderland, South Shields, Blaydon, Houghton, Fatfield, and Gateshead. Most travelled in by bus, although a significant minority walked or cycled, presumably to save costs.[125]

Even if miners and their families did move around the county, they might not imagine the various villages and towns as part of one cohesive region. In 1915, when Edward Farbridge moved from Hamsterley Colliery to Stanley at the age of 16, he was shocked by the difference between the small, isolated pit village and the more-populated urban district, later describing 'a feeling of being thrust into a land of bricks and mortar enveloped with a pungent smell of coal-fired soot and smoke, with people seemingly everywhere'.[126] Thirty years later, the contrast between different parts of the county remained. The civil

[125] The Brandon results are somewhat of an anomaly as Brandon Colliery was situated over two miles away from Brandon village, and most of its 'commuters' travelled in from there.

[126] E. Farbridge, *Recollections of Stanley* (DRO typescript, 1973).

servant in Mark Benney's semi-fictionalized wartime account of the coalfield, *Charity Main*, commented on the striking difference between the 'remote and self-contained . . . stable, hard-working' communities in the old pits of the west, and the situation on the coast, where 'the miners' cottages stretched in long unbroken miles along the roads linking the collieries . . . [Nothing could] mitigate the unvarying meanness and joylessness of the view'.[127]

During the strike, local differences were exacerbated owing to variations in the degree of deprivation. Although the strike meant a universal cessation of labour, some colliery villages had already suffered during the early years of the 1920s owing to a greater incidence of unemployment or strikes, and their families had both fewer savings to draw upon and a lower standard of physical and mental health to start with. Chopwell Colliery had been on strike for nearly a year when the lockout began owing to a localized dispute over pay. Meanwhile, the pits of the eastern coast had enjoyed relative prosperity. The continued expansion of Londonderry's Seaham Colliery had been the single main development in the Durham coalfield in 1925.[128] Intra-regional divisions were further emphasized by differences in relief rates. Despite centralized guidelines for scales of outdoor relief, boards of guardians differed in their generosity.[129] In May, the *Durham Chronicle* reported on the sense of dissatisfaction in Coxhoe, where one side of the main street fell under the jurisdiction of the Durham union and the other side under that of Sedgefield union. The residents relieved by Sedgefield were receiving a much higher rate of benefit.[130]

An attachment to the individual locality could therefore be more important than an attachment to the county as a whole. Sid Chaplin, the child who had lived in seven villages and eleven pit rows, described his memories of the 1920s:

> The village was the real concrete thing and it meant something and one clung to it . . . One was hardly aware of the villages next door. There was a tremendous sense of insularity . . . If any Byers Green boys ventured into our village we'd stone them till they took to their heels. And likewise if we went to the pictures in Byers Green we'd more likely than not be stoned on the way through . . . We used to sneak away round the hedges at the back of the village and nip into the

[127] M. Benney, *Charity Main: A Coalfield Chronicle* (1946), 61, 82.

[128] TNA:PRO, POWE7/59, Annual Report of the Secretary of Mines, 1925, 15.

[129] Dependants of strikers were permitted to apply for poor relief, although strikers themselves were not.

[130] *DC*, 22 May 1926.

pictures as quickly as possible and get lost among the audience of about two or three hundred people—tremendously cosmopolitan.[131]

This degree of geographical insularity may have been particularly keenly felt in children, who were likely to be more focused on their own individual experiences and for whom childhood games could articulate local rivalries unrestrained. Other groups were susceptible as well, and mobility was often more restricted for women, especially when money was short. During the deprived years of the 1930s, John Newsom interviewed a miner's wife who told him that she had married at 18 and had never been out of County Durham in her life. In the previous four years she had not been more than five miles from the village.[132] But, the miners themselves also might be discouraged from identification with a wider community. In some collieries, for example, tight rules governed the employment of men arriving from elsewhere. In 1921, an agreement was made between the owners and workmen of Chopwell and Whittonstall Collieries regarding a man's advance from putting to hewing. Hewing was the most prestigious job of the underground worker, and was typically the natural progression up the pit hierarchy for a putter. Under the new regulations, such a promotion would only be given to a man who had first proved himself at Chopwell pit, even if he had long experience of putting elsewhere. Only once a man had put specifically for a Chopwell seam, it was stated, would a 'stranger's' claim for hewing be considered.[133]

A local sense of identity could be intensified through sport. The region possessed a vibrant playing culture, and in the 1925–6 season Durham Football Association was comprised of 703 clubs, forty-nine leagues, and sixty-eight charity competitions.[134] 'Every town or village in the county boasts its football club or clubs,' commented the *Stanley News and Consett Chronicle*, adding that 'in many of the smaller villages there is an intensity of local patriotism by no means confined to the menfolk'.[135] These clubs inevitably felt the effects of the strike, and although newspapers reported on large crowds at local football matches, it was also noted that the gate receipts were not correspondingly large, as many spectators took up vantage points outside the grounds.[136] Some

131 S. Chaplin, 'Durham Mining Villages', in M. Bulmer (ed.), *Mining and Social Change* (1978), 62.

132 Newsom, *Out of the Pit*, 42.

133 DRO, D/X1005/29, Memorandum of agreement between owners and workmen of Chopwell and Whittonstall collieries, 17 Nov. 1921.

134 *DCA*, 4 June 1926.

135 *SNCC*, 27 May 1926.

136 *DC*, 22 May 1926.

teams folded as their miner players were unable to continue to pay their subscriptions or travel to fixtures; Langley Park Football Club were forced to withdraw from the season after being unable to afford a ball.[137] However, efforts were made to maintain such activities during the strike, and many sporting clubs relaxed their rules. Chester-le-Street Amateur Swimming Club, for example, decided to waive its fee to those out of employment for the 1926 season.[138] In addition to official sporting activities, unofficial sports were organized by men with long, empty days to fill. It is possible that such local teams became more important during the lockout, with miners less able to travel to see bigger teams in action. One miner remembered that at Stanley 'games of football and cricket were contested almost every day and there was no lack of spectators and partisanship ran high according to the locality in which they lived'.[139] Another spoke of the cricket competitions organized at Murton—'but you could only play for the street you lived in'.[140]

Such local divisions, which centred on streets rather than villages, were closely tied up with concepts of respectability. One old man interviewed by Mark Hudson in Horden in the 1990s remembered a strict hierarchy. It also hints at ethnic differences, one of the very few comments in oral or written testimony to do so:

> Where we lived in Oak Terrace was considered one of the more desirable places to live . . . Further down it got a bit rougher. Our house faced the end of Sixth Street, and we were absolutely forbidden by my father to go beyond Fifth Street. The very bottom, First, Second and Third Streets were known as China Town. That was where the really rough families, the real hard men, lived.[141]

The tension between the 'respectables' and the 'roughs' (and all the varying grades in between) that could fracture a working-class community has been a recurring theme in historical scholarship, with reference both to Durham miners and other regions and groups.[142] There is little to add that is new here. What is worth comment, however, is the effect that the strike had on such divisions.

137 *DC*, 18 Sept. 1926.

138 *CC*, 16 July 1926.

139 Farbridge, *Recollections*. See also Chap. 5, Sect. V, for further comment on sport as a means of social mobility.

140 GCLOT, ii (Mr H.).

141 M. Hudson, *Coming Back Brockens* (1994), 120.

142 See, for example, R. Moore, *Pitmen, Preachers and Politics: The Effects of Methodism in a Durham Mining Community* (Cambridge, 1974); Griffiths, *Lancashire Working Classes*; P. Johnson, *Saving and Spending: The Working-Class Economy in Britain 1870–1939* (Oxford, 1985).

In his study of the financial habits of the working classes in Britain before 1939, Paul Johnson argued that although the need to use credit arrangements, pawnshops, and hire-purchase affected a family's status within the working-class community, the shame associated with such practices diminished in the event of a strike, lockout, or depression because almost everyone was forced to do the same.[143] Yet, in his study of Lumphinnans in West Fife, Stuart Macintyre suggested that the effect of widespread poverty was more subtle. He agreed that the increasing scale of distress altered attitudes towards the unemployed, because it became impossible to stigmatize those out of work when they constituted a majority of the inhabitants. But, he argued that while 'general poverty was manageable . . . individual poverty could be a source of shame,' and noted that when the Scottish miners' leader Abe Moffat and his wife moved house in the 1920s, they left at night so that their meagre possessions could not be surveyed by the neighbours.[144]

In Durham, too, it is clear that even during a lockout in which to remain on strike and forgo earnings was the 'respectable' position to take, issues of respectability amongst the strikers themselves did not cease to matter. Recourse to charitable handouts was still resisted by some, not just because of the shame that might make them hesitant to take advantage of such provision, but also because of their reluctance to associate with others who might attend. One old woman interviewed in the 1970s remembered visiting the soup kitchens where a bathtub full of broth had been prepared; whether accidentally, or on purpose, a girl had fallen in. She couldn't recall if the children had then been expected to eat the broth, but remembered going home in tears. Her grandmother was outraged: 'If we have to eat green grass, the bairn is not going back to those soup kitchens anymore.' She never did.[145]

However, if respectability continued to matter, the widespread hardship did mute its effects. This is particularly noticeable amongst evidence gathered from those who were children at the time, and may be indicative of a generational divide: children, then as now, were perhaps keener than their parents to conform to the dominant culture of their peers. For one girl, therefore, the fact that her parents had savings in the Co-op was a source of deep regret as it rendered her ineligible for the free boots

[143] Johnson, *Saving and Spending*, 185.

[144] S. Macintyre, *Little Moscows: Community and Working-Class Militancy in Interwar Britain* (1980), 120–1.

[145] GCLOT, i (A.C.).

distributed in her area.[146] Another child already had some good boots, but most of his friends played barefoot. As an adult he remembered that 'when I used to go out to play with my mates, I used to hide my boots in the toilets so I wasn't ostracized from my friends . . . otherwise you were classed as a snob, and if you wanted to make friends with the rest of the boys, you just had to hide your boots and be the same as them'.[147] But it may be that even children were aware of a status hierarchy. Two years after the lockout, a school inspector observed that many girls in the area wore light, inadequate shoes, usually necessitated by the inability of their parents to purchase boots. He regretted that 'out of this element of necessity a fashion is growing up in the girls' schools for wearing light shoes even though they possess good boots at home'.[148] It is possible that the wearing of poorer quality shoes acted as an insurance against a potential loss of status in the event of future hardship: if children wore light shoes even if they did not have to, then the change from not *wanting* to wear boots, to *having* to wear plimsolls could not be monitored by others.

## VII

Against the local loyalties and rivalries of county, village, neighbourhood, and street, competed the ties of national belonging. That the strikers might have any national loyalties at all was fiercely denied by their opponents, keen to discredit them in the eyes of the wider public. On 4 May, one Northern newspaper praised the character of those who had stood up to the strikers, drawing upon the imagery of war: 'Our people faced the situation yesterday in the spirit that is typical of our island race,' it declared, '[that] spirit . . . with which our nation took up the challenge on a summer evening in August 1914'.[149] Such comments largely ceased with the calling off of the general strike, but some continued to view the ongoing miners' struggle in the same vein. One Conservative peer wondered why poor relief continued to be permitted to families of the locked-out miners: 'We did not feed the Germans.'[150] It would take a later leader of a later struggle to coin the phrase, but to some in 1926, the miners had already become the enemy within.

146 GCLOT, i (Mrs P.).
147 Ibid., iii (L.A.).
148 TNA:PRO, ED50/7, 'Investigation into the Physique of School Children in Durham Coal Mining Areas', Apr. 1928.
149 *North Mail*, 4 May 1926.
150 Lord Hunsdon, quoted in *HPD*(C), 197, c. 1501.

Fears that a miner's ties were to his class rather than his country were exacerbated as statements of support and financial backing began to arrive from trade union organizations abroad. During the lockout, several MFGB delegations were dispatched to Europe, Russia, and America to raise funds. By the end of the dispute, over £1.25 million had been donated to the MFGB's relief fund by the workers of other countries. The vast majority of this came from the All-Russian Trade Union Council, but other substantial donations were also received from Germany, Holland, and the United States.[151] In June, *The Miner* printed a telegram received from Friedrich Husemann, the German miners' leader, promising that the German miners would 'render every help in their power'.[152] Admittedly, international assistance of other kinds was harder to secure. Attempts to persuade the International Federation of Miners to implement a more powerful sympathetic strike were unsuccessful, while Samuel Cohn has noted that a one-day French strike called in solidarity received 'only tepid backing'.[153] The MFGB leaders did at least secure a promise that efforts would be made to limit the amount of foreign coal arriving in Britain, but, imported coal, reduced to 600,000 tons per week in June, was back to four million tons by October, compared to usual pre-stoppage imports of five million.

Within the historiography of the British coal industry, it is the South Wales coalfield that has been credited with a reputation for international solidarity, notably in the work of Hywel Francis and Dai Smith. They describe the SWMF as characterized by a 'proletarian internationalism', influenced by the pacifism and radical Nonconformity present in Wales since the eighteenth century, given its distinctive character by the waves of immigrants into the coalfield in the early twentieth century, and manifest through support for Irish independence, the young Soviet Union, and, most clearly, for the Spanish republicans.[154] More recently, Lewis Mates has argued that Durham also deserves credit: that although not many more than a dozen miners volunteered to fight in Spain compared to around a hundred Welsh miners, in financial terms, the money donated by the DMA 'is likely to have been close to the total given by

[151] Ibid., 196, c. 2471; R. Page Arnot, *The Miners; Years of Struggle: A History of the Miners' Federation of Great Britain (from 1910 onwards)* (1953), 470, 472.

[152] *Miner*, 26 June 1926.

[153] S. Cohn, *When Strikes Make Sense—and Why: Lessons from Third Republic French Coal Miners* (New York, 1993), 215.

[154] Francis and Smith, *The Fed*, esp. 10–13, 52–4; H. Francis, *Miners Against Fascism: Wales and the Spanish Civil War* (1984), esp. 29–41.

the SWMF and it *may* even have exceeded it'.[155] In 1926, of course, the miners were seeking international solidarity, rather than providing it, but there is some evidence that even amidst the disruption of the lockout, men and women retained a concern with wider international issues. In June, for example, Brancepeth colliery band lent its backing to a public meeting held at Willington in support of the principles of the League of Nations Union; a similar gathering was held in Ryton in October and was reportedly well attended.[156]

However, whether expressing gratitude or criticism, the evidence tends towards the conclusion that the miners' conception of the geography of the strike was a limited one. None of the oral or written reminiscences of the strike that I have come across make any reference to aid given from abroad, with the exception of that provided by Russia, which is noted for different reasons, and still only occasionally. In contrast, men and women often give fulsome accounts describing the kindnesses of shopkeepers and other local neighbours. Nor have I come across criticism that international assistance was not more generous. It is significant that, in his official history of the MFGB, Robin Page Arnot was later unwilling to blame foreign workers for the continued importation of coal during 1926. He suggested that it had been very difficult for foreign workers to discover the destination of coal cargoes and preferred to suggest that the real blame lay with the British leaders of the seamen and railwaymen's unions, and their adamant refusal to implement an embargo on the handling of foreign coal.[157]

In the House of Commons, too, the miners and their supporters were more likely to employ a patriotic rhetoric than to praise international class solidarity. The Labour MP David Kirkwood was not from a mining background himself but often became passionately involved in the miners' debates. During the general strike, this veteran of 'Red Clydeside' claimed in the House of Commons that:

> We [the Labour movement] are standing for something upon which the very fate of the British Empire depends. We are standing here today defending the British Empire [Hon. Members: Oh!] Standing here today as we have stood before, not to smash the British Empire. The ruling class of Britain at the moment are doing their utmost to smash the British Empire. It is impossible

155 L. Mates, 'Durham and South Wales Miners and the Spanish Civil War', *Twentieth Century British History*, 17 (2006), 389. Original emphasis.

156 *DC*, 7 Aug. 1926; *BC*, 16 Oct. 1926.

157 Arnot, *Years of Struggle*, 499–500.

for us to continue a British Empire if our race submit willingly to conditions such as these.[158]

The following month, one of his mining counterparts, Duncan Graham, stood up to address the House. Before launching into his attack of the government's position, he set out his credentials as a British citizen: 'I am a miner, a native of this country, with no foreign blood in my veins. In that respect I do not know either Germans or Russians. I was born the son of a miner, and since 1878 I have seen every strike and lockout that has taken place in Scotland.'[159]

Such rhetoric was a judicious line of argument to employ against a government fond of accusing the strikers of national betrayal. But the speeches made in the Commons during the lockout reveal a genuine sense of national pride amongst the Labour leaders. In June, William Whiteley, the miners' MP for Blaydon, worried that an increase in hours would persuade European employers to do likewise, 'bring[ing] down the general standard of leisure in the whole of the mining community right throughout Europe'. They would have no choice, he explained with some pride, for the output of the British miner per man was already higher than that of his continental counterpart, meaning European collieries would have to work longer hours in order to remain competitive.[160] When the debate continued the following day, the Labour cotton MP Tom Shaw, a multilinguist who was heavily involved in the international trade union movement, was still more defensive of the status and the rights of a specifically British mining community:

imagine the humiliation which it is to our miners to be asked to work longer hours than the Germans, the French, or the Belgians. Can Hon. Members imagine what that must be to the British miner, proud as he has been all his working life that, whatever were his conditions, at any rate he worked shorter hours and more efficiently during those shorter hours than any other miner in Europe. Now the British miner, proud of his craft, his strength and his skill, is to be told that he must work longer than the German, the Belgian or the Frenchman.[161]

158 *HPD*(C), 195, c. 478.

159 Ibid., 197, c. 455.

160 Ibid., 197, cc. 886–7. William Whiteley (1881–1955) initially worked underground before becoming a clerk in the DMA offices. He followed in the footsteps of his father when he became involved in union politics, and was elected agent to the DMA in 1912. He was Labour MP for Blaydon 1922–31 and again 1935–55. He held a variety of junior ministerial posts before being appointed Parliamentary Secretary to the Treasury and Chief Labour Whip in Attlee's 1945 government.

161 *HPD*(C), 197, c. 1245.

His words were echoed again and again by other Labour MPs. 'Our miners are better than any other miners in any European country', stated G. H. Hall, while his fellow miner David Grenfell made the same claim, explaining simply, 'we work harder'.[162] Such sentiments, expressed in Parliament, are obviously easier to document than rank-and-file opinion in the coalfield itself. However, one indication that the Durham miners were receptive to patriotic callings—or at least that they had been—can be seen from their response to the outbreak of the First World War. By mid-1915, over 230,000 British miners had enlisted, representing around one-quarter of the pre-war workforce. Among them the Durham miners were well-represented, and 24.9 per cent of the coalfield's workforce joined the services between August 1914 and August 1915. In England, only neighbouring Northumberland and the small coalfields of Kent and North Staffordshire could boast a higher proportion.[163] Just over a decade later the shared memory of the war continued to contribute towards a national awareness, with its dissemination of national rather than class symbols. At a Durham miners' demonstration in August 1926, for example, one man noticed that a red flag had been hoisted from the local memorial institute. He wrote to his local newspaper in disgust: 'This institute was built as a memorial to the men of Tanfield Lea who gave their lives for their country, serving under the Union Jack. Perhaps the men of Tanfield Lea responsible for the displaying of the Red Flag have forgotten this.'[164]

But, it is possible that those men who had been responsible for flying the red flag had chosen its position with care: by doing so they appropriated the memory of the war for themselves. Certainly, the comparison between the sacrifices made by the miners during the war and the position they currently occupied was alluded to time and time again by Labour MPs during parliamentary debates.[165] Not only had miners fought and died, but it was argued that an invaluable contribution had been made even by those who had stayed at home during 1914–18, when the MFGB had agreed to the price-fixing of coal. A decade later, Robert Smillie, angry at the government's refusal to renew the 1925

[162] *HPD*(C), 195, c. 137; 196, c. 698.

[163] The English coalfield average stood at 22.9% and the UK coalfield average at 23.3%. See J. M. Winter, *The Great War and the British People* (1986), 35; *Parl. Papers, 1914–16*, xxviii (307), Second General Report of the Departmental Committee Appointed to Inquire into the Conditions Prevailing in the Coal Mining Industry due to the War (1916), 14–31.

[164] *SNCC*, 26 Aug. 1926.

[165] See, for example, *HPD*(C), 197, cc. 938–9.

subsidy, argued that even if it had been renewed for years, 'the nation would not have repaid the miners for the subsidy which they gave to the nation in 1916 and 1917'.[166] As another Labour MP claimed, some miners in his Lancashire constituency had been asking whether it had been worth winning the war at all: 'Would it not have been better digging reparation coal rather than accept the conditions imposed upon us under this Bill [sanctioning an eight-hour day]?'[167] It was a patriotic legacy that would remain attached to the miners. 'It breaks my heart to see what is happening in our country today,' Harold Macmillan told the House of Lords in November 1984, 'A terrible strike is being carried on by the best men in the world. They beat the Kaiser's army and they beat Hitler's army. They never gave in.'[168]

In 1926, such feelings were not confined to parliamentary debates. In the coalfields the memory of the war interplayed with the strike: special memorial services were held in July, for example, to mark the tenth anniversary of the Battle of the Somme. Four months later, in the lead-up to Armistice Day, a feeling of betrayal surfaced. The *Durham County Advertiser* published one letter received from a 'Miner's Wife'. The miners had been the heroes of England during the war, she wrote, 'Now they are the scum of the earth'.[169] Just over a week later another miner's wife penned a similar letter to the organizers of Poppy Day at Blackhill:

> I hope you think you are doing good with having a poppy day for all ex-servicemen. 'What the Hell' does you or any other rotten patriot care about any such men. My husband is an ex-serviceman of four years fighting, and the same bloody dirty rotten government are starving us today . . . I wish every poppy you sell would mean a bullet through Stanley Baldwin's Body, you class of Bloody Traitors, but our day will come and all your bloody heads will be marked.[170]

This letter was also published in the local press, and drew a rush of angry responses. One came from a miner in the Swalwell branch of the British Legion. He claimed that well over three-quarters of his branch members were miners, 'and we protest most emphatically against so disgusting an epistle'.[171] Indeed, the letters of the two miners' wives seem to be exceptional, for there is little other suggestion that memories

166 Ibid., 196, cc. 2225–6.

167 Ibid., 197, c. 1211.

168 *Hansard Parliamentary Debates* (Lords), fifth ser., 457, c. 240.

169 *DCA*, 5 Nov. 1926.

170 *BC*, 13 Nov. 1926. Grammar as in original.

171 Ibid., 20 Nov. 1926.

of the war became more bitter during the lockout. In the Consett and Blackhill districts the funds raised by the sale of poppies in November 1926 were nearly double that of the previous year; in the Brandon area, the amount raised exceeded that of 1925 by £8.[172] On Armistice Day itself, Eppleton church reached its capacity of 1,000 and some could not gain admittance to hear the service, which concluded with the National Anthem.[173] Even at Birtley, where preparations for Armistice Day had engendered some controversy, a gathering of well over 2,000 people attended a service at the war memorial, where a Union Flag had been draped over a raised platform.[174] Less than a month earlier, Birtley parish council had taken advantage of the absence of some of its more moderate members to force through a ban on any war memorial service because of its role in fostering militarism. The ban had been rescinded a week later.[175]

Although memorial services were inevitably surrounded with patriotic imagery, this does not necessarily mean that national symbols were uppermost in the minds of those who attended them. As Adrian Gregory has discussed, the two minutes of silence around which Armistice Day centred were an intensely private as well as a public ritual, and the legacy of the war was contested by both right and left.[176] A. J. Cook, the man whose name would come to be most associated with the lockout, had actively opposed the war and spent time in prison under the Defence of the Realm Act. However, in many places in the Durham coalfield the symbols of the war became integrated with the symbols of the mining community. At least three of the Durham miners' banners of the 1920s featured the local village war memorial.[177] In 1926, several Armistice Day processions departed from miners' halls and were led by colliery bands,[178] while at Easington the collection taken during the service by the British Legion was handed straight over to the local boot fund for miners' children.[179] Meanwhile, many of the Durham leaders were in London, attending the special MFGB conference called to discuss the latest progress of the lockout. Opening the day's proceedings on 11 November, Herbert Smith announced the intention to observe

[172] *BC* and *DC*, 20 Nov. 1926.

[173] *DC*, 20 Nov. 1926.

[174] *DCA*, 12 Nov. 1926.

[175] *DC*, 16 and 23 Oct. 1926.

[176] A. Gregory, *The Silence of Memory: Armistice Day, 1919–1946* (Oxford, 1994).

[177] Namely Elemore, Murton, and Tursdale. See N. Emery, *Banners of the Durham Coalfield* (Stroud, 1998), 78–9.

[178] For example, at Silksworth and Sherburn Hill. See *DC*, 16 Oct. 1926; *DCA*, 12 Nov. 1926.

[179] *DCA*, 12 Nov. 1926.

the silence: 'I expect that there are no conscientious objectors. Whilst looking after the living, we have equally as much regard for the dead.'[180]

Perhaps the most striking example of the way in which identities could overlap is provided by another complaint sent to the local press after the publication of the letter from the miner's wife cited above. It came from a representative of the Flint Hill branch of the British Legion, all of whose members (the letter claimed) were miners. But, its author did not feel that the active commemoration of the war was at all contradictory to political support for the strike. An ex-serviceman and a miner himself, he recalled his mates who had died in France. 'Would that they had all been on strike now,' he remarked poignantly.[181]

## VIII

During the months of the lockout, tensions between the local and the national were particularly apparent in the relationship between the DMA and the MFGB. Durham had come late to the national union, permanently joining the MFGB only in 1908, after earlier affiliations in the 1890s had collapsed in acrimony over the eight hours question. Yet a historically ambivalent attitude to the national union vied with a proud tradition of regional solidarity. Durham had been one of the first coalfields to attempt systematic union organization in the 1830s and it was to this earlier tradition that the national leaders appealed in 1926. In July, A. J. Cook addressed the unofficial gala held at Burnhope.[182] 'I want to remind the miners of Durham they have a record to maintain,' he called out to the crowds, holding up a copy of Richard Fynes' history of the Northumberland and Durham miners' unions, and urging every man to read it. Fynes had been a Durham miner himself and his book recounted the story of the North-Eastern miners and their struggles to create an effective union, claiming that a collective consciousness could be detected amongst them from as early as the 1660s. 'The book told of the old struggles of the miners in the Northern Counties,' Cook explained, 'In those days the miners of Durham and Northumberland made short work of blacklegs, and never hesitated to put up a fight against greater odds than had to be faced today.'[183] The low rate of blacklegging

180 MFGB, *Proceedings for 1926*, 943.

181 *BC*, 27 Nov. 1926.

182 For discussion of the gala, see Chap. 6, Sect. III.

183 *SNCC*, 22 July 1926; R. Fynes, *The Miners of Northumberland and Durham: A History of their Social and Political Progress* (Sunderland, 1923 edn.; first pub. 1873).

in the coalfield in 1926 provided another reason for Durham to be proud of its solidarity. One delegate boasted to the MFGB conference in October: 'I think that we in Durham can stand for ever.'[184]

A national strike might have been expected to have reversed any regional insularity, as the MFGB embarked on a nationwide propaganda campaign. In June, the union launched its first national paper, *The Miner*, which was published weekly and was selling over 100,000 copies by the end of its first month.[185] Meanwhile, the miners' leaders toured the coalfields ceaselessly as part of a new strategy initiated by Cook: such activities had not been known in the days of Frank Hodges and Robert Smillie, the predecessors of Cook and Smith, even during the strikes of the early 1920s.[186] Indeed, one of the top four officials of the MFGB was already present in Durham: W. P. Richardson, MFGB Treasurer, was also General Secretary of the DMA.[187] As fundraising took place throughout Britain, coalfield groups overlapped. In August, a visiting Welsh miners' choir gave concerts at Willington and South Dene; in September, another group from South Wales sang at Langley Moor, reminding their North-Eastern audiences that this was a national struggle.[188]

Furthermore, one of the fundamental issues over which the strike was fought was the rejection of district settlements. Its lack of rhyming potential made it less memorable to posterity, but on the miners' banners and placards of 1926, the slogan 'No district settlements' was often allied with the more famous one, 'Not a minute on the day, not a penny off the pay.'[189] The fight for a national wage agreement had long been a major aim of the MFGB, and, in 1919, Robert Shirkie, the secretary of the National Federation of Colliery Enginemen and Boilermen argued for such, explaining to the Sankey Commission that the lack of uniformity in the wages paid in different districts had caused considerable trouble in the coal industry.[190] During the lockout itself,

184 MFGB, *Proceedings for 1926*, 891.

185 P. Davies, *A. J. Cook* (Manchester, 1987), 105.

186 J. Lawson, *The Man in the Cap: The Life of Herbert Smith* (1941), 216.

187 W. P. Richardson (1873–1930) was appointed Secretary of Usworth Colliery in 1898, the same pit in which he had started work and where his father had been killed in the explosion of 1885. In 1912 he was elected to the executive of the DMA and he became general secretary in 1924. He was treasurer of the MFGB 1921–30.

188 *DC*, 21 Aug. 1926; *DCA*, 3 Sept. 1926.

189 See, for example, the report of a meeting addressed by Peter Lee. *DCA*, 4 June 1926.

190 *Parl. Papers, 1919*, xi (373), Coal Industry Commission (1919), i, Reports and Minutes, March 1919, q. 9536.

after the Coal Mines Act added minutes to the day, and as it became increasingly likely that pennies were going to be taken from the pay, the miners found that the only issue left to fight for was the promise of a national settlement.

The primary reason for the emphasis on a national settlement was the wide regional variation in the profitability of the coalfields. The historian might expect, therefore, that the strike would have made the Durham miners more grateful to the policy of the MFGB, as Durham was one of the coalfields which stood to benefit most by its success. As an exporting coalfield it had been particularly hard hit by the return to the gold standard in 1925 and both employment and wages had suffered. In the final three months of 1925, 97 per cent of coal cut in Durham was raised at a loss: in 1926, it was therefore in Durham that some of the most savage wage cuts were proposed by the owners seeking a solution to their difficulties, even though the miner's wage in Durham was already lower than the national average (see Tables 1.4 and 1.5).[191]

Such facts support John McIlroy's defence of the traditionally maligned Nottinghamshire miners. He observed that significant problems did not emerge in their coalfield until the end of August, 'and then in the context of a marathon struggle from which many Nottinghamshire miners would derive little future advantage but faced immediate costs'.[192]

However, whether or not the Nottinghamshire miners felt that they were fighting a battle on behalf of other more stricken coalfields, the rhetoric heard in Durham in 1926 was very different. Rather than simply appealing for help, the Durham miners gave the impression that they were fighting for Durham very much on their own terms. This was particularly the case over the hours question. In the early 1920s, Durham's hewers worked days of between six and six and three-quarter hours, a tradition untouched since a district agreement of 1890; their example had been important in convincing the Sankey Commission of the viability of a seven-hour day. Such a history meant that the coal owners' proposals, rather than being seen as the exploitation of a desperate population, were interpreted as an attempt to dislodge the Durham miners from a privileged position. Speaking during the parliamentary debates over the eight hours' legislation, William Whiteley stood up to declare his opposition. 'I come from the North of England,

191 TNA:PRO, LAB27/4, PRO 30/69/1274.

192 J. McIlroy, 'Nottinghamshire', in McIlroy et al., *Industrial Politics*, 212.

**Table 1.4** *Average earnings per shift, January 1922 (figures provided by the Miners' Federation of Great Britain)*

| Coalfield | Wage per shift (skilled coal-getter) | Wage per shift (underground labourer) |
|---|---|---|
| Nottingham | 17*s* 3.76*d* | 13*s* 7.69*d* |
| Derbyshire | 16*s* 3.17*d* | 11*s* 6.5*d* |
| South Yorkshire | 15*s* 2.57*d* | 11*s* 6.5*d* |
| Leicestershire | 14*s* 11.4*d* | 10*s* 8.11*d* |
| Cannock Chase | 14*s* 11.4*d* | 11*s* 2.31*d* |
| East Yorkshire | 14*s* 8.28*d* | 11*s* 2.31*d* |
| Warwickshire | 14*s* 8.28*d* | 11*s* 9.21*d* |
| South Derbyshire | 13*s* 7.69*d* | 9*s* 5.22*d* |
| West Yorkshire | 13*s* 7.69*d* | 10*s* 10.11*d* |
| Lancashire and Cheshire | 12*s* 11.0*d* | 9*s* 0.51*d* |
| North Staffordshire | 11*s* 1.35*d* | 9*s* 7.35*d* |
| Kent | 10*s* 5.0*d* | 7*s* 11.0*d* |
| Northumberland | 9*s* 7.17*d* | 5*s* 6.87*d* |
| North Wales | 9*s* 5.86*d* | 7*s* 4.13*d* |
| Scotland | 9*s* 3.72*d* | 7*s* 6.77*d* |
| South Wales | 9*s* 0.24*d* | 6*s* 4.6*d* |
| Somerset | 8*s* 11.96*d* | 7*s* 9.3*d* |
| Cumberland | 8*s* 3.45*d* | 7*s* 9.6*d* |
| **Durham** | **8*s* 0.93*d*** | **5*s* 10.63*d*** |
| Forest of Dean | 7*s* 5.08*d* | 5*s* 11.22*d* |
| Bristol | 7*s* 4.45*d* | 6*s* 4.74*d* |

*Source*: G. D. H. Cole, *Labour in the Coal Mining Industry (1914–1921)* (Oxford, 1923), 241.

where we have the reputation of being the pioneers of shorter hours for mine-workers,' he began.[193]

The objectives of the Durham miners could also be in direct competition with other coalfields. In June, James Robson spoke at Chester-le-Street, explaining the importance of the lockout to his county organization: 'If the uneconomic pits are stopped in Durham, half the collieries in the county will close down . . . If he could help it, Durham was not going out of production for the purpose of aggrandising other counties.'[194] Others echoed his concern that the owners' proposals would disadvantage Durham disproportionately. At a meeting in August, Joshua Ritson admitted that he might have accepted an eight-hour day had it been confined to Durham, as it would have given the coalfield a chance to compete against Yorkshire. But, if both

[193] *HPD*(C), 197, c. 885.

[194] *DC*, 5 June 1926.

**Table 1.5** *Average earnings per shift, May 1925 (figures provided by the Mineowners' Association of Great Britain)*

| Coalfield | Wage per shift |
|---|---|
| Kent | 13*s* 0.47*d* |
| Eastern Area | 11*s* 7.12*d* |
| South Wales | 10*s* 8.84*d* |
| Lancashire, Cheshire, North Staffordshire | 10*s* 1.37*d* |
| Scotland | 10*s* 2.89*d* |
| **Durham** | **9*s* 11.51*d*** |
| Northumberland | 9*s* 5.67*d* |
| North Wales | 9*s* 4.74*d* |
| Forest of Dean | 8*s* 11.82*d* |
| Somerset | 8*s* 11.19*d* |
| South Staffordshire, Salop | 8*s* 9.38*d* |
| GREAT BRITAIN | 10*s* 7.87*d* |

*Notes*: Neither the MAGB nor the MFGB explained how they reached their figures. However, the MAGB were unlikely to have taken account of short-time working and the MFGB were almost certain to have done so, which is a possible explanation for the discrepancy in the position of Durham (and the other exporting coalfields) in their tables.
*Source*: TNA:PRO LAB27/7.

Durham and Yorkshire could work eight hours, he continued, then the latter county could beat Durham, because their coal was richer, they had higher seams, were better organized, and had newer pits: 'Durham would be left in the lurch again.'[195]

Even if the miners' leaders acknowledged that the Durham coalfield was economically disadvantaged, they continually emphasized that the strength of the area union provided benefits that were not enjoyed elsewhere. In June, Ellen Wilkinson described the appalling conditions in the Somerset mines to the House of Commons; in response, Jack Lawson expressed his shock that such conditions still existed.[196] A month later Lawson made an appeal against the continuation of the Emergency Regulations not on behalf of the miners of his own county, but on behalf of those of Warwickshire. He had recently visited that county, and observed the population still to be in a state of development left behind by Durham over a century before. 'When I saw them I understood something of what my own

[195] *SNCC*, 26 Aug. 1926. [196] *HPD*(C), 197, c. 1033.

people went through in the early part of the nineteenth century,' he declared.[197]

Such comments had little effect on the government benches opposite, but were perhaps more valuable as propaganda amongst Lawson's constituents. As well as providing yet another example of the benefits of a strong union, the reminder that other counties were suffering as much or more than their own was also intended to quell mutterings of discontent within the coalfield. Throughout the dispute the MFGB insisted that all donations from foreign and domestic sources were administered through its central relief fund, allowing it to control how much each coalfield received. In June, a DMA circular was sent around the lodges, reminding members why this was necessary—namely that while Durham miners benefited from Labour majorities on the county council and boards of guardians, not all counties were similarly provisioned, so 'our obligation is to see that whatever money is raised is equally divided as far as possible . . . We must not overlook the fact that the miners in the United Kingdom are involved in the struggle and together we stand or fall.'[198] A couple of months later, Ritson used the same argument to soothe the potential objections of a crowd at Kelloe:

> He could not understand why some sections objected to the general secretary [of the MFGB] having a free hand to conduct negotiations, as by standing aloof they would get nowhere. They had also to consider the conditions and oppression that existed in other counties where they had not Labour Boards of Guardians and County Councils to feed the children the same as in Durham county. In other areas they were finding the full weight of the oppression of the squire, capitalists and others, while Durham was fortunate.[199]

Despite such attempts by the local leaders to persuade their members to think within a national context, a few remained hesitant. When Whiteley addressed a meeting at Rowlands Gill in August he acknowledged that there were still some amongst the older generation in particular who felt that the DMA should have remained aloof from the MFGB: 'These people did not understand all the situation or they could not talk like that . . . without the unity of the Miners' Federation today their position would not be worth calling anything but a slavish position under conditions which the owners were prepared to offer them.'[200] Alan Campbell has documented similar suspicions about the

[197] *HPD*(C), 198, c. 2557.
[198] DRO, D/DMA108, DMA circular, 14 June 1926.
[199] *DC*, 21 Aug. 1926.
[200] *BC*, 21 Aug. 1926.

national union in the Scottish coalfields of the interwar years, manifest in support for the non-political trade union movement. Campbell, like Whiteley, suggested that this tended to be concentrated amongst the older generation, for whom the philosophy of the non-political trade union correlated more readily with the earlier tradition of the independent collier.[201] In Durham, older people were also more likely to remember the years before affiliation in 1908 with affection, when a different economic climate had rendered the coal industry more prosperous. For those who looked back upon this time with nostalgia, it also represented an age before the national politics of the MFGB came to disrupt those of Durham. In June 1926, one miner interviewed anonymously in a local paper blamed the national union for the desperate state of the Durham industry:

> He said Durham and Northumberland collieries, if unhampered, could resume work tomorrow without a reduction . . . the great majority of the miners in those two counties would never have been asked to suffer either a reduction in pay or an increase in hours if they had not been partners in the great national organisation. The effect of the Federation had been to reduce the miners of Durham and Northumberland to the level of the worst-paid districts in the country.[202]

A similar complaint was sent to the *Blaydon Courier* a couple of months later: 'If the Federation will not settle, why not our county leaders?'[203] Another grumbled: 'If we miners in Durham and Northumberland would only trust our own leaders, both counties would be at work in less than a week. Why not leave Mr Cook and his brass hats and place ourselves in the hands of our local leaders?'[204] Distrust of the MFGB was not therefore incompatible with loyalty to the DMA.

Yet, if a national lockout failed to engender a commitment to the MFGB amongst a minority of the coalfield's population, others simply fitted its concerns into a frame of reference that was already highly localized. One old man remembered fifty years later that Cook had frequently visited County Durham to address the miners. But, rather than remembering him as a representative of the MFGB (or even of South Wales), he explained to his interviewer that Cook had been 'the miners' man from Durham'.[205] Such opinions were familiar enough to be caricatured even at the time. In the 'local anecdotes' column of

201 Campbell, *Scottish Miners*, ii, *Trade Unions and Politics* (Aldershot, 2000), 242.
202 *DCA*, 18 June 1926.
203 *BC*, 25 Sept. 1926.
204 Ibid., 26 June 1926.
205 GCLOT, ii (Mr P.).

the *Durham Chronicle*, one entry was submitted from Pittington in September:

In a local village, a party of miners were talking of their leaders. One of them said, 'Cook's the best man we ivvor [ever] had or ivvor will have.' 'Aa divvent [don't] agree wi' that,' remarked another. 'Where's there a bettor?' asked the first one. 'Wey, that fellow that comes te the Store Hall te pay the relief oot ivvory week!'[206]

## IX

Finally, it is worth exploring in more detail the miners' interaction with the police as an example of a relationship in which issues of both regional and occupational identities came into play. At first, it had been the state of relations between the strikers and the army that had worried official opinion, when, during the general strike, explicit appeals had been made to class loyalties and placards raised at demonstrations and rallies pleading to soldiers and sailors not to shoot their working comrades.[207] Nevertheless, the government remained confident, reassured by reports that, although units recruited from mining areas were often sympathetic to the miners, 'their general tone and spirit was excellent'.[208] However, if the army was briefly responsible for maintaining the peace during the abnormal circumstances of the general strike, it was the police who had to deal with the more prolonged tensions of a seven-month lockout. 'We must remember that [the police] . . . had a very unpleasant job,' remembered one who had been a young striker in 1926:

In many instances they had fathers and brothers and uncles and cousins on strike from the pits, and yet they, the police, would be escorting men to the job that their relatives should have been doing . . . And we must not forget that police donations also helped to provide aid for the miners, and on many occasions, whether acting under orders or individual understanding, they the police turned a blind eye and saw nowt.[209]

This particular man had been involved in the Territorial Army during the 1920s and had received his own share of hostile comments, which

206 *DC*, 4 Sept. 1926.
207 TNA:PRO, ADM 116/2439, diary of Commander H. Tafrell Dorling, 2 May 1926.
208 TNA:PRO, WO 30/143, War Office bulletin, 11 May 1926.
209 GCLOT, ii (T.L.).

may have coloured his attitude. But, other sources give some impression of a police force integrated into the local community. It is a tiny sample, but to return briefly to Easington Deanery, the marriage registers record five policemen who married in its churches over the two years 1925–6. Every one of them was the son of a miner.[210] Other ties can also be discovered: the miners' MP Joshua Ritson had served eight years in the Sunderland constabulary in between stints in the Durham coalfield.[211] Such loyalties must have encouraged sympathy towards the strikers, and when a series of sports for the unemployed were organized at Seaham Harbour over the summer the local police force contributed money.[212] The miners' propagandists actively stressed the common interests of the two groups and emphasized that police action against the miners would amount to class betrayal. In July, a Labour county councillor faced the Consett magistrates after allegedly denouncing the police force as traitors to their class and announcing that he would see that their wages were reduced in turn.[213]

If common ground could often be found between miners and local policemen, the arrival of policemen from other regions proved much more divisive. During the period of the general strike, police were drafted to Durham from the West Riding, Lincolnshire, and Norfolk; towards the end of the lockout, the Durham constabulary also received reinforcements from Hull.[214] This was a deliberate strategy on the part of the government, who were aware that local policemen were likely to have loyalties to those involved in the dispute, and the policy worked. One man interviewed in the 1970s was keen to point out that 'there was no trouble with the policemen here, none you know'. But, then he spoke of those imported from elsewhere and his attitude changed as he described their provocative behaviour, the baton charges, and the men who were hurt.[215] Such complaints were taken to the highest level, and Whiteley explained in the House of Commons: 'We have no fault to find with the Durham county policemen, who have been there all the time. It is with these imported people, who are sent there with the idea that there is going to be a row, and if they cannot find one, they are determined to make one.' In words which anticipated the complaints

210 DRO, marriage registers, var.

211 'Ritson, Joshua (Josh)', in Bellamy et al. (eds), *Dictionary of Labour Biography*, 12 vols. (1972–2005), ii, 322.

212 *SWN*, 9 July 1926.

213 *DCA*, 23 July 1926. The case was dismissed due to lack of evidence.

214 TNA:PRO, HO144/6902, Home Office report, 10 May 1926; *DC*, 30 Oct. 1926.

215 GCLOT, iii (J.R.).

made during the miners' strike sixty years later, he told of the influx of large numbers of men who had previously been connected to the armed forces: 'They are put into policemen's clothes and sent out with no numbers, so no-one can report them. These people are creating disaffection. They are using provocative, filthy and foul language, and irritating the people, making even the most mild men turn round and take action from time to time.'[216]

A few months earlier, Lawson had spoken in similar terms in the aftermath of the general strike. In answer to Home Office figures which noted an extremely high number of prosecutions in the Durham coalfield, Lawson suggested that this was because so many policemen had been rushed into Durham, adding an extra edge to the atmosphere. Usually the Durham constabulary mixed with the communities they policed, he argued, but those who arrived from elsewhere knew nothing of the people they were dealing with. He called for an investigation into events at Birtley, where a group of policemen had led a baton-charge on a crowd. 'One of the things people were indignant about,' Lawson explained, 'was not merely that they were batoned, but that a sergeant of police used the vilest language some of the men had ever heard'.[217] It is striking that Lawson, like Whiteley, specifically referred to the language used by the police as a point of contention. It is unlikely that Durham pitmen had never heard or indeed used whatever words they were that the imported policemen employed so freely, but it is commonly attested that such language was reserved for the pit.[218] By using such language above ground, within earshot of women, such policemen offended not only because of the angry sentiments expressed, but also because they upset local norms of behaviour.

This is not to suggest that miners did not sometimes resent the power and presence of the Durham constabulary, whose local knowledge could prove just as much of a deterrent to young miners seeking to break the law. One old man remembered that he and his brother used to go to Burnopfield to seek coal from the heap. But 'you couldn't go during the day, because the policeman had left our village for Marley Hill and he knew everybody . . . the ones who Joe Wallis knew their names, it was no good running away'.[219] But, again, local policemen belonged

[216] *HPD*(C), 199, c. 2072.

[217] Ibid., 196, cc. 839–40.

[218] For the persistence of such attitudes well into the twentieth century, see, for example, N. Dennis, F. Henriques, and C. Slaughter, *Coal is our Life: An Analysis of a Yorkshire Mining Community* (1979 edn.; first pub. 1956), 213–16.

[219] GCLOT, i (J.R.D.).

to some kind of wider geographical community and to that end they were susceptible to local sanctions. In July, when Will Lawther and Harry Bolton, checkweighman and chairman of Blaydon urban district council, respectively, were released from jail and made a triumphant return to Chopwell, one local constable was accused of exaggerating the threat posed by the welcoming crowd and of ringing headquarters to request reinforcements. Police arrived with batons, and violence ensued. Soon afterwards he found his garden vandalized and eventually the hostility forced him to leave the area.[220]

## X

On 6 May 1924, two years before the wheels of the coal shafts ground to a halt, Stanley Baldwin had made an after-dinner speech to the Royal Society of St George. 'To me, England is the country, and the country is England . . . ', he famously began.[221] As David Gilbert has pointed out, the Baldwinian evocation of England, with its sounds of the blacksmith's anvil and sight of the farmer's plough, was never one that encompassed the miners, despite the fact that they accounted for one-tenth of the male working population.[222] Yet, mining communities could be similarly forgetful of the outside world, and by 1926 an emotional attachment to the locality had become a recognizable phenomenon. In *Clash*, Ellen Wilkinson's semi-autobiographical novel set during the general strike, Joan questioned her friends about why negotiations were breaking down with the government as the strike deadline approached:

> And to make things more difficult, the Miners' Executive had all taken the first trains home after the conference.
>
> What on earth for? Are they as trustful as all that? . . .

[220] Ibid., i (Mrs C.). Will Lawther (1889–1976) initially worked in the Northumberland coalfield before his family moved to Chopwell in 1907. He soon became involved in union and Labour Party politics, along with several of his brothers: militant sons of a village which was dubbed a 'Little Moscow' (see also Chap. 2, Sect. III). Lawther was elected Labour MP for Barnard Castle in 1929 but returned to union politics after defeat in 1931, becoming President of the MFGB in 1939. His political views mellowed with age and he would eventually turn against his former allies on the left within both the TUC and the Labour Party.

[221] S. Baldwin, *On England* (1927 edn.), 6–7.

[222] D. Gilbert, 'Imagined Communities and Mining Communities', *Labour History Review*, 60 (1995), 47.

'The miners are always homing pigeons,' Royd explained. 'They won't stay in London one second after a meeting is over.'[223]

The North East in particular was geographically more remote than many other coalfields. Local newspapers were parochial in their coverage; only a couple contained even a limited amount of national and international news.[224] In 1930, only 10.9 per cent of Durham families held radio licences, the lowest figure for any English county.[225] Stereotypes of the Durham miner as tightly bound to his homeplace therefore became useful currency in humour as well as literature. Even when asked about the war, which brought men into contact with those of other regions and might have fostered a patriotic, national identity, one old man chose to tell an interviewer only one anecdote, about his own Durham countrymen:

> These two chaps they were from Sacriston. They were sappers . . . They come to the mouth of the drift with their clay pipes and they were having a smoke. While they were having a smoke two officers passed. And these two fellas out for a smoke didn't lift an eyelid. So one of the officers comes back and he says to these two chaps, 'Don't you know who I am?' Well one of them says, 'Are you from Sacriston?' . . . Did you get that, they asked these two officers if they were from Sacriston![226]

However, even if regional horizons remained closely defined during the strike, cooperation across occupational (if not class) boundaries was more fluid, as others of the working-class within County Durham tended to lend their support to the miners. This is not to dilute the importance of a strictly occupational consciousness amongst the miners themselves, who still identified themselves first and foremost as miners rather than members of a wider working class. Rather, the dominance of the mining industry within local society was such that even those not associated with coal were drawn in. As early as 1880, a local reporter had commented that 'our whole community is so essentially and totally a coal-mining one, that the sympathy of those who are not actually engaged with those who are is but little else, perhaps, than the instinct of selfishness'.[227] His words would not have been out of place almost fifty years later. However, sympathy was not just a result of

[223] E. Wilkinson, *Clash*, with a new introduction by B. Vernon (1989 edn.; first pub. 1929), 89.

[224] For example, the *Seaham Weekly News* and the *Chester-le-Street Chronicle*.

[225] *Daily Telegraph*, 3 Sept. 1930.

[226] BMOA, 1999/2.

[227] *DCA*, 10 Sept. 1880, cited in Lee, *Church of England*, 6.

cynical self-interest, although cooperation had clear pragmatic benefits. Many not associated with the mines felt genuine sympathy for mining families, as the Gateshead oral history project illustrates. But, miners were in a numerical majority in many urban and rural boroughs, mining officials dominated local authorities, and miners' propaganda could achieve a wider distribution than propaganda from other sources. The mining community therefore continued to define the boundaries of local culture. In November, Bowburn Football Club chose to debar any miner who had returned to work from playing in either their senior or junior teams. Bowburn was not a miners' club, and presumably open to any young man in the village whatever his occupation, but the morals by which it abided were those of the mining community.[228] In 1975, one old miner thought back to the strike:

> Few people in the mining areas escaped the hardships, the shopkeepers especially. I am quite certain that these people had never realised, until now, just how dependent they were on the mining community for their living. I attended one meeting . . . where a decision was being taken whether or not to return to work. When a show of hands was called for I noticed that quite a few people who were not miners, but railwaymen and shopkeepers, had their hands up.[229]

These people, suffering because of the lockout, were clearly prepared to make their feelings known—they wanted to see the miners return to work. But, it is significant that they chose to register their discontent in this way, by nothing more than a show of hands, and as part of a miners' meeting.

[228] *DCA.*, 12 Nov. 1926.

[229] R. W. Morris, 'The General Strike in the North East', *North East Group for the Study of Labour History Bulletin*, 10 (1976), 5.

# 2

# The Testing of Political and Union Loyalties

On 25 November 1926, nearly seven months after the lockout began and a few days before it ended, a confidential government bulletin estimated that just over 14,000 Durham colliery employees were back at work, corresponding to 8.3 per cent of the usual workforce, and representing a lower proportion than in any other county in England or Wales. A month earlier, the proportion had been just over 4 per cent (see Table 2.1). The DMA could boast exceptional loyalty in this respect, and in normal times, too, it remained the beneficiary of an unusually high union density (see Table 2.2). Even in the aftermath of 1926, the Northumberland and Durham Non-Political Trade Union was unable to make inroads and had not attracted many more than 4,000 members by the end of 1928.[1] Jack Lawson would later comment that the union 'is an integral part of the life of the Northern miner; in truth, it is in the texture of his thought even when he is not conscious of it'.[2]

With regard to the valleys of South Wales, Chris Williams has commented on the social centrality of union and Labour officials, whose role remained as critical in peaceful times as it did in times of conflict. 'They were the advice bureaux of their localities, a role for which they were occasionally named: Will "Knowledge" Hughes, Dafydd Hughes "Income Tax".'[3] The union fulfilled a similar function in Durham, and lodge minute books and correspondence are filled with the mundane details of unemployment benefit, compensation, housing conditions, and safety issues. Furthermore, 34,000 miners belonged to the Approved Society attached to the DMA in 1931.[4] Lawson portrayed the union

[1] The National Archives: Public Record Office (henceforth TNA:PRO), FS11/411, returns of the Northumberland and Durham Miners' Industrial Non-Political: Trade Union, 1927–8.

[2] J. Lawson, *A Man's Life* (rev. 2nd edn., 1944),136.

[3] C. Williams, *Democratic Rhondda: Politics and Society, 1885–1951* (Cardiff, 1996), 207.

[4] *Parl. Papers, 1931–2*, xiv (879), Report by the Government Actuary on the Third Valuation of the Assets and Liabilities of Approved Societies, 76.

**Table 2.1** *Numbers of men returning to work (Home Office figures)*

| County | Approximate number usually working | Number at work, 25 Oct., and as percentage of usual total | | Number at work, 15 Nov., and as percentage of usual total | | Number at work, 25 Nov., and as percentage of usual total | |
|---|---|---|---|---|---|---|---|
| Durham | 170,000 | 7,142 | 4.2% | 9,765 | 5.7% | 14,042 | 8.3% |
| Carmarthen | 16,100 | unrecorded | – | unrecorded | – | 1,788 | 11.1% |
| Glamorgan | 161,000 | 12,058 | 7.5% | 14,430 | 9% | 22,737 | 14.1% |
| Cumberland | 11,700 | unrecorded | – | 1,549 | 13.2% | 1,646 | 14.1% |
| Monmouth | 54,000 | 3,934 | 7.3% | 6,674 | 12.4% | 8,040 | 14.9% |
| Northumberland | 57,600 | 6,339 | 11% | 8,700 | 15.1% | 11,847 | 20.6% |
| Yorks./West Riding | 181,000 | 24,947 | 13.8% | 35,497 | 19.6% | 48,889 | 27% |
| Somerset | 5,200 | 498 | 9.6% | 1,542 | 29.7% | 1,683 | 32.4% |
| Flint | 2,800 | 969 | 34.6% | 1,230 | 43.9% | 1,441 | 51.5% |
| Denbigh | 13,900 | 4,082 | 29.4% | 5,869 | 42.2% | 7,914 | 56.9% |
| Lancashire | 75,700 | 14,258 | 18.8% | 25,487 | 33.7% | 43,391 | 57.3% |
| Staffordshire | 42,500 | 24,901 | 58.6% | 27,591 | 64.9% | 31,529 | 74.2% |
| Gloucestershire | 8,500 | 5,858 | 68.9% | 6,034 | 71% | 6,647 | 78.2% |
| Salop | 4,400 | 3,109 | 70.7% | 3,123 | 71% | 3,473 | 78.9% |
| Nottinghamshire | 53,300 | 34,553 | 64.8% | 42,819 | 80.3% | 44,945 | 84.3% |
| Leicestershire | 12,000 | 8,196 | 68.3% | 9,507 | 79.2% | 10,157 | 84.6% |
| Warwickshire | 20,100 | 16,035 | 79.8% | 15,913 | 79.2% | 16,998 | 84.6% |
| Derbyshire | 64,500 | 41,828 | 64.8% | 52,223 | 81% | 55,393 | 85.9% |
| Worcestershire | 1,100 | 1,026 | 93.3% | unrecorded | – | 1,082 | 98.4% |
| ENGLAND AND WALES | 954,300 | 208,707 | 21.9% | 267,943 | 28.1% | 332,560 | 34.8% |

*Source*: TNA:PRO, HO144/6902.

**Table 2.2** *Mining union membership, January 1926 (major districts)*

| Coalfield | Men employed | Union membership | Union density (%) |
|---|---|---|---|
| Durham | 153,000 | 155,773 | 101.8%* |
| Cumberland | 11,300 | 10,036 | 88.8% |
| Yorkshire | 188,600 | 164,196 | 87.1% |
| North Wales | 17,600 | 14,224 | 80.8% |
| Northumberland | 56,700 | 43,482 | 76.7% |
| Lancashire and Cheshire | 97,100 | 72,902 | 75.1% |
| Derbyshire | 58,700 | 43,000 | 73.3% |
| Nottingham | 56,000 | 38,767 | 69.2% |
| South Wales | 211,200 | 129,155 | 61.2% |
| Warwickshire | 20,000 | 11,500 | 57.5% |
| North Staffordshire | 35,400 | 10,679 | 30.2% |

* In Durham the practice of registering men as union members even if they were unemployed meant that union density exceeded 100%.

*Source*: Numbers of men employed taken from Royal Commission on the Coal Industry (1925), iii, Appendices (1926), 179; union membership taken from *Parl. Papers, 1928*, ix (781), Return Showing Details of Membership, Income, Expenditure and Funds of Registered Trade Unions With 10,000 or More Members in the Years 1925 and 1926, 4–5.

official as the lynchpin of his community: 'People come to his home for advice; they meet him on the road, in the pit, at the hall of the meeting, everywhere; at all hours of the day they come—and he is ever the patient listener and advisor.'[5]

Durham's place in the popular imagination as a union stronghold was complemented by the strength of its Labour allegiance. By 1926, the Labour Party held nine of Durham's eleven county seats; three years later it overturned slim Conservative margins in Barnard Castle and Sedgefield to take them all. The party was similarly dominant in local politics and in 1919 Durham became the first English county to boast a Labour majority, briefly lost in 1922, but regained in 1925 and held for the rest of the century.[6] Within the DMA itself, the strength of Labour support could be seen in the high rate of subscribers to the political fund. Approximately 120,000 members contributed to this fund in 1926, but the passing of the 1927 Trade Disputes and Trade Unions Act required union members to opt in to their political fund rather than have to opt out. Even then, 113,287 Durham miners had subscribed by the end of 1928. Twelve months later, this number had increased to 119,086.[7]

[5] Lawson, *Man's Life*, 137.

[6] Monmouthshire elected a Labour council in the same year. See B. Keith-Lucas and P. Richards, *A History of Local Government in the Twentieth Century* (1978), 114.

[7] TNA:PRO, FS12/10, DMA returns, 1926–9.

However, other influences might divert attention away from the dominant ideologies of the DMA and the Labour Party. The Tory, the coal owner, and the blackleg have long made up an unholy trinity within union mythology and too often this has influenced their portrayal within historical writing—if indeed they are acknowledged at all. Within an academic context, coal miners have been particularly resistant to sociological attempts to explain the existence of some kind of alternative working-class consciousness. In his famous 1966 study, David Lockwood described the miner as a typical 'traditional proletarian worker', with a strong awareness of an 'us' and a 'them'. In opposition was the 'traditional deferential worker', who was exposed to paternalistic authority and deferred to his superiors.[8] Very crudely, the implication could be drawn that a worker was more likely to vote Conservative in proportion to how far he deviated from the profile of a miner. However, Durham's election statistics point to a significant proportion of men and women who opted out of a Labour identity (see Table 2.3). Indeed—unlike in South Wales—the fragility of Labour support was

**Table 2.3** *General Election results, 1924: County constituencies with over 20 per cent of the male population over the age of 12 engaged in the coal industry*

| Constituency | Percentage of males aged 12 + engaged in the coal industry | Result | Labour votes | Conservative votes | Turnout |
|---|---|---|---|---|---|
| Barnard Castle | 44.6% | Con. gain | 49.2% | 50.8% | 84.9% |
| Bishop Auckland | 50.7% | Lab. held | 55.1% | 44.9% | 80.9% |
| Blaydon | 53.6% | Lab. held | 62.6% | 37.4% | 77.0% |
| Chester-le-Street | 56.6% | Lab. held | 71.0% | 29.0% | 78.7% |
| Consett | 55.4% | Lab. held | 55.9% | 44.1% | 83.4% |
| Durham* | 58.5% | Lab. held | 54.9% | 35.1% | 85.2% |
| Houghton | 53.6% | Lab. held | 57.8% | 42.2% | 79.6% |
| Seaham | 71.4% | Lab. held | 65.5% | 34.5% | 78.8% |
| Sedgefield | 33.7% | Con. held | 47.3% | 52.7% | 85.4% |
| Spennymoor | 62.2% | Lab. held | 63.0% | 37.0% | 78.3% |

* In Durham, a Liberal candidate stood and polled 10 per cent of the vote. In every other constituency there were only two candidates.

*Source*: F. W. S. Craig, *British Parliamentary Election Results, 1918-49* (Glasgow, 1969), 338–48; M. Kinnear, *The British Voter: An Atlas and Survey since 1885* (1968), 116.

[8] D. Lockwood, 'Sources of Variation in Working-Class Images of Society', *Sociological Review*, 14 (1966), 249–67.

revealed in the 1931 election, when only Jack Lawson and Joe Batey held their seats for Labour.[9] The rest were claimed by the Conservatives, the Liberals, or, in the case of Seaham, by the man freshly castigated as the Judas of the Labour movement, Ramsay MacDonald.

## I

In the autumn of 1926, the Conservative MP for Sedgefield, Major Ropner, admitted to a grudging respect for the solidarity he had witnessed amongst his mining constituents, telling the House of Commons that 'I cannot refrain from paying a tribute to the magnificent loyalty and patience with which over a million men continue largely to support leaders in whom, for the most part, they have entirely lost faith and trust'.[10] But, his words hardly amounted to unqualified admiration, and his implication that the rank and file had been misled by manipulative leaders was in tune with his government's attempts to express criticism of the strike without blaming the miners themselves too directly. To those who stood in political opposition to the strike, a blind loyalty to the union seemed a convenient explanation for the fact that the strike remained solid.

In fact, in Durham the radicalism of the rank and file might frequently outdo the moderation of district leaders who were far from comfortable with the implications of the dispute. In a confidential letter to Ramsay MacDonald, Peter Lee revealed his fears as early as 24 May. After condemning the attitudes of both government and coal owners, the miner's leader criticized his own side:

> it will be a crime against the women and children if we men do not get together and establish a peace which will once more set the wheels of industry moving. It ought to be clear now to the minds of all thinking men, who have no knowledge of past strikes and lockouts, that the longer we stand apart the greater will be the loss to all concerned and greater also the bitterness arising from the dispute.[11]

[9] Joe Batey (1867–1949) worked in the Northumberland coalfield before moving to Durham at the age of 18. He was elected president of his local lodge in 1888, to the executive of the DMA in 1901, and first sat on the executive of the MFGB in 1917. He was Labour MP for Spennymoor 1922–42.

[10] *Hansard Parliamentary Debates* (Commons), fifth ser. (henceforth *HPD*(C)), 199, c. 605.

[11] John Rylands University Library, Manchester (henceforth JRUL), RMD/1/4/81, P. Lee to J. Ramsay MacDonald, 24 May 1926.

Lee was also prepared to voice his reservations more publicly, and at a mass meeting in June he made a veiled attack on the national leadership. A report was provided by the *Durham County Advertiser*:

> 'What we want,' he [Lee] added, 'Is men who have thought out the problem. Anyone can shout this motto,'—pointing to a banner with the words, 'No reductions, no increased hours, no district settlements' . . . This observation was met with considerable interruption . . . Mr Lee remarking: 'If this is a sample of the new democracy, God help Britain, and God help the people.'[12]

Conflict between the Durham leadership and the local lodges frequently spilled over into animosity. A damaging challenge to district unity came in July, when the DMA Executive recommended cancellation of the annual gala, citing lack of funds. This was duly endorsed in a coalfield ballot, but a significant minority of lodges disagreed and an unofficial gala was organized at Burnhope. The result was a bitter and public exchange of views between the DMA Executive and the officials of Burnhope lodge.[13]

The most striking example of rank-and-file militancy came at the end of November, when votes were taken across the coalfields on whether or not to resume work on district terms. The leaders of both the DMA and the MFGB recommended acceptance of the proposals, but only 41 per cent of the Durham workforce chose to follow their lead; 59 per cent voted for rejection in a ballot in which 64 per cent of the workforce voted.[14] The lack of a two-thirds majority meant that the result was taken as a mandate for the return to work, but it was not the result that the Durham leaders expected; nor, presumably, were they happy to receive a number of resolutions from local lodges demanding their resignation. Sidney Webb wrote despairingly to Beatrice: 'It is clear that the rank and file are absolutely stubborn, and will not encourage or allow any leadership in a rational direction.'[15]

The local leaders were as keen to ensure a strict model of union conduct as they were to prevent unauthorized radicalism. Across the coalfield, accusations spread that Labour- and miner-dominated boards of guardians did not always treat financial and unfinancial [men in

12 *Durham County Advertiser* (henceforth *DCA*), 4 June 1926.

13 Ibid., 23 and 30 July 1926.

14 Durham County Record Office (henceforth DRO), D/DMA109, result of ballot on owners' proposals, 29 Nov. 1926.

15 London School of Economics (henceforth LSE), Passfield 2/3/1, S. Webb to B. Webb, 18 Nov. 1926.

arrears] union members equally when allocating relief,[16] while in Birtley the meeting place of the local Labour Party was co-opted as the distribution centre for boots and shoes. One resident, himself a Labour voter, believed that many Conservatives would therefore simply not have gone.[17] At Eden lodge, the sporadic payments of lockout benefit were withheld from any man who disappointed his union's expectations, and committee meetings were increasing taken up with questions of eligibility. Blacklegs were struck off, but so also were those whose actions damaged the miners' cause in other ways. In October, for example, benefit was withheld from two men after they were found to have been selling coal.[18]

The Eden minutes also provide a striking example of the difference between the union's vision of control and the muddle that often proved to be the reality. In November, with the strike at an end, the Eden miners were told to take their place in the queue behind their lodge officials when they signed on for work, after a warning that 'anyone who signs on before the Secretary shall forfeit his lockout benefit this weekend'.[19] The insistence that lodge officials should lead, even in defeat, could have been a powerful symbol of union discipline. But, the restrained record of the lodge minutes stands in marked contrast to the later recollections of one of the Eden miners:

> I remember the final meeting prior to signing on for the pit at the miners' hall and everybody was that eager in case they didn't get a job that when it came to the time when they were released from the meeting to fall in and march away and go down to the pit in an orderly manner, well they smashed the doors off the meeting room in their hurry to get out.[20]

Of course, divisions between the leaders and the rank and file were not unique to the months of the lockout, and discontent with the union leadership was a feature of normal times too. The self-aggrandizing leader who reached the top only to forget the men who had put him there was a common stereotype in Durham as elsewhere. 'Oh I believe in the Union, but I didn't believe in the class of the men that we had here. Oh, dear me terrible,' commented one ex-miner in his old

16 *Stanley News and Consett Chronicle* (henceforth *SNCC*), 22 July 1926.

17 Gateshead Central Library, Oral History Transcripts (henceforth GCLOT), ii (T.L.).

18 DRO, D/DMA 334/5, Eden lodge minutes, 8 Oct. 1926.

19 Ibid., 2 Dec. 1926.

20 F. Whitfield, 'And of Course, I've got some Dust', in K. Armstrong and H. Beynon (eds), *Hello, Are You Working? Memories of the 'Thirties in the North-East of England* (Whitley Bay, 1977), 61.

age.[21] He was particularly scathing when asked about Lawson: 'divent mention Jack Lawson to me.'[22] Suspicions that the DMA leaders could and did negotiate special treatment for themselves continued during the lockout, and at the end of May, W. P. Richardson was stung into issuing a circular repudiating claims that the DMA was still paying its officials their full salary.[23] In one staggering case of misremembering, one old woman managed to reverse the intentions of the union and the government entirely when asked years later about the provision of relief:

> Well, we got it just from, err, well, the government, it must have been, yes the government, must have been . . . Well because, what was like the miners' lodge, really, the miners' lodge, they supplied the children, the women and children but not the men because they were on strike—they made it as if they were, as if it was, their own fault.[24]

Lawson suggested that such criticism was healthy, and speaking of the opposition that Herbert Smith had often attracted, he explained:

> miners make full use of their right to attack a leader, but it almost seems as though their criticism is a measure of their loyalty. For the leader who bears the brunt of most attacks is ever highly regarded—even with affection. Which waywardness often deceives the stranger.[25]

Even with Lawson's rather hopeful corollary, the district leaders in County Durham were clearly not the beneficiaries of blind loyalty. There are also problems with analyses based on too strict a division between leaders and rank and file. With relation to the South Wales coalfield, Chris Williams has argued that such a cleavage has been over-exaggerated in recent writing. Rather, he suggested that the terms 'activist' and 'rank and file' conceal 'a multitude of different standpoints . . . and it must be remembered that, outside the pit, many "leaders" and "activists" shared homes, pubs, clubs, chapels, relatives, perhaps even lovers'.[26]

However, those who claimed in 1926 that ordinary miners had been misled by manipulative leaders might have thought they had more of a case with regard to the national leadership. A. J. Cook, in particular, was seen as the demagogic face of MFGB radicalism, and Conservative commentators worried over his heroic reputation in the coalfields. Durham

21 Beamish Museum, Oral History Archive (henceforth BMOA), 1991/82. See also GCLOT, ii (T.L.).

22 BMOA, 1991/82.

23 *DCA*, 28 May 1926.

24 BMOA, 1976/125.

25 J. Lawson, *The Man in the Cap: The Life of Herbert Smith* (1941), 156.

26 C. Williams, *Capitalism, Community and Conflict: The South Wales Coalfield, 1898–1947* (Cardiff, 1998), 47–8.

**2.** A. J. Cook Addressing a Meeting in 1926 (location unspecified)
Reproduced by permission of the People's History Museum.

was no exception. In June, Cook received 'a wild and tumultuous welcome' at a meeting of nearly 10,000 people at Wardley; in August, a crowd of around 20,000 congregated at Murton, where he was carried 'shoulder-high through the cheering multitude'; a couple of months later, he visited Horden where about 8,000 attended and again lifted him shoulder-high amidst cheers and a rendition of 'He's a jolly good fellow'.[27] Years later, Will Lawther remembered the crowds that Cook had been able to attract: 'Oh my God!—when he got onto the platform—there he just went at it—imagine, where is the man today, I don't care who he is, what sphere of life—political, economic, trade union, religious—could draw the crowds like Cook did'.[28] Another old miner was more succinct: 'Cook at that time, he was the miners' pin-up.'[29]

If Cook's name became shorthand for all that was wrong with the strike within Conservative circles, within the Durham pit villages it

[27] *SNCC*, 24 June 1926; *Durham Chronicle* (henceforth *DC*), 14 Aug. 1926, 23 Oct. 1926.

[28] 'Transcripts from an Interview with Will Lawther', *North East Group for the Study of Labour History Bulletin*, 2 (1968), 10–11.

[29] GCLOT, iv (Mr S.).

represented all that was right. When Cuthbert Headlam, the Conservative MP for Barnard Castle, addressed a meeting in Lanchester in December, he was continually interrupted with cries of 'Good old Cook'. 'You can still say, Good old Cook!' retorted Headlam angrily. 'If it had not been for Mr Cook you would have been working today'. He was unable to silence his hecklers, however, and they continued: 'Good old Cook! Give him the VC!'[30] Even those who were not so impressed with the message espoused by the miners' secretary recognized his magnetic qualities. R. W. Morris, a former colliery mechanic highly critical of Cook as 'the chief culprit in the whole sad business', remembered Cook's oratory with admiration: 'He peeled off his coat and rolled up his sleeves as he warmed to his task. It was a masterly performance as he carried his audience along with him in his mesmeric peroration. You could not help being carried away. He was cheered to the echo.'[31]

In September 1926, Beatrice Webb described Cook as 'a quivering mass of emotions—a mediumistic magnetic sort of creature—not without personal attractiveness—an inspired idiot, drunk with his own words, dominated by his own slogans'. Six weeks later she had further hardened her opinion, damning him as being like 'the gangrenous gas of a badly wounded body'.[32] Of course, Webb differed from the miners' leader in both politics and style and she was unlikely to be sympathetic. As a young man, Cook had absorbed a Marxist education at the Central Labour College (CLC). He left the Communist Party in 1921, but remained sympathetic to its aims and particularly to those of the CPGB-inspired Minority Movement, breaking with them only in 1928 after their instruction of 'class against class'. Recent scholarship has been kinder and has shown the miners' secretary to be rather more pragmatic than his reputation as a wild demagogue suggests. In private (and, if the occasion allowed, in public as well), Cook displayed a considerable amount of flexibility during the negotiations of 1926, particularly in contrast to an intransigent Herbert Smith.[33] But, if Beatrice's words underestimated the miners' leader, they also underestimated his

30 *Blaydon Courier* (henceforth *BC*), 4 Dec. 1926. Cuthbert Headlam (1876–1964) was Conservative MP for Barnard Castle 1924–9 and 1931–5, and then for Newcastle North 1940–51. He held a couple of junior government offices in the 1920s and 1930s.

31 R. W. Morris, 'The General Strike in the North East', *North East Group for the Study of Labour History Bulletin*, 10 (1976), 4, 12.

32 Diary entries, 10 Sept. and 24 Oct. 1926 in M. Cole (ed.), *Beatrice Webb's Diaries*, ii, *1924–32* (1956), 116, 124.

33 See, for example, P. Davies, *A. J. Cook* (Manchester, 1987), 111–33, esp. 117–18; J. McIlroy, A. Campbell, and K. Gildart (eds), *Industrial Politics and the 1926 Mining Lockout*

followers. For his was not a temporary appeal, and Cook continued to be one of the most popular faces portrayed on the banners of the Durham miners' lodges, not only in the 1920s but throughout the twentieth century. As a member of the Bishops' deputation reminded Baldwin in July, 'they are wonderfully solid behind him . . . They could chuck Cook tomorrow, and they have not done it'.[34]

For, although Cook's charisma undoubtedly widened his support, his popularity was also entwined with the slogan from which he became inseparable. 'Not a minute on the day, not a penny off the pay' became the focus on which solidarity was centred. According to Peter Lee, this was all that was needed to secure the steadfast resolve of the Durham miners during the lockout. As he told the MFGB conference:

> I come from a county where you don't need any intensive propaganda. All you have to do in Durham is to put the owners' proposals on a piece of paper, and circulate them amongst the people—eight hours a day where we have been working six and a half. Twenty-one per cent off the wages. You need no more propaganda. This is sufficient in Durham.[35]

This was not simply rhetorical bluster, and most ordinary miners and their families saw the struggle for a decent wage and working day as the sole reason for fighting. Paul Jeremy has suggested of South Wales that the reason why destitution did not force many back sooner was the draconian surrender terms insisted upon by the owners.[36] Similarly in Durham, throughout the autumn those pits that reopened did so on harsh terms: eight-hour days and lower wages, conditions that would mirror the final district settlement. Men longing for the end of the strike knew what the outcome would be and, in August, Will Lawther noted that 'the men realise that to submit would be a living death'.[37] The strike 'was over the fact that the owners wanted us, the miners, to work for less pay and longer hours, and we were definitely against it,' explained one old miner fifty years later.[38] Furthermore, it was argued that it was the owners, not the miners, who had moved the

(Cardiff, 2004), 57, 271. Although traditional images of Cook still linger. See A. Perkins, *A Very British Strike: 3 May–12 May 1926* (2006), 80–1, 259.

34 TNA:PRO, CAB21/296, Minutes of Bishops' meeting with the Prime Minister, July 1926.

35 MFGB, *Annual Volume of Proceedings for 1926* (1927), 652.

36 P. Jeremy, 'Life on Circular 703: The Crisis of Destitution in the South Wales Coalfield during the Lockout of 1926', *Llafur*, 2 (1977), 65.

37 Cited in W. R. Garside, *The Durham Miners, 1919–1960* (1971), 212.

38 GCLOT, ii (Mr A.).

goalposts and precipitated the struggle; that the miners were simply waging a defensive battle to maintain the status quo. 'Why were the miners on the stones today?' asked James Gilliland in the House of Commons, 'The world said that they refused to work. There was no truth in that. They were locked-out. They never put their notices in. The owners had determined that unless the miners accepted what they termed economic conditions of work they would keep them out'.[39] Many of those later interviewed for the Gateshead oral history project disputed the very use of the term 'strike'. 'The 1926 strike was really a lockout, it was a massive lockout by employers and the issue was very simple. You must work longer hours for less pay. It was a condition that trade unions couldn't accept,' explained one.[40] It was a quarrel over vocabulary that was fought throughout, and amongst the pages of *Hansard* the historian imagines the chamber noisily erupting whenever a government speaker refers (after 12 May) to 'the strike': '[Hon. Members: "Lockout!"].'[41]

Wages in the coal industry varied enormously in the 1920s, between regions and within individual collieries, and the propaganda battle over the 'average' wage of a mineworker was fierce. A number of variables could be altered to produce a higher or lower figure, and calculations inevitably differed (see Table 2.4 for MFGB estimates). At the beginning of May, Baldwin complained in the House of Commons

> of the mysteries of minimum percentages and datum lines, bonus terms, ascertainment of allowances, additions and subtractions going to two places of decimals, and then you have . . . to check and alter and regulate them by a dozen district agreements in which old practices and calculations may upset everything that you have hitherto understood.

He suggested that this was one of the difficulties facing any resolution to the dispute: 'You can never get an agreed amount of what a miner is earning. Neither owner nor miner will ever agree on a figure.'[42]

[39] *SNCC*, 29 July 1926. James Gilliland (1866–1952) worked in the Durham coalfield all his life. In 1925 he was appointed DMA agent and elected as representative of Birtley on Durham County Council. In later years (1935–45) he would be President of the DMA.

[40] M. Ridley, 'Making a Contribution', in T. Austrin et al. (eds), *But the World Goes on the Same: Changing Times in Durham Pit Villages* (Whitley Bay, 1979), 59.

[41] Robert Smillie argued in Parliament that by refusing to work the eight-hour day demanded by the lockout notices, the miners were simply refusing to break the law, which limited underground work to seven hours. In fact, under the law as it stood in May 1926, there was a sixty-day grace period during which, in an emergency, an extra hour could be worked. The miners argued that no such emergency existed. See *HPD*(C), 196, cc. 2227, 2234.

[42] Ibid., 195, cc. 58–9.

**Table 2.4** *Day wages in Durham, selected grades**

| Class of Labour | Basis wage per shift (fixed 1879) | Wage per shift, July 1914 (basis + 57.5%) | Wage per shift, October 1925 (basis + 110%; minimum of *7s 6½d*) | 1914 wages, plus 76% (rise in cost of living, 1914–25)** | Wage under owners' proposals, July 1926 (basis + 89%; minimum of *6s 8½d*) |
|---|---|---|---|---|---|
| UNDERGROUND WORKERS | | | | | |
| Deputies | *4s 8.5d* | *7s 4.99d* | *11s 6.6d* | *13s 0.62d* | *8s 10.79d* |
| Hand-Putters (piece) | *4s 2d* | *6s 6.75d* | *9s 8d* | *11s 6.6d* | *7s 10.5d* |
| Hewers (piece) | *4s 2d* | *6s 6.75d* | *9s 8d* | *11s 6.6d* | *7s 10.5d* |
| Horsekeepers | *3s 0d* | *4s 8.7d* | *7s 6.5d* | *8s 3.79d* | *6s 8.5d* |
| Pony-Putters (piece) over 21 | *4s 2d* | *6s 6.75d* | *9s 8d* | *11s 6.6d* | *7s 10.5d* |
| Pony-Putters (piece) under 21 | *4s 0d* | *6s 3.6d* | *8s 4.8d* | *11s 1.06d* | *7s 6.72d* |
| Shifters | *3s 0.75d* | *4s 9.88d* | *7s 6.5d* | *8s 5.71d* | *6s 8.5d* |
| Sinkers | *3s 11.5d* | *6s 2.81d* | *8s 3.75d* | *10s 11.67d* | *7s 5.78d* |
| Stonemen (piece) | *4s 4d* | *6s 9.9d* | *9s 1.2d* | *12s 0.14d* | *8s 2.28d* |
| Stonemen (datal) | *3s 7.5d* | *5s 8.31d* | *7s 9.5d* | *10s 0.58d* | *6s 10.22d* |
| Timber Drawers | *4s 3d* | *6s 8.33d* | *9s 11.7d* | *11s 9.38d* | *8s 0.39d* |
| Wastemen | *2s 11d* | *4s 7.13d* | *7s 6.5d* | *8s 1.03d* | *6s 8.5d* |
| SURFACE WORKERS | | | | | |
| Banksmen (piece) | *4s 2d* | *6s 6.75d* | *8s 9d* | *11s 6.6d* | *7s 10.5d* |
| Banksmen (datal) | *3s 5.5d* | *5s 5.36d* | *7s 6.5d* | *9s 7.03d* | *6s 8.5d* |
| Blacksmiths | *3s 10d* | *6s 0.45d* | *8s 4.8d* | *10s 7.51d* | *7s 2.94d* |
| Electricians | *3s 10d* | *5s 10.88d* | *8s 4.8d* | *10s 4.75d* | *7s 2.94d* |
| Labourers | *3s 1d* | *4s 10.28d* | *7s 6.5d* | *8s 6.57d* | *6s 8.5d* |

* Wages for pieceworkers are approximate only: for pieceworkers, basis rates differed from pit to pit and even from stall to stall according to geological conditions; the final wage then varied according to output. Also note that the figures are indicative calculations rather than actual wages; in the case of some grades the sums vary slightly owing to minor changes in the basis rate between 1914 and 1925.

** The Samuel Report disputed the comparison between the situation in 1914 and 1925, suggesting that if 1925 was compared to the years 1909–13, increases in wages and the cost of living were more evenly matched.

*Notes*: Day wages were calculated on an agreed basis rate plus a percentage, which was regulated according to net proceeds (divided 87% to wages and 13% to the owners). The existing percentage was not, however, allowed to fall below 110%: under the lockout notices posted on 30 April this was to be reduced to 80%; by July, the owners had amended their offer to 89%. See MFGB, *Proceedings for 1926*, 1170. 'Average' wages also varied depending on which classes of worker were surveyed, whether the provision of free housing was taken into account, whether or not a full week's work was assumed, etc. Nor were different figures always deliberately misleading: even the statistics produced by the MFGB were sometimes inconsistent. Cf. Tables 1.2, 1.4, and 1.5.

*Source*: Royal Commission on the Coal Industry (1925), iii, 56–7. Figures submitted by the Miners' Federation of Great Britain.

However, even claims of a relatively high average wage could not disguise the fact that the lockout notices posted on 30 April necessitated a wage cut. This amounted to a reduction of the percentage wage from 110 per cent to 89 per cent and a reduction of the subsistence wage (which guaranteed minimum earnings) from 7*s*. 6 1/2*d*. to 6*s*. 8 1/2*d*. per shift.[43] Just under two months earlier, the Samuel Report had observed that 'any material fall in wages will . . . bring real wages at the present cost of living, below pre-war level for a large proportion of the miners'.[44]

Nevertheless, the Samuel Report had recommended wage cuts as the lesser of two evils. Several Labour MPs voiced a similar preference in the Commons, arguing, like Samuel, that a wage cut would at least be easier to reverse than lengthened hours when prosperous times returned.[45] Opposition to an increase in hours was particularly strong in Durham, owing to the traditionally short shifts of its hewers, who worked fewer hours than any other mineworker in Britain. In June, Cuthbert Headlam bumped into Stanley Baldwin in the smoking room of the House of Commons: 'I told him that I did not expect that the 8 Hours Bill would be accepted in Durham. He said that other people had told him the same thing, but there were hopes that it might be acceptable in some of the other coalfields.' Five years later, fighting to regain his Barnard Castle seat, Headlam would remember the bitter feeling provoked by the Act, confessing in his diary that his record of support for it continued to prove a handicap: 'it is quite extraordinary how that vote counts against me.'[46]

An increase in hours also raised the spectre of unemployment, and although this had not yet assumed the terrible proportions to come, it was still known and feared. In the months between May 1924 and the end of March 1925, thirty-eight pits closed in Durham, affecting 19,000 men.[47] A year later, by April 1926, the number of insured persons registered as unemployed in the county exceeded 28,000, accounting for almost a third of the 98,000 registered as such in England and Wales.[48] Unemployment in the British coal industry as a whole stood at 15.8 per cent

[43] Ibid., 198, c. 21.

[44] *Parl. Papers, 1926*, xiv (1), Report of the Royal Commission on the Coal Industry (1925), i, Report, 229.

[45] *HPD*(C), 197, cc. 803, 1196, 1254; 198, c. 1730; Royal Commission on the Coal Industry, i, 175.

[46] Diary entries, 28 June 1926 and 15 Oct. 1931, in S. Ball (ed.), *Parliament and Politics in the Age of Baldwin and MacDonald: The Headlam Diaries, 1923–35* (1992), 93, 219.

[47] TNA:PRO, LAB 27/7, report submitted to Ministry of Labour by MAGB.

[48] Ibid.

in May 1925, a figure only trumped by the shipbuilding industry—upon which the North East also relied—in which unemployment was 31.5 per cent.[49] The Samuel Report rejected an increase in the miner's working day mindful of precisely such concerns, suggesting that if output remained the same, employment in the industry would fall by a further 130,000.[50] It was to prove a close estimate and, by September 1927, the total number of people employed in and about British mines had fallen from 1,107,100 to 982,600, a decrease of 124,500. The shorter shifts of Durham meant that the eight-hour legislation entailed a greater degree of reorganization than elsewhere, and Durham and Northumberland accounted for over one-third of the total loss.[51]

The miners also fought against a lengthening of the working day owing to the nature of their occupation. The miners' witnesses called to give evidence to the Sankey Commission in 1919 had stressed the savage conditions of underground work, and their concerns were reiterated in Parliament as the government debated the Coal Mines Bill in 1926.[52] The hardship most frequently referred to was the workplace accident rate, with which many in the union hierarchy were familiar. W. P. Richardson's father had been amongst the forty-one men and boys killed in the Usworth Colliery explosion of 1885 when his son was 12; Herbert Smith's father had been killed in a solitary accident a few days before his son was born; while the first day that the young A. J. Cook had spent underground had been marred by the death of the man working next to him. Durham and Northumberland had a better accident record than most, partly due to geological conditions which rendered roof falls less common.[53] Even so, the DMA's memorial record is reminiscent of the lists of names on war memorials; most poignant when the same names repeat themselves and the observer is left to guess at the broken families that resulted.[54] During the lockout, accident statistics became a constant reference point and were used precisely to oppose an increase in the length of the working day. One letter published in the *Chester-le-Street Chronicle* drew attention to government statistics which counted

[49] TNA:PRO, LAB 27/7.

[50] Royal Commission on the Coal Industry, i. 173.

[51] Garside, *Durham Miners*, 273.

[52] For example, *Parl. Papers, 1919*, xi (1), Coal Industry Commission, ii, *Minutes of Evidence*, q. 9279; *HPD*(C), 197, c. 1053.

[53] In 1925 the death rate due to accidents on the surface and underground was 1 per 1,000 in the UK, compared to 0.9 in Durham and Northumberland and 1.3 in South Wales. See F. A. Gibson, *The Coal Mining Industry: Supplement with Statistics for the Year 1925* (Cardiff, 1927), 32–3.

[54] E. Hall (ed.), *DMA Fatal Accidents Book, 1920–50* (Durham, 1995).

an average of seven miners killed every day. An addition of one hour to the working day would mean five more deaths each week; over five years that would mean 1,300 more miners 'killed by Act of Parliament'.[55]

The primary focus of the strikers rested on wages, hours, and conditions of work, but there was some attempt to acquaint the miners with wider political issues. Labour intellectuals made occasional appearances in the Durham villages and, in August, G. D. H. Cole visited Chester-le-Street to give a talk entitled 'The Coal Problem', hosted by the Workers' Educational Association.[56] A month earlier, a crowded meeting in Sherburn Hill had heard a former local miner, now teaching at the CLC, discuss 'Economics as Applied to the Present Crisis'.[57] On at least one occasion, A. J. Cook himself also sought to locate the strike within a wider context when he spoke to Durham miners about the problems caused to the industry by the Dawes scheme and the return to the gold standard.[58] In her letters to the Seaham women, Beatrice Webb had long been referring to international issues. In September 1923, she wrote of the French occupation of the Ruhr, the iniquity of the Treaty of Versailles, Italy's aggressive behaviour towards Greece, and the terrible conditions of life in Germany. 'These terrible and perplexing problems of Foreign Affairs seem far removed from the Durham pit villages,' she explained, 'But, I need not remind you how vitally they are affecting our own prosperity, and even our daily bread.'[59]

Despite such efforts, it is unlikely that many miners were familiar with, for example, the ins and outs of government monetary policy, and for many the slogan sufficed. One ex-miner, writing at the end of the twentieth century, had been a young man of 24 in 1926: 'I didn't know much about objectives, concepts, or what it was all about except what the slogan said.'[60] It is nevertheless surprising that the issue of nationalization was not mentioned in any of the oral or written memoirs that I have come across, though it remained an ongoing dream of the miners' leaders. This may partly be due to the fact that most memories were recorded at least twenty or thirty years after nationalization had been achieved, by which time many men and women had lost their enthusiasm for it. In 1919, the

55 *Chester-le-Street Chronicle* (henceforth *CC*), 23 July 1926.
56 Ibid., 27 Aug. 1926. 57 *DC*, 17 July 1926. 58 Ibid., 14 Aug. 1926.
59 LSE, Passfield 4/15, B. Webb to Seaham women, 20 Sept. 1923.
60 F. Proctor, *I was There: An Autobiography* (Gibsons, British Columbia, 1999), 46.

Sankey Commission had been conducted in a blaze of publicity and optimism, and the Chairman's decision to recommend nationalization had provided Sacriston lodge with a hero for its new banner.[61] But, it may also be because such hopes had been so clearly shattered by 1926. Proposals for nationalization put to Parliament by the miners' MPs in 1924 and 1925 had been easily voted down, and the Samuel Commission, deliberating in private, made no reference to the issue. In any case, by 1925, yet another government report seemed an empty promise. 'Commissions will not settle the problem of today,' declared Cook, 'Commissions may sit as long as they like but there is no solution for the miners under the capitalist system. The issue is one of bread and cheese.'[62]

A commitment to the principles for which the miners were fighting does not mean that every man threw himself into the strike with enthusiasm. Even during the excitement of the general strike, many local newspapers emphasized the apparent passivity of the Durham miner. One described the 'utmost quietude' prevailing in Brandon, Littleburn, and Browney, wondering 'whether the average miner took any interest . . . except to echo the fervent wish that it would soon be over'.[63] Gratified government officials noticed similar apathy, recording that an anti-blackleg march at Stanley had attracted only a few protestors 'owing to the wet'. It was postponed, but two days later had to be cancelled altogether when 'no-one turned up for it'.[64] Poor weather (this time the cold) was also blamed in October when a mass meeting convened by Houghton miners' lodge attracted fewer than one thousand people.[65] In any case, attendance at meetings did not necessarily indicate political commitment. One old miner's wife remembered that 'when the General Strike was on at Spen and Greenside, and they had nothing else to do, they just used to get away to the meetings to put the time in'.[66]

Nor was apathy synonymous with opposition to the strike. On several occasions, when Hensley Henson chatted to miners in Auckland Park, they expressed a desire that the strike could be over and even voiced regrets that it had ever happened. However, this does not mean that they opposed its aims, or even the way it was being conducted

[61] Unfortunately, Lord Sankey joined the National Government only a month after the banner's first gala outing.

[62] Davies, *A. J. Cook*, 88.

[63] *DC*, 22 May 1926. For similar reports, see also *DCA*, 21 May 1926.

[64] TNA:PRO, HO144/6902, Home Office reports, 6 July 1926; 8 July 1926.

[65] *DC*, 16 Oct. 1926.

[66] GCLOT, ii (Mrs F.).

by the MFGB leadership (as Henson interpreted such comments).[67] In November, when the miners were balloted over the acceptance of the owners' proposals, one local newspaper reported the results with surprise: 'Miners' Amazing Apathy. 52,603 fail to vote on vital question,' announced the headline. The DMA calculated a lower figure of non-voters (46,857), although even this smaller number still represented 36 per cent of those union members eligible to vote.[68] But, this does not necessarily mean that these men were apathetic and it seems unlikely that men who had endured months of hardship (and only men who were still on strike were qualified to vote) could have viewed with nonchalance the possibility of its end. Presumably the non-voters also encompassed those wracked with indecision, those desperate to return to work but unwilling to commit such a betrayal to paper, or those who could not bear to vote for a return to work and see the struggle of seven months nullified.

Varying levels of enthusiasm amongst the miners may also have had some relation to generational differences. Nearly ten years later, a 40-year-old miner described with painful candour the inertia induced by getting older. He had not worked since 1928: 'at first I used to feel bitter and want to do something violent . . . I read a lot about Russia and Communism and joined some demonstrations. But, it leads to violence and you can't take risks with the authorities when you've a wife and kids. That's what makes a lot of us only armchair revolutionaries.'[69]

Headlam believed that such attitudes were also apparent in 1926. After talking to a group of older men in his constituency, he commented that many were keen to get back to work. They had worked through the prosperous war years, had put money aside for their old age, and were now seeing their savings eroded.[70] It is impossible to know how far those to whom Headlam spoke were representative of a wider feeling of discontent, but other evidence backs up his suggestion. In July, the *Durham County Advertiser* recorded the words of a miner disgruntled with the policy of the MFGB. Asked why he did not make his opposition known at lodge meetings, he replied that 'if I were to speak in these terms . . . I would be pitched out. All the young men

[67] Dean and Chapter Library, Durham Cathedral (henceforth DCL), diary of H. Hensley Henson, 31 Aug. and 7 Sept. 1926.

[68] *SNCC*, 4 Dec. 1926; DRO, D/DMA109, result of ballot on owners' proposals, 29 Nov. 1926.

[69] J. Newsom, *Out of the Pit: A Challenge to the Comfortable* (Oxford, 1936), 20.

[70] *HPD*(C), 196, c. 711.

back Cook . . . The middle-aged miners are in the minority and are outvoted every time by the younger element'.[71] Younger men were also more likely to have spent the war in uniform, with the radicalization that might have ensued.

Yet, some sources give the opposite impression. During the general strike, Robin Page Arnot, then only 35 himself, had been active in the Northumberland and Durham joint strike committee. Fifty years later he remembered being thrilled when some young men had offered their services. 'Well, there couldn't have been anything better, these were the people that some of the checkweighers were complaining about: "The younger people today, they're nothing like they used to be—they don't come to the lodge meetings, and if we make appeals to them they pay little attention." '[72] But, the 1920s was neither the first nor the last decade in which an older generation griped about the apathy of the young. In fact, while some labour leaders bemoaned the lack of interest shown by some young people to Labour and the union, in part this was the fault of such institutions actively to engage with them. In September 1926, the Labour Party divisional conference urged local parties to take steps to form youth sections in their representative areas.[73] Dipton was one of many villages that had no such provision. Maurice Ridley was born there in 1911 and came to political awareness in the late 1920s:

> If there had been the possibility within the local mining area of getting into a youth section of the Labour Party I would automatically have been in at sixteen. But, membership of the Labour Party was not for kids . . . in County Durham the local Labour Party set-up had little to do with young people at all . . . I began to associate with people in the Young Communist League and they said 'Well, if you haven't got an organized set-up in your area, why not join the YCL?'[74]

Amongst the young, therefore, alienation from the established Labour traditions of their elders did not preclude a radical political consciousness. They may also have felt a closer identification with A. J. Cook (aged 42), than with their own district leaders and the more moderate rhetoric they espoused. Of the six highest officials in the DMA, J. E. Swan (aged 48), was the only man under 50.[75] James Robson, the

71 *DCA*, 30 July 1926.

72 R. Page Arnot, '1926 Remembered and Revealed', *Llafur*, 2 (1977), 14.

73 DRO, D/Sho94, agenda for the Divisional Labour Party conference, 4 Sept. 1926.

74 Ridley, 'Making a Contribution', 64.

75 John E. Swan (1877–1956) worked in the Durham coalfield all his life. He was Labour MP for Barnard Castle 1918–22 and held various DMA offices. In 1926 he was the DMA's compensation secretary.

President, was 66, and Peter Lee, the most outspoken critic of a radical MFGB policy, was 62.

## II

If the miners' loyalties to the lockout and their union were more complex than that assumed by traditional historiography, so too was their attitude to the Labour Party. The miners had been latecomers to Labour politics and the MFGB had been the last major union to affiliate to the Labour Party in 1909. The North East in particular had been a stronghold of Liberalism due to the prevalence of Nonconformity and an attachment to free trade, and the Lib–Lab leaders William Crawford and John Wilson had dominated the early DMA. However, the politics of the coalfield had begun to change after the war. Pockets of Liberalism remained in County Durham into the 1920s, but in the predominantly mining constituencies the reaction against Liberalism had become acute. In December 1923, Beatrice Webb observed that 'on the North East coast Liberalism has disappeared, the turnover of the miners being complete and the disaffected trade unionist of the Labour Party being Conservative when he is not Liberal'.[76] One old miner put it more simply: 'if you were in the union you weren't allowed to vote Liberal. That would be a crime.'[77] In the general election of 1924, when 339 Liberal candidates were fielded across the United Kingdom, the party chose to risk just one man amongst the nine county seats of Durham; he received 10 per cent of the vote.[78]

Even the local organizations of the Liberal Party had little impact on the conduct of the lockout. During the months of the strike, local newspapers detailed hundreds of social events organized by the DMA or the Labour Party and dozens organized by Conservative or Conservative-affiliated groups such as the Mothers' Union. However, only a few solitary references to anything Liberal are recorded: the existence of a Durham division Young Liberal football club; a social evening held by the Men's Liberal Association in South Moor (although about 250 people were present); a meeting of the National League of

[76] Diary entry, 3 Dec. 1923, in M. Cole (ed.), *Beatrice Webb's Diaries*, i, *1912–24* (1952), 253–4.

[77] British Library Sound Archive (henceforth BLSA), C707/155/1–3C1.

[78] F. W. S. Craig, *British Parliamentary Election Results, 1918–49* (Glasgow, 1969), 343.

Liberal Trade Unionists in Stanley, at which the speaker regretted the state of Liberal affairs and noted that the League was £1 million in debt.[79]

In contrast, by 1926 it was the Labour Party which enjoyed the security delivered by the mining seats and commanded the allegiance of almost all within the union. Indeed, to many observers, the Labour Party and the DMA could appear one and the same. The Durham divisional Labour Party had been inaugurated in 1918; appropriately enough, as Matthew Worley has pointed out, in the miners' hall.[80] Since then, an extensive patronage network had developed, with union support often critical in the nomination or appointment of local and national political candidates. One old miner remembered with enthusiasm that 'to be elected onto the [Blaydon] Urban District Council depended on one thing, that you got the nomination of the working-class organizations, namely the miners' lodge'.[81] With rather less approval, Lord Londonderry had complained in 1929 that 'the Urban District Council is the Dawdon Lodge, and the Dawdon Lodge is the Urban District Council'.[82] Comprehensive records are difficult to find, but of interest here are the reports of elections held for parish councillors in Chester-le-Street rural district in 1925. The figures contained in the papers of the five miscellaneous parishes for which records survive show that miners dominated the lists of both nominated and successful candidates. Only in Great Lumley did a significant number of nominated miners fail to achieve election, and this was because of the restricted number of posts available. Also of note is the low proportion of colliery supervisory staff endorsed (see Table 2.5).

The dominance of the DMA was not always absolute. In South Shields, which had been slower than most to break with its Liberal heritage, lodge officials complained of Labour men 'pouring contemptuous and scurrilous odium upon Trade Unionists and particularly miners'; in Bishop Auckland, miners faced competition from railwaymen and teachers in their attempts to control the local party; even in the concentrated mining constituency of Seaham, the divisional Labour Party

79 *DC*, 18 Sept. 1926; *SNCC*, 18 Nov. 1926.

80 M. Worley, *Labour inside the Gate: A History of the British Labour Party between the Wars* (London and New York, 2005), 49.

81 GCLOT, iv (Mr S.).

82 JRUL, RMD/1/5/14, Londonderry to J. Ramsay MacDonald, 19 Mar. 1930.

**Table 2.5** *Nominations and elections to Chester-le-Street Rural District Council, 1925*

| | Great Lumley | Little Lumley | Ouston | Urpeth | Witton Gilbert | Total |
|---|---|---|---|---|---|---|
| Miners nominated | 14 | 9 | 5 | 10 | 10 | 48 |
| (and elected) | (7) | (8) | (5) | (10) | (9) | (39) |
| Colliery craftsmen nominated | | 3 | 1 | 1 | 2 | 7 |
| (and elected) | | (0) | (1) | (1) | (2) | (4) |
| Colliery supervisory staff nominated | | 5 | 7 | 2 | 3 | 17 |
| (and elected) | | (0) | (1) | (0) | (1) | (2) |
| Other manual workers nominated | 1 | 1 | | 6 | 2 | 10 |
| (and elected) | (0) | (1) | | (0) | (0) | (1) |
| Non-manual workers nominated | 4 | 1 | | 2 | 6 | 13 |
| (and elected) | (2) | (0) | | (0) | (2) | (4) |
| Retired/women nominated | | | | 1 | 2 | 3 |
| (and elected) | | | | (0) | (1) | (1) |
| **Total number of nominations** | 19 | 19 | 13 | 22 | 25 | 98 |
| **(total positions available)** | (9) | (9) | (7) | (11) | (15) | (51) |

*Source*: DRO, RD/CS502.

snubbed the DMA Executive in July 1920 in its nomination of Sidney Webb as its parliamentary candidate.[83]

However, despite such tensions, the influence of the Labour Party and DMA within local government proved critical in 1926. Education committees were swiftly authorized to commence school feeding arrangements, for example, and, by the end of the dispute, Durham (still one of only a very few county councils to boast a Labour majority) had provided several million more free meals to school children than any

[83] D. Tanner, 'The Labour Party and Electoral Politics in the Coalfields', in A. Campbell, N. Fishman, and D. Howell (eds), *Miners, Unions and Politics, 1910–47* (Aldershot, 1996), 80; Worley, *Labour*, 88, 49.

other county in Britain, at an estimated total cost of nearly £300,000.[84] Jack Lawson later waxed lyrical:

> As on the former occasion [1921] there was one gleam of sunshine in the gloom of the 1926 lock-out. The bairns were fed in schools, and well fed, too . . . The County Council saw that nothing was lacking, and Peter Lee and his colleagues in the midst of all their troubles and anxieties had the satisfaction of knowing that they at least saved children from the carnage of industrial war.[85]

Labour-dominated boards of guardians were able to provide further practical aid to the strikers and, despite strict Ministry of Health guidelines, many boards attempted to bend the rules in order to maximize the help they could give. The amount of relief provided again exceeded that of most other coalfields. Five of the Poor Law boards in County Durham were relieving more than one in three of their populations by mid-June.[86] The most striking example of this was at Chester-le-Street, where 42,722 people out of a population of about 86,000 were receiving relief when numbers peaked at the beginning of July. Forty-seven of Chester-le-Street's fifty-nine guardians belonged to the Labour Party and thirty-nine were miners' officials, miners, or miners' wives. Indeed, members were acutely aware of the need to preserve the numerical domination of those sympathetic to the strikers. Guardians were not legally permitted to be reliant on the poor law themselves, so, when money was paid out at Birtley, recipients were asked to donate anything they could spare to ensure that no striking guardian lost his status by being forced to apply for relief.[87]

In fact, the elimination of the Chester-le-Street guardians would come by other means. By the beginning of August, a rift had opened between the elected Labour guardians who wanted to give as much support to the miners as they could, and the professional relieving officers, who had begun to disregard the instructions of the guardians and instead distribute relief according to Ministry of Health guidelines. On 28 August, the board suspended the relieving officers, 'because they refuse to give relief to people who are suffering from privation and

[84] *HPD*(C), 204, cc. 583–6. Durham provided 19,387,504 meals to school children, May–Dec. 1926; Glamorgan, which provided the second highest number, only 6,468,043, spending £72,500. School feeding is further discussed in Chap. 5, Sect. II.

[85] J. Lawson, *Peter Lee* (1936), 256.

[86] Namely, Houghton-le-Spring, Easington, Chester-le-Street, Sedgefield, and Lanchester. TNA:PRO, MH132/8, Ministry of Health annual report, 1926–7, 127.

[87] *Parl. Papers 1927*, xi (1123), Chester-le-Street Union. Report of the Board of Guardians on the Administration for the Period 30 Aug. 1926 to 31 Dec. 1926, 4–8.

want through destitution'.[88] A furious Neville Chamberlain was swift to respond and, under the Board of Guardians (Default) Act 1926, which had received Royal Assent only weeks earlier, as Minister of Health he authorized the mass sacking of the board, replaced them with three of his own appointees, and reinstated the compliant relieving officers.[89]

Active and visible attempts at assistance by local Labour politicians could reinforce support for the Labour Party within the coalfield. Referring to the Ministry of Health appointees who replaced the Chester-le-Street guardians, one old miner remarked wryly, 'Well, I can tell you, they didn't belong to the Labour Party.'[90] Another explained years later that at least he and his fellow-miners had had 'one thing in our favour, and it was this, the Chester-le-Street Rural District Council was controlled by Labour . . . and they were in there to seek to help those that couldn't help themselves'.[91] Such evidence is not limited to oral history respondents. In August 1926, one George Suggett secured first prize in a fancy dress competition. He had dressed as a miner and carried the motto 'The Guardian is my shepherd, I shall not want.'[92] There is no way of knowing whether George Suggett opposed an indiscriminate distribution of relief and so meant his costume to be seen ironically, whether he was a miner glad of the help, or perhaps even a guardian himself. None the less, his costume reveals the widespread awareness that partisan guardians could and were actively providing support.

Labour politicians deliberately contrasted such benefits with the political alternatives. In 1925, W. H. Handley stood as Labour candidate for Rainton in the council elections. His campaign propaganda reminded voters of the free meals provided for school children by Labour councillors during the 1921 lockout. He won the seat, reversing the defeat he had suffered to the Conservative candidate three years earlier.[93] In the autumn of 1926, W. P. Richardson again paid tribute to the good work being done in feeding the school children: 'Workers in Durham will now realise the wisdom of electing their own members to the County Council. No County Council which had not a Labour majority on it would have treated the workers in this way.'[94]

88 DRO, U/CS14/1, Chester-le-Street Poor Law Union minutes, 28 Aug. 1926.

89 Similar proceedings took place against the guardians of West Ham (July 1926) and Bedwellty (Feb. 1927). See S. Webb and B. Webb, *English Poor Law History; Part II: The Last Hundred Years* (1963 edn.; first pub. 1929), ii. 925–34.

90 GCLOT, iii (R.E.). 91 Ibid., i (N.C.). 92 *DC*, 7 Aug. 1926.

93 DRO, D/Sho 129/41, local election campaign material, 1925.

94 *Miner*, 25 Sept. 1926.

The dominance of the Labour Party within local government therefore provided an opportunity to consolidate support by a practical demonstration of its benefits. But, it also made it vulnerable to blame when hardship intensified. Boards of guardians in particular were subject to intense criticism, for, however sympathetic their members might be, they could not always be as generous as they might like, and Poor Law Union minute books reveal an almost tangible sense of frustration. Guardians struggled to reconcile their desire to give more help to the miners with an awareness that to provoke the wrath of the Ministry of Health would ultimately prevent them from giving any at all, for, even if not threatened with dismissal, the continued provision of loans from central funds depended on an adherence to Ministry guidelines. The most contentious issue was over the payment of relief to strikers themselves (as opposed to their dependants), prohibited since the Merthyr Tydfil judgement of 1900. This stated that poor relief could not be given to an able-bodied man who simply refused to work unless he became so reduced by want as to be unable to do so, in which case he could be relieved temporarily until his condition improved.[95] In June, the clerk to the Easington guardians warned his colleagues that not only were they personally liable to be surcharged if they departed from this principle (the Chester-le-Street guardians would receive notice to pay such charges later in the year), but such actions jeopardized the continued relief of women and children. Instead, he put forward an alternative proposal:

> providing the Relieving Officer is satisfied that any single man has been without food for twenty-four hours and has no means to obtain food, it may be assumed that he is no longer able-bodied and may be relieved *in kind and on loan*, to the value of a day's ration. Thus to keep him from starvation, he would receive one day's ration every alternate day. This may be stretching the law, but I think it may be assumed that although a man may not have reached the point of actual starvation on three days' rations a week, he cannot be said to be able-bodied for work.[96]

The guardians adopted his suggestion, but three days' rations per week did not satisfy hungry men. A couple of weeks later the clerk regretted the criticism since received, noting that 'I feel sure that if the men

[95] See P. Ryan, 'The Poor Law in 1926', in M. Morris (ed.), *The General Strike* (Middlesex, 1976), 361–2.

[96] DRO, U/Ea17, Easington Poor Law Union minutes, 27 May, 7 June 1926. Original emphasis.

realised how anxious this Board were to assist them, they would accept this relief in the spirit with which it is offered.'[97]

Across the county, Labour guardians faced similar dilemmas and similar discontent when they attempted such compromises. In November, two Labour members of the Lanchester board were assaulted by a hostile crowd a few weeks after reluctantly conceding to the Ministry's demand that the scales of relief be reduced.[98] At Houghton, guardians initially reacted to the dispute by openly relieving 'destitute single adults,' but by September they had been forced to cut back. Under central pressure, they agreed to implement a system whereby single men would receive relief only after medical examination. At the first such inspection, the vast majority of applicants were rejected, the atmosphere turned aggressive, and one doctor was 'slightly mauled'.[99] Even the doomed Chester-le-Street guardians did not escape criticism: in June, 200 unemployed men marched to their offices to demand more generous relief.[100]

## III

Despite the fact that the miners' concerns revolved around industrial questions of wages and hours, bolstered by moderate Labour and union leaders who sought to retain the focus on these issues, not all viewed the strike as devoid of a more political threat. Allegations that the strike was prompted by a revolutionary agenda had plagued the miners since the first heady days of May, when images of the lockout and those of the general strike became intertwined. The decidedly unrevolutionary aims of the TUC leaders during the general strike have been demonstrated thoroughly elsewhere and do not need further evidence here,[101] but such large-scale action inevitably became constitutional and political issue. Only a decade earlier, millions of working-class men had been taught to use guns, and with the USSR only nine years old the spectre of Bolshevism seemed real enough to some. One troubled member of the public wrote to

97 Ibid., 7 June 1926.
98 *BC*, 20 Nov. 1926.
99 TNA:PRO, HO144/6902, Home Office report, 11 Sept. 1926.
100 *CC*, 25 June 1926.
101 See, for example, G. Phillips, *The General Strike: The Politics of Industrial Conflict* (1976), 220–41.

Ramsay MacDonald imploring him to speak out for constitutional government: 'Don't be like Kerensky.'[102] Another wrote to Baldwin with relief following the end of the general strike: 'If they happened to have had a few men of the type of Lenin, they would have grasped the power, and this Country would have been flooded with blood.'[103] A couple of years later, Walter Citrine and Winston Churchill sat next to each other at a dinner party. 'He and I fell to talking quietly about the strike,' recalled Citrine in his memoirs, 'I said, "You didn't really believe all that stuff about our making an attack on the Constitution?" "Oh yes, I did," he replied. "I was in the country at the time and I saw Red." '[104]

In County Durham itself, Headlam also worried that the TUC 'mean revolution'. A few days after the general strike was called off, he visited Burnhope and spoke to several men in the village. 'They were truculent and inclined to be nasty,' he noted, adding, with a touch of melodrama, 'but they made no attempt on our lives'.[105] Meanwhile, the Councils of Action set up by the strikers, and their success in temporarily commanding the normal business of the region, continued to attract Soviet-inspired comparisons. On 20 May, Sir Alfred Palmer, as the Chairman of the Gateshead magistrates, found Edward Wilson of Chopwell guilty of having circulated literature likely to cause disaffection. Sentencing him, he announced:

> the manner in which Chopwell has been governed for some time past is a scandal . . . If you think that the Council of Action can hold up the inhabitants in a state of tyranny you are very much mistaken. Why you and those associated with you don't go to Russia, I don't know. I am sure the Government, and I personally, would subscribe willingly to get rid of the whole lot of you and let you go and live in that country where everything is so blissful and happy.[106]

To men such as Palmer, there seemed real cause for concern. The period of the general strike in the North East saw intermittent outbursts of violence and Durham provided more prosecutions and convictions during the nine days of May than any other county in England, with

[102] TNA:PRO, PRO 30/69/1436, Capt. R. Boumphrey to Ramsay MacDonald, 7 May 1926.

[103] TNA:PRO, CAB21/296, T. Evans to T. Jones, Cabinet Secretary, 5 July 1926.

[104] W. Citrine, *Men and Work: An Autobiography* (1964), 217.

[105] Diary entry, 4 May 1926, in Ball, *The Headlam Diaries*, 86; DRO, D/He22, Headlam's diary, 16 May 1926.

[106] *Newcastle Chronicle*, 21 May 1926. Cited in A. Mason, *The General Strike in the North East* (Hull, 1970), 72 n.

183 cases of violence and disorder out of a total of 583.[107] Tony Mason has commented on the difference between the North East and the rest of the country. In Plymouth, for example, football matches famously characterized relations between police and strikers; in the North East, there were serious clashes with police as a consequence of picketing and attacks on road traffic.[108] Admittedly, those involved were not necessarily miners. In the blunt assessment of one naval officer posted on the Tyne, the vast majority of agitators during the general strike were 'not genuine workers but excellent examples of unemployable hooligans, many of them unwashed, collarless and extremely oderiferous [*sic*]'.[109] Nor did events in Durham hit the headlines like those in Newcastle, where several serious incidents occurred, or in Northumberland, where the *Flying Scotsman* was derailed. Nevertheless, there is evidence to suggest that at least some Durham miners chased a revolutionary agenda. One was jailed for three months after he had allegedly warned:

> If Stanley Baldwin uses his forces—police force, Air Force, Army and Navy—we will meet them and they will go to the bottom. I make no bones about it. I am corresponding with a friend of mine in the Army and another in the Navy and they say that a good number will follow the Red Flag when the opportunity arises.[110]

However, the Durham leaders constantly sought to separate the wider issues of the miners' lockout from the show of solidarity in May and its more revolutionary implications. It was a distinction that sometimes the government itself seemed reluctant to make, as month after month it continued to renew its Emergency Powers legislation (initially introduced as a response to the general strike), until the very end of the mining dispute in December. Arguing passionately against the continuance of such regulations, Robert Richardson, the miners' MP for Houghton-le-Spring, protested in July that 'there is a great difference between a general strike and a dispute in the mines, and I want the Government to keep that in mind . . . I have seen many disputes, and I have to say that in 95 per cent of them we have been defending, and not attacking . . . '.[111]

107 *HPD*(C), 196, c. 825. The West Riding provided 110 and Northumberland 103.

108 Mason, *General Strike*, 102–3.

109 TNA:PRO, ADM116/2439, diary of Commander H. Tafrell Dorling, 2 May 1926.

110 *DCA*, 23 July 1926.

111 *HPD*(C), 197, cc. 1526–7. Robert Richardson (1862–1943) began work in the Durham coalfield at the age of 9. He was appointed secretary of Ryhope lodge in 1887 and

This was not just rhetoric to reassure a jittery House of Commons. From the outset, the DMA leaders had seemed consciously to distance themselves from the general strike. When a regional strike committee was set up in Newcastle at the beginning of May, Will Lawther was one of the few miners to get involved, and then not as an official representative of the DMA. Indeed, apart from a belated official visit paid by Peter Lee, the DMA stalled on giving any positive commitment at all.[112] During the general strike the DCOA also remained detached from the regional branch of the Organization for the Maintenance of Supplies (OMS), declining either to donate funds or advise colliery managements to support the organization; Tony Mason has argued that this was a severe blow to the OMS in Durham.[113] It also suggests that the coal owners, like the miners' leaders, sought to contain the strike within industrial, and even regional boundaries, and did not see it as an overtly political dispute.

At least some rank-and-file members of the union also feared the political implications of the events of May. Some served the government even while they were locked out. In June, the miners' MP for Pontypridd, T. I. Mardy Jones, wondered whether it had been government policy 'to recruit a goodly number of young miners as special constables during the period of the general strike? . . . my information at the moment is that there are in the coalfields, and particularly in South Wales, a number of young miners recruited as special constables for the police force, and paid £3 a week for their job'.[114] His counterparts in the North East shared similar concerns. On 7 May, the Chairman of the Durham branch of the OMS declared himself delighted with the number of applications he had received, remarking that 'many applicants were from the ranks of the strikers who intimated that they were prepared to sink their individual views in the dispute in order to render all the assistance within their power on behalf of the community'.[115] When those who were miners remained on strike after 12 May, the DMA found itself in a dilemma over whether to pay lockout benefit to men regarded by many as irredeemable blacklegs. At the beginning of June it supported West Bitchburn lodge, insisting that the men there 'who

then to the DMA Executive ten years later. He was Labour MP for Houghton-le-Spring 1918–31.

[112] Mason, *General Strike*, 17–20.

[113] The OMS was a volunteer group formed in Sept. 1925 to ensure the maintenance of essential services in the event of a general strike. In May 1926 it put itself under the control of the government. For details of its development in the North East, see ibid., 50–3.

[114] *HPD*(C), 196, c. 860.

[115] *DCA*, 7 May 1926.

joined the Civil Constabulary [during the general strike] cannot be paid lockout benefit'.[116] Some had done so covertly and by the end of May Langley Park lodge was still attempting to confirm the names of its members who had been special constables.[117] The question was finally resolved at the end of June, when the DMA was advised that under the Emergency Regulations it was likely to lose any court case brought against it by a striker refused lockout benefit on such grounds. Without the money to risk such a confrontation, the recommendation was made to local lodges to pay those men who claimed.[118]

Even lodge officials could be guilty of such apparently inconsistent positions. After the collapse of the general strike, Eden lodge debated the conduct of William Jackson, its Financial Secretary, for 'being a Special Constable, and also other Lodge officials absenting themselves from the Strike Committee'. Following 'heated discussion', Jackson and another official were asked to resign. However, Jackson's decision to act as a special constable was not enough to lose him the goodwill of the wider lodge membership. On 4 June, when fresh elections were held for the vacancy, Jackson stood again against six other contenders and won back his position with ease, before resigning again on 1 July, under increased pressure from his colleagues.[119] For at least some workers, therefore, the general strike and the coal dispute were seen not only as separate issues but as requiring different responses. On the one hand, the coal dispute was a purely industrial battle conducted against the owners' attack on wages and hours, in which loyalty lay with workmates. On the other, men such as Jackson may have feared the political implications of a general strike, in which loyalty lay with the government.

The calling off of the general strike put an end to fears of a constitutional crisis, but the lockout continued to provoke political concerns. The TUC rejected Russian roubles during the general strike, but the money was then offered to and accepted by the miners, an Anglo-Russian Miners' Committee was set up, and by the end of the dispute the Russian Trade Union Council had donated over £1 million to the MFGB relief fund, far more than any other foreign or domestic

116 DRO, D/DMA (Acc. 2157(D)), 208 (box)/3, DMA executive committee minutes, 5 June 1926.

117 DRO, D/DMA (Acc. 2157(D)), 187 (vol.), Langley Park lodge minutes, 27 May 1926.

118 DRO, D/DMA (Acc. 2157(D)), 208 (box)/3, DMA executive committee minutes, 21 June 1926.

119 DRO, D/DMA334/5, Eden lodge minutes, 13 May, 4 June, 1 July 1926.

source.[120] Henson expressed his disgust in his diary: 'It is now evident that the Russian money was no voluntary contribution of the Russian Trade Unionists, but the calculated gift of the Soviet Government, designed to promote revolution in Great Britain.' He was still upset the following day: 'It becomes difficult to believe that Cook and Smith are free agents. Have they given any kind of pledge to their Russian paymasters?'[121] Such opinions were not confined to bishops. After the close of the dispute, the Non-Political Miners' Union would invoke similar sentiments when its journal claimed that, in May 1926, 'the agents of an alien and revolutionary junta, mad with the blood of innocent victims dripping from their hands, plunged Industrial England into chaos and turmoil, drove over a million brave men out of employment, [and] kept them and their wives and little children on the brink of starvation for many months'.[122]

In fact, violence in the Durham coalfield had largely ceased by mid-May, to return only in the autumn as the beginning of a noticeable back-to-work movement exacerbated tensions, a pattern common throughout the British coalfields.[123] Towards the end of May, the *Durham County Advertiser*, which had denounced the general strike as illegal earlier that month, could find nothing but praise for the miners, remarking that the colliery villages exuded the atmosphere of a peaceful Sunday afternoon.[124] In December, Malcolm Dillon, Chairman of Seaham Harbour Police Court (and chief agent to Lord Londonderry) expressed satisfaction at the way in which law and order had been maintained. His police superintendent agreed: 'He could say that the people of the Division, as a whole, had conducted themselves in an orderly way. They had had exceptions, but there had been nothing serious, and in the Seaham quarter they had had no trouble whatever.'[125] It was a record frequently cited in Parliament by the miners' leaders as they argued against the Emergency Regulations. Robert Richardson thought the proposed legislation absurd: 'In my own town I expect that the chief mischief into which the miners will get will be the playing of

[120] DRO, D/DMA (Acc. 2157(D)), 194 (box), MFGB statement of accounts for the fourteen months ending 30 June 1927; All-Russian Council of Trade Unions, *Red Money* (1926), 26–7.

[121] DCL, Henson's diary, 8–9 June 1926.

[122] DRO, D/DMA (Sam Watson), 41 (box)/3, *Non-Political Miners' Journal*, June 1927.

[123] See S. Catterall, 'Police', in McIlroy et al., *Industrial Politics*, 261. Also A. Campbell, *The Scottish Miners, 1874–1939*, ii, *Trade Unions and Politics* (Aldershot, 2000), 226.

[124] *DCA*, 14 May 1926.

[125] *Seaham Weekly News* (henceforth *SWN*), 3 Dec. 1926.

marbles.'[126] Indeed, by the time the Emergency Powers' Act was being debated for the seventh (and still not the last) time in October, even Headlam thought some of its provisions to be excessive. Announcing his opposition to the banning of public meetings, he declared that 'in Durham . . . the relations between the police and the miners are as good as they can be'.[127]

Membership figures suggest, however, that the strike did increase the appeal of Communism, at least temporarily. In a recent reassessment of Communist Party membership, based on access to newly opened Russian archives, Andrew Thorpe has calculated that the party's national membership doubled from 6,000 on the outbreak of the strike to a peak of 12,000 in October, an increase that 'was not merely dramatic, but sensational'.[128] In Tyneside, which had been a particularly weak area of Communist support before 1926, the rush of miners and their wives to the party during the strike briefly transformed it into the largest section of the CPGB, with 2,600 members, most of whom were miners.[129] Decline was as swift. National membership was back to just over 6,000 within a year, and Tyneside returned to being a concern to the Communist leadership, with only 200 members left by 1933.[130]

An increased enthusiasm for more radical political organizations benefited other groups as well. In November, it was reported from Wingate that during a three-month propaganda campaign launched by the ILP, packed weekly meetings had been held in the miners' hall and various speakers invited. Sixty-nine new members had been enrolled, compared to the previous year, when the Wingate membership had remained stubbornly at six.[131] Such reports are not necessarily symptomatic of a search for alternative politics, and are also reflective of a coalfield in which 150,000 men found themselves with empty days to fill. However, the strike provided a cause for Communist leaders to focus on, and Communist propaganda could also take advantage of the complaints regarding the Labour-dominated councils. In June, one local newspaper printed a letter from a Communist Party member from Stanley. Aggrieved at the recent

126 *HPD*(C), 195, c. 396.

127 Ibid., 199, c. 767.

128 A. Thorpe, 'The Membership of the Communist Party of Great Britain, 1920–45', *Historical Journal*, 43 (2000), 780.

129 M. Worley, *Class against Class: The Communist Party between the Wars* (2002), 34; Thorpe, 'Membership', 787.

130 Thorpe, 'Membership', 790–2.

131 *Miner*, 20 Nov. 1926.

decision of the Lanchester Guardians to cut the relief scales, he complained:

Now I, my wife and four kiddies, have got to live on a dole of 31*s.* a week, and pay 13*s.* 9*d.* a week rent. The same crowd of class-conscious brethren are holding an eviction order over my head, in respect of my rent, which is for a Council house. And yet we have a Labour Council and a Labour Board of Guardians, as well as a Labour majority on the Durham County Council. It seems to me that some of these people try to punish the workers even more than the boss has attempted.[132]

Any increase in Communist support was achieved despite fierce opposition from the regional mining leadership. In October, Peter Lee gave a stark warning of the danger of following a revolutionary course: 'They would be met by machine guns . . . planes would crush them down. They would be over the precipice, and it might be impossible for them to get back again.'[133] Indeed, the strike hardened DMA opposition to the Communist Party and the Minority Movement. W. P. Richardson blamed the existence of such extremes for the weakening of trade union ranks that allowed Spencerism to flourish.[134] Local lodges could sometimes be more receptive, and when in July the Communist MP Shapurji Saklatvala addressed 5,000 people at New Kyo, the Morrison lodge banner flew alongside that of the local Communist branch.[135]

But, for all the claims of Communist involvement in the dispute, its relevance always remained overshadowed by the more fundamental Labour loyalties of the coalfield. 'The miners are a hard-headed lot, and though they have always been in the forefront of Labour politics . . . they have always refused to be carried away by this and that new craze,' Beatrice Webb told the Seaham women in 1923 (a few years before she herself became an ardent admirer of the Soviet system).[136] Certainly some miners, while remaining loyal to the principles of the union, continued to be suspicious of Communist and particularly Soviet overtones to the struggle. At one meeting of the Enginemen, Boilermen and Firemen's Association, an enquiry was discussed from Hetton lodge

[132] *Workers' Weekly* (henceforth *WW*), 4 June 1926. [133] *DCA*, 29 Oct. 1926.

[134] R. Martin, *Communism and the British Trade Unions, 1924–1933: A Study of the National Minority Movement* (Oxford, 1969), 90. Spencerism was the derogatory name given to the breakaway non-political union movement, named after George Spencer, the Nottinghamshire Miners' Association official and Labour MP who negotiated the return to work of the Nottinghamshire miners in Oct. 1926. [135] *SNCC*, 12 Aug. 1926.

[136] LSE, Passfield 4/5, B. Webb to the Seaham Women, 23 July 1923.

asking if the relief money received from the MFGB had been supplied by the Russian government, as one member was refusing to accept it if so. Somewhat disingenuously, the reply was given that lodges should work on the general assumption that Co-operative Societies and British trade unions were the chief benefactors.[137] For another miner, Soviet involvement became a reason to disown the strike altogether and he explained that it was why he had decided to blackleg: 'There are very few of us like this tainted money from Russia, because we believe the Russians will be on our necks before long.'[138]

In fact, where some kind of allegiance to Communism existed, it frequently revolved around a more fluid and less theoretical sense of what Communism meant. One miner explained years later that 'there are two ways of making a Communist; one is reading Karl Marx and the other is by being kicked around by the employers. Consett Iron Company made far more Communists than ever Karl Marx made.'[139] Another had been a Communist Party member but left after his colliery manager expressed displeasure. Still, he did not see it as a betrayal of his beliefs: 'You can be a fighter without being a member of the Communist Party, and I could cause as much trouble as anybody.'[140] Indeed, the miners could draw upon older traditions of radicalism that had existed in the coalfield long before Marx and Lenin were recognized names. Several historians have documented a Chartist presence in the coalfield in the 1830s and 1840s, particularly amongst the miners' leaders but also, to a greater or lesser extent, amongst the pitmen themselves.[141]

Across the Durham coalfield, it was the small town of Chopwell, with about 10,000 inhabitants, which gained particular notoriety as an area of Communist subterfuge. At the end of May 1926, the *Newcastle Chronicle* revealed what it imagined to be the true horror of the place. As Stuart Macintyre has described, ' "UNDER THE RED BANNER. CLUTCHING HAND OF COMMUNISM. SPECTRE OF A

137 DRO, D/DMA (Acc. 2157(D)), 258 (box)/7, Enginemen's Association, executive committee minutes, 8 June 1926.

138 *DC*, 19 June 1926.

139 J. Stephenson, 'A Comment by James Stephenson of Winlaton', *North East Group for the Study of Labour History Bulletin*, 4 (1970), 29. Consett Iron Company owned Chopwell Colliery.

140 GCLOT, i (N.C.).

141 K. Wilson, 'Chartism and the North East Miners: A Reappraisal', in R. W. Sturgess (ed.), *Pitmen, Viewers and Coalmasters: Essays on North East Coalmining in the Nineteenth Century* (Newcastle-upon-Tyne, 1986), 81–104; R. Colls, *The Pitmen of the Northern Coalfield: Work, Culture and Protest, 1790–1850* (Manchester, 1987), 267–301; M. Chase, *Chartism: A New History* (Manchester, 2007), 102–3.

3. Chopwell Lodge Officials with Banner, 1926. Will Lawther stands with his hands in his pockets, second from left. The banner depicts Marx, Lenin, and Keir Hardie.
Photograph courtesy of Gateshead Council, Libraries & Arts.

MINIATURE RUSSIA," ran its headline above a sensational account of sedition, economic sabotage and Communist Sunday Schools'.[142] In fact, Chopwell had had a radical reputation before the First World War. By 1926, its Communist Sunday School had already waxed and waned, its famous Lenin, Marx, and Owen Terraces had already been christened and an attempt by its footballers to register as the Chopwell Soviets had been denied by the Durham FA.[143] However, even in 'the reddest village in England'[144] men might be uncertain about what such allegiances meant. When Lenin Terrace was built, a member of Blaydon council explained that it was so named 'after the greatest and noblest of trade union leaders that ever lived'.[145] Another old miner had lived

142 S. Macintyre, *Little Moscows: Communism and Working-Class Militancy in Interwar Britain* (1980), 13.

143 The 'Chopwell Reds' was then suggested but again permission was refused; eventually the club was registered as Chopwell White Star. See GCLOT, i (A.W.).

144 *Morning Post*, 15 June 1926.

145 *BC*, 20 June 1925, cited in Macintyre, *Little Moscows*, 14.

in Marx Terrace: 'Well, I never worried about it. At the time it was a home.'[146] Even someone such as Harry Bolton, a Chopwell militant who served a jail term during the lockout, might be influenced by apparently conflicting heroes. According to his grandson's memories, works by Trotsky and Lenin sat next to those by Gladstone and Ramsay MacDonald on his bookcase.[147] One Labour councillor for the area, and brother of DMA agent James Gilliland, was asked years later about the 'Little Moscow' tag awarded to Chopwell by the press. 'We sort of gloried in it . . . it was very Red,' he explained, 'But, as far as the Communist Party was concerned, I think it was just a spirit of fighting the boss, you see, that was the idea, fighting the boss. Well, if you fought the boss, you were a Communist, you see.'[148]

In his study of the 'Little Moscow' community of Mardy in South Wales, Stuart Macintyre discovered a hesitancy to embrace the label even amongst Communists: 'Paradoxically, when some of the residents of Mardy were interviewed in their old age, a sense of pride about Mardy's radical identity was expressed by the less political informants, while leading Communists were apt to dwell on its disadvantages.'[149] In Chopwell the reaction against the 'Little Moscow' epithet was perhaps even more widespread. Many denied that an important Communist influence had existed at all. Jack Parks, another Chopwell militant, shrugged off the label: 'Oh well it was just wide talk you see, and it was in the very early days [that it was given]. Lawther was making himself a bit of a nuisance.'[150] Meanwhile, one old miner who had not been involved in the politics of the period later wondered how the term had come about at all: 'I can't understand it really . . . There never was any trouble in this village other than blacklegs, and of course we wouldn't stand for that. But, apart from that, this had been a very, very peaceful village.'[151] His comment is indicative of the fact that, for many, pride in the solidarity of the union and the radicalism that might accompany it was no indication of a more aggressive politics. Another old miner agreed:

> No, it ['Little Moscow'] was never justified. We never knew anybody that was Reds, in the village, nobody preaching communism or anything like that . . . It was really because Chopwell lodge was a lodge which fought for every privilege they could get in the pits . . . They were the best lodge in the county. But, you see, we had loyal pioneers that fought for these and we were the people that got

146 GCLOT, iv (Mr S.). 147 Ibid. (J.F.). 148 Ibid. (Mr G.).
149 Macintyre, *Little Moscows*, 16. 150 GCLOT, iv (J.P.). 151 Ibid. (B.M.).

it. Well, it made a lot of people jealous. Well, in any other collieries or anything that couldn't get it, they would say, 'Oh, little Moscow's got it.'[152]

## IV

If the union and Labourist ideology which dominated the coalfield usually prevailed over its more radical rivals, it also faced competition from more moderate philosophies. One of the most important was the continuing influence of paternalism. The model of the 'deferential' worker imagined by a rash of studies in the 1960s is one that has rarely been applied to coal miners.[153] Perhaps the most successful examination of employer control over a British mining workforce is Robert Waller's study of the Dukeries coalfield in Nottinghamshire, in which he emphasized the dominance of the colliery management over the occupational, religious, educational, and social lives of the inhabitants of the company villages. He observed that such control severely hindered the strength of the Nottinghamshire Miners' Association and prevented the development of an independent Labour politics until after the Second World War, although he remained reluctant to ascribe this purely to deferential attitudes.[154] He found that, in contrast to the Dukeries, a Labour and trade union presence played a much more important role in the older settlements to the west of the county, where the coal companies did not enjoy such a monopoly on the provision of goods and services, and where other occupational groups existed who were able to sustain Labour politics themselves.[155]

By contrast, the issue of paternalism in the Durham coalfield has been largely neglected, particularly for the interwar period. This is perhaps surprising given the ubiquity of tied housing, which has been cited by several mining historians as an explanation for non-militancy. With regard to the miners of the Ruhr, for example, Dick Geary has written of the 'sinister and authoritarian side of paternalism,' whereby those living in company housing faced rigorous control and the possibility of eviction if they came into conflict with their employers. Twenty-two

152 Ibid. (Mr S.).

153 See Lockwood, 'Sources of Variation'; E. Nordlinger, *The Working-Class Tories: Authority, Deference and Stable Democracy* (1967); F. Parkin, 'Working-Class Conservatives: A Theory of Political Deviance', *British Journal of Sociology*, 18 (1967), 278–90.

154 R. Waller, *The Dukeries Transformed: The Social and Political Development of a Twentieth-Century Coalfield* (Oxford, 1983), 108–63.

155 Ibid., 291–2.

per cent of all Ruhr miners lived in such housing in 1914; Geary argued that this acted as a significant bulwark against militancy and communal culture.[156] In the company towns of West Virginia, Roger Fagge suggested that coal operators 'used the company houses as a method of controlling the labour force'; Waller also mentioned it as one of the factors which limited solidarity in Nottinghamshire.[157]

The number of Durham miners living in tied housing was significantly higher than the 22 per cent of the Ruhr, and Robert Moore's study of Durham's Deerness Valley placed emphasis on its symbolic importance. He described villages made up of colliery houses in which every brick was stamped with the name of the colliery company. But, he suggested that the paternalistic ethos of the Deerness coal companies declined from the turn of the century, due both to the harder attitude of the managerial class and a more politicized workforce, to be finally destroyed by the bitterness of 1926 itself.[158]

However, in the mid-1920s, the adoption of paternalistic techniques was still being touted as good management practice. In 1925, a revised edition of *Colliery Working and Management* was published. It served as a handbook for the colliery hierarchy, and amidst sections on wage costs, mechanization, and various methods of working, it asked, 'What manner of man ought a colliery manager to be?' As well as being a well-trained mining engineer and a good man of business, the authors argued, a manager should also be capable of dealing effectively with his men and 'should cultivate pleasant relationships with them':

> This he may do by taking a personal interest in their reading-rooms and institutes, their athletic clubs, their musical bands, or in some of the various institutions which usually exist in colliery villages—in short, by taking advantage of opportunities for personal contact with them in circumstances favourable to friendliness and goodwill.[159]

In fact, the DMA and DCOA could already boast a long and unique history of mutual cooperation. In 1898, following the passing of

[156] D. Geary, 'The Myth of the Radical Miner', in S. Berger, A. Croll, and N. LaPorte (eds), *Towards a Comparative History of Coalfield Societies* (Aldershot, 2005), 57–8.

[157] R. Fagge, *Power, Culture and Conflict in the Coalfields: West Virginia and South Wales, 1900–1922* (Manchester, 1996), 40; Waller, *Dukeries Transformed*, 78–9.

[158] R. Moore, *Pitmen, Preachers and Politics: The Effects of Methodism in a Durham Mining Community* (Cambridge, 1974), 81, 91–2. For further comment on the system of tied housing see Chap. 1, Sect. IV.

[159] H. F. Bulman and R. A. S. Redmayne, *Colliery Working and Management* (rev. 4th edn., 1925), 63–5.

the first Workmen's Compensation Act the previous year, a joint committee had been established by the DMA and DCOA to hear and resolve compensation disputes. It was still in operation in the 1920s, one of only two such initiatives in any industry across the country.[160] It consisted of five men from both sides amongst whom a unanimous decision had to be reached, otherwise the claim was referred, as elsewhere, to the courts. A 1920 government report hailed it as a shining example of what could be achieved and regretted that such committees were so rare, largely due (unlike in Durham, it implied) to the 'want of confidence and cooperation between employers and workmen'.[161] Called to give evidence, Robert Cooper, the solicitor who sat on the committee on behalf of the Durham owners, described a model of mutual helpfulness.[162]

Many Durham coal owners maintained such a philosophy even during the strike, despite its massive financial implications. Men of the colliery hierarchy continued to involve themselves in the social events of the mining communities: Elemore Colliery's officials organized a fancy dress parade for their workmen, judged by the wife of the colliery manager; at Bewicke Main, the colliery manager opened the local leek and vegetable club's annual show; at Eppleton, the colliery manager presented the prizes at another fancy dress parade.[163] Various colliery managements also gave their blessing to events organized by the union or Labour Party specifically to raise funds or sustain the morale of the strikers. Of all the many and varied entertainments, one of the most popular was pit pony racing, which could only take place with the consent of the manager. One race meeting was held at Spennymoor in July. Bookmakers attended, twenty-one ponies from neighbouring collieries were ridden by pit boys, and over 5,000 spectators attended. The proceeds were donated to the district distress fund.[164]

Some of the most remarkable scenes of cooperation between management and union took place in association with the provision of aged miners' homes. One of the greatest boasts of local lodges, such houses were funded partly by the union itself and partly by the levy on royalties imposed by the Sankey Commission. Employers often also chose to

[160] The Cumberland miners operated a similar system.

[161] *Parl. Papers 1920*, xxvi (1), Report of the Departmental Committee on Workmen's Compensation, 59.

[162] *Parl. Papers 1920*, xxvi (87), Report of the Departmental Committee on Workmen's Compensation, Minutes of Evidence, i, qq. 5933–6175.

[163] *DC*, 10 July 1926; *CC*, 8 Oct. 1926; *SWN*, 15 Oct. 1926;.

[164] *DC*, 3 July 1926.

contribute voluntarily. In August 1926, a new batch of homes was opened at Brancepeth, and Viscount Boyne, Sir Hugh Bell, and Captain Paddon—royalty owner, colliery owner, and manager of Browney Colliery respectively—were invited to the ceremony. Three months into the strike, the speeches given to the assembled crowd by both trade unionists and employers were astonishing for their good humour. Bell's words were recorded by the *Durham Advertiser*:

> [He] said it was sixty-two years since he began to have anything to do with coal mining . . . It was a troublesome job. If anyone killed him it would be Mr Batey [miners' MP for Spennymoor] (laughter)—but he did not want to; in fact he thought Mr Batey rather liked him (laughter, and Mr Batey: 'Oh yes'). Mr Batey has had his knife into me as I have had my knife into Mr Batey (laughter) . . . To the best of his ability he had behaved justly and uprightly to those with whom he had been brought into contact, and he would further say, on the other hand, he had been treated well and justly by all whom he met (hear, hear). He would not even exclude Mr Batey—(laughter)—but would include all the members of the association of which Mr Batey was a very prominent person (hear, hear) . . . They had differed but had not borne any animosity towards each other . . . On the one side he had tried to persuade his opponents that his view was not altogether wrong, just as he knew that his opponents thought that his view was not altogether right (laughter).[165]

During the dispute many colliery owners were prepared to go even further than usual in their support of miners' welfare. The threat of eviction from colliery houses, for example, could have been used to put pressure on strikers. Instead, as early as 7 May, Londonderry Collieries became one of the first of several to recommence the supply of free weekly loads of coal to its workmen.[166] A couple of weeks later, the Charlaw and Sacriston Colliery Company not only gave permission to their workmen to work various outcrop seams but supplied the necessary timber, tubs, and materials for the purpose.[167] Even when coal was not given as a free gift but on the proviso that allowances received would be deducted after the stoppage,[168] such concessions were enormously helpful to families trying to survive without a wage. Indeed, to the outside world, itself suffering from a shortage of coal, it appeared that the colliery companies were positively discriminating in favour of those workmen who had caused the strike in the first place. When, therefore,

165 *DCA*, 6 Aug. 1926. 166 *SWN*, 7 May 1926. 167 *CC*, 21 May 1926.

168 As at South Moor Colliery. DRO, NCB/1/X124, South Moor Colliery Company minutes, 10 May 1926.

Straker and Love provided their employees with a large quantity of coal in June, others could only buy surplus stock, and then only with official permission.[169] Later in the year, two men not connected to the industry stole into Shotton Colliery to help themselves. They were subsequently charged with theft. The police inspector explained that the colliery company allowed its own workmen to take the duff [waste] coal, but it was forbidden to 'outsiders'.[170]

Amongst at least some of the colliery hierarchy this resulted from a genuine belief in their paternalistic duty. In April 1925, Lord Londonderry had written an angry letter to Sidney Webb referring to a speech made by the latter during the county council election campaign:

> I cannot allow statements of this kind to pass unchallenged . . . [In 1921] my agents took an active and leading part in the feeding of the children, and I subscribed money and found the coal, the boilers and every resource which would permit for the purpose of feeding the children, and in many cases not only the children but the old people and invalids. I am not aware that you gave anything in money or in personal help towards the provision of food . . . the allegation that the colliery owners were averse to spending money on feeding the children comes very badly from you.[171]

Webb wrote an apologetic letter in reply, claiming that his speech had been misrepresented by the press. Yet, Londonderry's accusation of hypocrisy rings true, if only to the historian: Beatrice's diary reveals that in 1926 the Webbs thought hard before donating money to the miners' relief fund, reluctant to endorse a stoppage that they did not believe in.[172] Londonderry on the other hand was seemingly generous to his workmen, and in January 1927 the secretary of Dawdon miners' lodge published his thanks (albeit after prompting):

> I have pleasure in acknowledging that during the lockout, whenever we asked for assistance [from Lord Londonderry] towards the feeding of school children it was always granted. We received an average fifteen tons of coal per week . . . and the Grangetown men also had a small supply of coal for the canteen during a period of the lockout. We also received from his lordship several cartloads of

[169] *DC*, 12 June 1926.

[170] Ibid., 23 Oct. 1926.

[171] DRO, D/X1268/38, correspondence between Londonderry and S. Webb, 8–29 Apr. 1925.

[172] Diary entry, 12 June 1926, in N. and J. MacKenzie (eds), *The Diary of Beatrice Webb*, iv, *1924–1943: The Wheel of Life* (1985), 85. Webb wrote: 'At the back of my mind is a certain personal discomfort about the miners' lockout: ought we or ought we not to give and ask others to give, to the fund for the miners' wives and children? Neither Sidney or I would have given a penny to it if no-one would have been the wiser.'

old timber every week, material to fix up sheds, and the use of field ovens and boilers for the kitchens.[173]

Such apparent concern on the part of an upper-class coal owner contrasted with the rather less effusive support of socialists such as the Webbs and inevitably affected their relative standing in the eyes of some miners. One man was quoted anonymously in the local press as describing Londonderry as 'the ideal coal owner', unlike his own misguided leaders who were damaging the living standards of Durham miners.[174]

Admittedly, the relationships between the colliery hierarchy and their employees were subject to regional variations. With regard to the Scottish coalfield, Alan Campbell has suggested that the more urbanized colliery settlements were better resistant to strategies of employer control.[175] Anecdotal reports indicate that the same may have been true of Durham. When in 1910 Jack Lawson moved from Boldon, one of the largest pits in the coalfield, to stand as checkweighman at the smaller Alma colliery, he found that 'it was no uncommon thing . . . to hear the workers' colliery representative and the company's chief agent address each other familiarly by their Christian names. They had gone to school and grown up together. Such a thing was impossible where I had come from'.[176] Thirty years later, such differences remained evident and when a character in Mark Benney's *Charity Main* visited the upland pits, he too found them very different to the large pits on the east coast, echoing Lawson's description of managers who were 'local men born and bred, speaking the same dialect as the men who worked under them'.[177] During the lockout the miners' leaders made some attempt to exploit such relationships, and when Peter Lee addressed a mass meeting at Stanley he appealed to the small colliery owners, 'probably the sons of working men', to stop their pits until the crisis ended.[178] However, there is little evidence that the size of the colliery made any substantial difference to the radicalism of its workforce. One possible measurement of militancy is the percentage of men who voted in November to reject the owners' proposals and continue the strike. This appears to have had little correlation with the size of pit. A tenuous line of best fit does suggest that the bigger collieries, which were likely to have a less-paternalistic

173 DRO, D/Lo/F613, newspaper cutting, source unknown, n.d., *c.*Jan 1927?

174 *DCA*, 18 June 1926.

175 Campbell, *Scottish Miners*, i, *Industry, Work and Community* (Aldershot, 2000), 296.

176 Lawson, *Man's Life*, 113–14.

177 M. Benney, *Charity Main: A Coalfield Chronicle* (1946), 85.

178 *BC*, 12 June 1926.

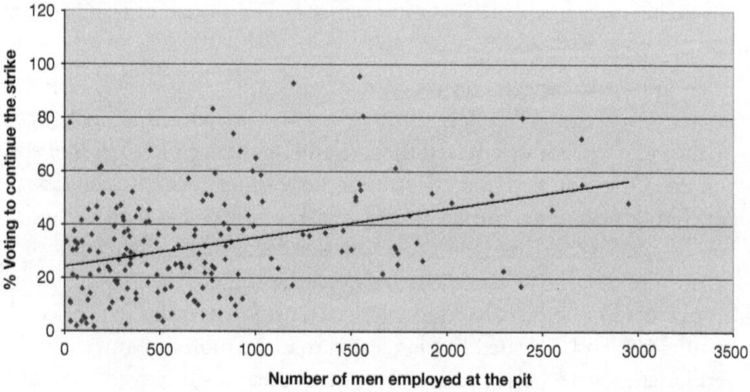


**Figure 2.1.** Results of Durham Miners' Association ballot, November 1926

ethos, did have a greater percentage of men voting against the proposals, but the scattering of results is considerable (see Figure 2.1).

In his essay on the relationship between the gentry and the labouring poor in eighteenth-century England, E. P. Thompson found 'a more active and reciprocal relationship than the one normally brought to mind under the formula "paternalism and deference"'. Not least important, he argued, was the difference between public actions and private ones: 'The same man who touches his forelock to the squire by day—and who goes down in history as an example of deference—may kill his sheep, snare his pheasant or poison his dogs at night.'[179] The Durham coalfield communities of the interwar years were of course very different from the agricultural villages of 200 years earlier. But, Thompson's suggestion that a simple delineation between paternalism and deference is misleading remains relevant. In any analysis of 1926, strong relationships between the union and management should not necessarily be interpreted as an indication of paternalistic attitudes amongst the employers, and still less of deferential attitudes amongst the workers. 'When one of our coalowners died the number of people who went to his funeral was amazing,' one old miner later remembered, 'I often wonder if it was respect or only to make sure they buried him.'[180]

In fact, a striking degree of collaboration between the two could be the result of other factors. The pragmatic colliery manager had

[179] E. P. Thompson, 'Patrician Society, Plebian Culture', *Journal of Social History*, 7 (1974), 396, 399.

[180] BMOA, 1991/82.

good reason to cooperate with the union. As Captain Brass, the sub-agent of Kimblesworth Colliery, declared, he 'did not think there was any colliery manager who would wish to return to the old time of struggling with individual workmen. Two or three non-unionists gave them more trouble than two or three thousand trade unionists properly organized'.[181] If anything, such practical considerations became more important during the lockout. Years later, one old miner laughed at the idea that the free provision of coal had been a product of generosity. He believed that it had been initiated only to limit the potential for trouble: 'so the miners wouldn't raid the two big coal heaps, because if they started raiding them they'd soon be joined by outsiders such as hawkers and tradesmen from the surrounding districts.'[182] Perhaps most importantly, both sides knew that sooner or later normal working relationships would have to be resumed. Asked fifty years later whether the owners had used their monopoly on housing to increase pressure on their workforce, one old miner put it bluntly: 'They never tried nowt of that, because you see, if they'd put all the men out of the house, where were they going to get anybody to start at the pit? They weren't that daft.'[183]

Public cooperation and generosity on the part of the coal owners were therefore no guarantee of private empathy or even sympathy for their mining workforce. Sir Hugh Bell, who had spoken with such warmth at the aged miners' homes' ceremony, remained a prominent member of the DCOA and provoked even Headlam's criticism when the Conservative MP heard that Bell was trying to exploit the opportunity provided by the strike to insist upon the abolition of the minimum wage.[184] Indeed, in December, Bell's name was brought up in Parliament by the Glaswegian MP George Hardie who furiously objected to the fact that a couple of months earlier Sir Hugh had allegedly suggested 'that the Prime Minister might want to do some shooting'.[185]

Mutual agreements were in the interests of the miners, too, and even the friendly relations suggested by the joint compensation committee were sustained by the desire of both parties to avoid costly court cases.

[181] *DCA*, 3 Aug. 1923. [182] GCLOT, ii (Mr H.). [183] Ibid., iii (J.R.).

[184] DRO, D/He22, Headlam's diary, 31 Oct. 1926.

[185] *HPD*(C), 200, c. 2424. Bell's stance is still more surprising given his family background. Lady Bell was still writing sympathetic social commentary about the working classes at this time, as demonstrated in a 1922 essay, 'Women at the Works—and Elsewhere', in her collection *Landmarks* (1929). The Bells' daughter, Molly, married the (then Liberal) MP C. P. Trevelyan in 1904. He would go on to be President of the Board of Education in the 1924 and 1929–31 Labour governments.

Between January and December 1926, over 1,000 compensation cases were heard regarding both fatal and non-fatal accidents: 434 were settled by committee; only fifty-eight were referred to the county court. These figures implied a little more discord than in 1925, when the committee had reached agreement over 498 cases and only forty-four had been referred, but they were certainly not unusual. Yet, rather than celebrating cooperation, John Swan, the DMA's Compensation Secretary, reviewed the year with a torrent of bitterness, declaring that 'the seven months' stoppage gave us no respite but added to our work. During that period the Owners saw to it that no truce existed. War was waged almost on every man in receipt of compensation.' Two years earlier, his predecessor James Robson had expressed similar sentiments, but reminded his readers that the committee still remained preferable to court cases, which were both cumbersome and costly.[186]

The hostility that many miners felt towards the colliery hierarchy has already been noted, and it appears that the strike did little to change this.[187] When Jack Lawson praised the ability of union and management to come together to honour an event such as the opening of aged miners' homes, therefore, he did not celebrate deference. Rather, he sought to stress the basic decency and *independence* of the miner:

> Would this House believe that in my own county . . . our people have built, out of their own wages, no fewer than 1,500 homes for their own people, and they are homes that are almost as beautiful as a fairy tale and that are a joy to see? The coalowners have given their contribution and done something towards that, and my friends and I here, during this conflict, with all its bitterness, have been time after time on the platform with coalowners and with managers . . . they have agreed to forget their feud for the moment, and some of the finest meetings I have ever addressed have been during this crisis. That is the miner, and I know there are Members opposite who know what he is and how intelligent he is.[188]

## V

Complementing the influence of paternalism was the effect of Conservative politics in the coalfield. Despite the importance of County Durham as a stronghold of Labour politics, those who identified with

[186] DRO, Library C47–8, DMA minutes of arbitration committee, 1924–5; D/DMA331/2, DMA minutes of arbitration committee, 1926.

[187] For attitudes towards the colliery hierarchy see Chap. 1, Sect. IV.

[188] *HPD*(C), 200, cc. 733–4.

alternative political identities were far from insignificant. When Headlam first visited Barnard Castle in June 1924 as its prospective MP he did not like what he saw: 'The gloominess of the pit manager . . . was very disheartening. They clearly think that it is a hopeless business to try and induce the miners to vote for me. In some places seemingly a man dare not admit that he is a Conservative.'[189] However, Headlam won his seat, and that year every county constituency saw at least one-quarter of votes cast for the Conservatives (see Table 2.3). Even in Seaham, which had the highest density of miners in the county, Sidney Webb's Conservative opponent won over one-third of the vote.

Five years and a lockout later, Labour took every seat at the general election, maintaining a fairly similar proportion of the vote. Conservative support was shattered, in part due to the resurgence of Liberal candidates. How far the strike was responsible for any weakening of Conservative support is difficult to ascertain, although Worley has suggested that its national effects could be seen in the gains made by the Labour Party in the municipal election contests of 1926–9.[190] Only one by-election was held in the North East (in Northumberland rather than Durham) during the months of May to November 1926. Touted in the press as an indication of popular reactions to the strike, it saw Margaret Bondfield increase the Labour majority over the Conservatives from 1,600 to 9,000 in Wallsend (the Liberal candidate forfeited his deposit).[191] Certainly, an increased hostility to Conservative politics is what might be expected as a result of the strike. The government strained to maintain an appearance of exasperated neutrality throughout the dispute, frequently denouncing the recalcitrance of both sides. To the miners, however, it appeared that the government's sympathies lay firmly with the colliery owners; the Baldwin family business even encompassed several collieries in South Wales. Accusations of prejudice became particularly bitter over the Coal Mines Act, by which the government effectively sanctioned the owners' demands for a longer working day. The Durham leaders were outraged: 'I can assure the government that they have never united the miners more than they have done by the introduction of this Bill,' warned Ritson.[192] Even the cautious Jack Lawson denounced the bill as nothing less than 'a declaration of war'.[193]

189 Cited in H. Beynon and T. Austrin, *Masters and Servants: Class and Patronage in the Making of a Labour Organisation* (1994), 321.

190 Worley, *Labour*, 114.

191 *Evening Chronicle*, 22 July 1926.

192 *HPD*(C), 197, c. 1237.

193 Ibid., 197, c. 1038.

If the Conservative government saw its limited popularity further shaken by the strike, local Conservative associations were much more successful at retaining support. Conservative social clubs were not unknown in the Durham coalfield, and where they existed they could boast memberships of a couple of hundred or more.[194] Many more men and women participated in social events organized by the Conservative Party, and continued to do so during the lockout itself. In October 1926, 100 people attended a whist drive and dance under the auspices of Barnard Castle Unionist Association; 200 attended a dance at Hamsterley organized by the Blaydon division; and a vegetable show put on by the Birtley Conservative Club was heralded 'in every way a decided success'.[195] Not all those who attended such events to enjoy the drink or entertainment necessarily adhered to the politics. Stefan Berger has noted that Conservative working men's clubs also existed in the South Wales coalfield, but that their customers were 'predominantly solid Labour voters who did not want to miss out on good and cheap beer'.[196]

However, support for more overtly political occasions was also forthcoming. At a meeting of the National Conservative League in September it was reported that the previous two years had seen the establishment of over 100 lodges in County Durham, with a membership of around 6,000.[197] The previous month, it was claimed that the annual rally of the Blaydon Divisional Unionist Association had attracted a crowd of nearly 5,000. One speaker remarked that 'it was certainly a stimulating sight to see a great gathering such as that in the heart of what one might call, from a political point of view, a somewhat enemy country'.[198]

Support for Conservative politics did not necessarily translate into union disloyalty. However often they were used as interchangeable insults, Tory and blackleg were not synonymous terms. It was a phenomenon that perplexed Beatrice Webb: 'How extraordinary it is,' she wrote, 'that there are still thousands of workmen who, in defence of their wages, are prepared, at the cost of much personal suffering, to strike against their employers, and yet who are not prepared to

[194] For example, in Birtley and Houghton-le-Spring. DRO, PS/CS44, register of clubs, Chester-le-Street division; Tyne and Wear Archives Service, MG.HS/91/1, register of clubs, Houghton-le-Spring division.

[195] *SNCC*, 21 Oct. 1926; *BC*, 30 Oct. 1926; *CC*, 1 Oct. 1926.

[196] S. Berger, 'Working-Class Culture and the Labour Movement in the South Wales and the Ruhr Coalfields, 1850–2000: A Comparison', *Llafur*, 8 (2001), 30.

[197] *BC*, 4 Sept. 1926.

[198] *SNCC*, 5 Aug. 1926; *BC*, 7 Aug. 1926.

vote against their employers!'[199] But, in fact, the local Conservative associations, badly organized and patchy as they were, never suggested that a trade union consciousness was incompatible with their politics. Rather than criticize the local union, they argued that the problem was its domination by the political organization of the MFGB. At a meeting of the Eden lodge of the National Conservative League during the strike, its secretary attempted to reason with the miners, arguing emotively that it was the socialists who posed the greatest threat to their welfare: 'If all the uneconomic pits were closed, as suggested by Herbert Smith, in order that better wages could be paid in those remaining, it might be sound economy, looked at from a cold-blooded point of view, but it meant that all those people engaged at uneconomic pits would be thrown onto the scrap heap.'[200] There was nothing wrong with district settlements, he argued, because 'the Durham Miners' Association was their Trade Union, not the Miners' Federation of Great Britain'.[201]

A decade or so earlier, the Liberal Party had sought to retain support by tailoring its politics to fit the interests of a specifically local mining culture.[202] In the 1920s, the Conservatives attempted to do the same: fighting for political support amongst a workforce already heavily influenced by the Labour traditions of its union they could not afford to ignore the dominant culture of the union altogether. Throughout the strike, local Conservative clubs demonstrated their active support of the miners and enthusiastically involved themselves in workers' sports and activities, often in aid of strikers' funds. In July, a race with prizes was organized at Easington by the local Conservative club, while subsequent months saw both Thornley and Sherburn Women's Unionist Associations host whist drives and dances in aid of their respective children's boot funds.[203] Direct financial help might also be given and the documents of one soup kitchen record a donation of £1 from the regional branch of the Local Conservative League.[204] Admittedly, some were reluctant to give preferential treatment to the strikers, and when the Pittington Women's Unionist Association held a whist drive and dance in November, they declared it to be in aid of 'a boot fund for children *of all classes*'.[205]

199 LSE, Passfield 4/15, B. Webb to Seaham women, 6 June 1925.

200 *DC*, 17 July 1926.

201 *DCA*, 16 July 1926.

202 See D. Tanner, *Political Change and the Labour Party* (Cambridge, 1990), 209–20.

203 *DC*, 10 July 1926; *DCA*, 13 Aug. 1926; *DC*, 2 Oct. 1926.

204 DRO, D/X411/172, financial report for Springwell and Mount's single men's soup kitchen, 1926.

205 *DC*, 20 Nov. 1926. Emphasis added.

The Conservatives could also appeal to voters through more traditional channels. Throughout the stoppage, the letter columns of local newspapers saw frequent complaints from Durham's middle class about local rates of taxation. Such letters were not unique to the months of the strike but, in 1926, the cost of school feeding and the enormous increase in relief payments accentuated such grievances. When new rates were set by the Chester-le-Street district council in October, for example, eight of its ten parishes saw an increase.[206] The anger that such measures provoked was reflected in the formation of a rash of new ratepayers' organizations.[207]

The assumption has often been made that such organizations excluded the miners. Norman McCord, for example, has suggested that while an indiscriminate provision of outdoor relief 'was perhaps an entirely intelligible standpoint on the part of those responsible for a prolonged coal strike . . . it was scarcely likely to recommend itself . . . to the ratepayers, whose money was being spent so freely'.[208] However, for many miners, interests overlapped. In 1924, 48,942 houses were provided free of rent to mineworkers in Durham, and local rates were paid on behalf of their inhabitants by the colliery companies. A further 54,639 mineworkers received an allowance in lieu of free housing; for these householders the increase in rates was therefore a cause for concern.[209] In June, when the Annfield Plain, Stanley, and Tanfield Ratepayers' Association protested to the Ministry of Health about the relief being given to strikers, the letter pointed out that the association's 1,500-strong membership was principally made up of miners.[210] Complaints were also heard at Ferryhill, where the local ratepayers' association passed a similar resolution, again adding that its members were mainly miners.[211] One frustrated ratepayer wrote to his local newspaper:

> I am one who has gone through a few strikes now and never asked for relief from rates, which I have contributed to for the last twenty years, and a 'coal hewer' at that too. Increased rates are the order of the day now every half year, and the mad-brained policy of Council and Guardians alike is driving hundreds of working men away from Labour. Until Labour asserts itself and

206 *CC*, 22 Oct. 1926.

207 For example, at Easington, Wheatley Hill, Hetton, and Ryhope. See *DCA*, 17 Sept. 1926; *DC*, 9 Oct., 16 Oct. and 6 Nov. 1926.

208 N. McCord, *North East England: An Economic and Social History* (Bristol, 1979), 245.

209 Royal Commission on the Coal Industry (1925), iii, Appendices (1926), 248–9.

210 *DCA*, 11 June 1926.

211 Ibid., 25 June 1926.

does something to relieve the responsibilities of men like myself, self-respecting ratepayers, we in turn say to Labour, 'You don't get our votes any more'.[212]

For the minority of miners who owned their own homes, the position was still worse, as they saw their rates rise while also finding that their possession of such an asset disqualified them from the relief given to their fellow strikers. Accurate figures of home ownership have proved impossible to find, but it appears that very few miners were in such a position, unlike in other coalfields such as South Wales where owner occupation was much more widespread.[213] However, amongst this minority, the sense of injustice was keenly felt. One retired miner from Pelton wrote to Neville Chamberlain to complain: 'I am an old miner who, through thrift and hard work, saved a little money, invested it in cottage property here, which now I am done work I expected to live on, but the rates are now so heavy that I cant [*sic*] live. I am worse off than those who are getting this relief.' 'PS,' he added, 'If the likes of me speaks to them we are insulted. We are not of the same Political Persuasion.' Sure enough, summoned to present his case before the Chester-le-Street Board of Guardians in person a couple of months later, he was accused of being a pawn of the local Conservative party and jeered out of the room.[214]

## VI

If support for the Labour Party could sometimes be lost in the Durham coalfield, it seems that a union loyalty remained much more resilient, surviving even in conjunction with Conservative allegiances, or to those susceptible to the paternalistic overtures of their colliery management. However, even the ties of the union were not unbreakable, as illustrated by the small minority of men who were prepared to defy the dominant ideology of their community and return to work as blacklegs.

Contemporary and later stereotypes of 'scabs' portrayed them as men who were distrusted and shunned within the community well before the strike began. After such men blacklegged, there would be a general nodding of heads and the suggestion that their deviation was only to have been expected. One man gave his opinion of blacklegs from the

212 *CC*, 23 July 1926.

213 Approximately two-thirds of miners in the Rhondda were estimated to own their own homes by 1914. See Williams, *Democratic Rhondda*, 18.

214 DRO, U/CS310, R. Bell to N. Chamberlain, 17 May 1926; *DC*, 17 July 1926.

vantage point of the 1970s. In 1926, he claimed, one of the first to blackleg in his village had been the boss's groom and odd job man, who had been known for taking any opportunity to ingratiate himself with management: 'Shall I sweep your yard, sir?' Another group of blacklegs he described with contempt as 'ale-drinkers from Staffordshire'. Various others were interlinked by blood or marriage either to other blacklegs or to management; two were men who 'would have sold their own wives for half a dollar'.[215] This particular commentator had served a jail sentence in 1926 for breaches of the Emergency Regulations and so might be expected to be a particularly unsympathetic observer, but his sentiments were echoed throughout the coalfield. In his written memoirs, Bill Carr wondered, 'What makes a Blackleg?' He explained: 'They vary in type. What they have in common is a shrinking away from any social responsibility, they can be expert at fawning on the boss, almost serf-like in their appreciation of authority.'[216]

Another common characteristic of blacklegs, according to popular opinion, was their terrible workmanship, and it was commonly heard that they were 'neither use nor ornament when they were in the pit'.[217] This not only lessened the blow felt by their defection, but the ridiculing of a blackleg's skill also struck at his masculinity. In November, the New Silksworth correspondent for the *Workers' Weekly* reported on the latest blacklegs in his area: 'One . . . is a scissor grinder from Sheffield, and will hardly know his way in-bye. The other too [*sic*], father and son, are as good as a blank file, so it will be some time before they make their presence felt.'[218] Blacklegs might even look different to the rest of the population. Given the absence of photographic evidence, the record of a children's fancy dress ball held at Murton on 1 December is desperately tantalizing. Amidst the usual variety of costumes—fairies, Christmas crackers, 'Red Indians', and so on—two little boys came dressed as 'pitmen'; another two dressed as 'scabs'.[219]

Such stereotypes may well have possessed some grain of truth. Poor workers were perhaps more likely to want or need to impress the management by showing loyalty and returning early. Indeed, a link between fickle union members and shoddy workmanship had already been noted the previous year in a DCOA survey which asked whether

[215] Durham University Library (henceforth DUL), misc. accession 1996/7:4, qualidata reference QD35/Moore, data gathered by Prof. Robert Moore.

[216] B. Carr, 'Memories of the General Strike', *North East Group for the Study of Labour History Bulletin*, 10 (1976), 18–19.

[217] GCLOT, iii (J.R.).

[218] *WW*, 5 Nov. 1926.

[219] *DC*, 4 Dec. 1926.

its members had ever dismissed men for being either non-unionists or in arrears. The Stella Coal Company found it a hard question to answer: 'It is a little difficult to express the exact position . . . In settling which men have to be dismissed it generally happens that "Unfinancial" men are also inefficient and bad in so many instances that they receive their notices in any case.'[220] Other scraps of evidence suggest that blacklegs were quickly replaced by the regular workforce once the dispute had ended. One oral account recalled, not without sympathy, that blacklegs were 'thrown to the wolves'.[221] James Stephenson, the secretary of Rowlands Gill lodge in 1926, later claimed to have been told by the colliery agent of the latter's reluctance to hire members of the Non-Political Union: 'I can't trust them,' he had explained, 'People who can't play fair and can't be trusted by their own kind are not likely to be very much use to me'.[222]

However, blacklegging was not a straightforward issue of solid, union men versus the weak, stupid, or greedy, and a significant proportion of those who blacklegged could not be dismissed as such. To walk the gauntlet of booing crowds must actually have required great courage and a strong character; far more so in Durham, where strike-breaking was infrequent, than in some of the Midland counties where it might be the strikers themselves who were in a minority. Within the pit villages they faced ostracism and alienation, and sometimes physical retaliation too. On a visit to Durham in August, A. J. Cook urged his listeners to think of the blacklegs as lepers.[223] The following month, the local press recorded ugly scenes at Fanny Pit when fifty working miners emerged to face several thousand men and women. Three who tried to escape were 'severely mauled'.[224]

Alan Burge has examined 'scabbing' in interwar South Wales and found that many of those involved in the region's breakaway union had once been liked, trusted, and respected by their neighbours and workmates.[225] His findings are echoed in Durham in 1926 where profiles of blacklegs also defy stereotyping. Indeed, some of those who returned to work not only held important positions within the community but played leading roles within collectivist institutions. In August, for example, a local newspaper reported that three men had started work

220 DRO, NCB1Co86(895), results of DCOA survey, May 1925.

221 For example, GCLOT, iii (H.M.).

222 Stephenson, 'A Comment', 27.

223 *SNCC*, 12 Aug. 1926.

224 *DCA*, 10 Sept. 1926.

225 A. Burge, 'In Search of Harry Blount: Scabbing between the Wars in One South Wales Community', *Llafur*, 6 (1994), 58–69.

in an unnamed district of Durham, all of whom were employees of the local Co-operative Society, and one of whom was on the management committee.[226] Even the lodge hierarchies were not immune to the pressures that could break a steadfast resolve and drive a man to blackleg. At Eden lodge, amidst rising anxiety over the numbers returning to work, a rota was drawn up organizing the union officials into teams to picket the colliery office where signing on took place. But, the task of watching other men going back proved too much for some and a couple of months later it was recorded 'that as W$^{m}$ Hughes signed on for Work whilst on duty as a Picket he be removed from Office as an Auditor at once'.[227]

A man's transformation from striker to 'scab' was rarely straightforward. Men who made the decision to return frequently then lost their nerve. On 17 June, 120 unionists held a meeting and decided to return to Frankland Colliery on the following Monday; yet, on the appointed day not one turned up. Eventually only forty-two restarted on the 28th.[228] Even once a miner was working again he might still change his mind, and government bulletins testify to agonies of indecision. On 6 September, Fanny Pit reopened with fifty-six men, albeit 'mainly strangers—not the regular men'; by the tenth the pit was silent again. The reason is unknown, but the bulletins of the intervening days record the gathering of hostile crowds around the pithead and the need for a strong police presence.[229] Decisions to return were also affected by local peculiarities. In July, a youth usually employed by Adventure Colliery was killed by a fall of stone whilst seeking coal, and 'the melancholy incident cast quite a gloom over the district, where the young man was a favourite'.[230] The funeral was held on a Saturday and may well have contributed to the return to work of a significant number of men the following week.

But, if the popular image of blacklegs demanded that they be written off in the most damning of terms, there is some evidence that, even at the time, men and women were capable of seeing subtleties. After writing about blacklegs as a group so vociferously in his memoirs as cited above, Bill Carr then immediately went on to describe one of his own friends who returned to work with more sympathy:

> He was not a rat; the sheer boredom of non-preoccupation with the pit that at least gave him, as a single man, a reasonable social life, was denied him, couple

[226] *DCA*, 3 Sept. 1926.

[227] DRO, D/DMA334/5, Eden lodge minutes, 2 Dec. 1926.

[228] TNA:PRO, HO144/6902, Home Office report, 17, 21, 28 June 1926.

[229] Ibid., 6–10 Sept. 1926.

[230] *DC*, 10 July 1926.

with family pressures and (father a non-miner), all this finally caused him to make the break. It was a deep, sad blow to me. He was blacklisted from our home for ever.[231]

Any analysis of the motives of blacklegs is difficult. Written sources are extremely patchy and oral sources, often illuminating in other areas, falter due to the small number of blacklegs in the first place and the reluctance of those few to talk with candour about their motivations. One stereotypical explanation is that of the nagging wife, but this will be deconstructed below.[232] Furthermore, despite an expectation of prejudice against the Irish as strike-breakers, there is little evidence of such. I have come across only one oral history respondent who suggested that Irish undercutting of native workers, particularly during strikes, contributed towards anti-Irish sentiment in the coalfield.[233] His solitary account can be set against the recollections of one of the Methodists interviewed by Robert Moore, who reeled off a long list of blacklegs and his opinion of them but stated that he 'never knew a Catholic who blacklegged'.[234]

More often, contemporaries simply stressed the fact that blacklegs came from outside the district union culture: they were either 'strangers' to the area and so lacked the geographical ties which made solidarity possible; or they were non-unionists—men who had already opted out of a union identity and by blacklegging simply emphasized a division that already existed.[235] When in June a small drift at Tanfield became one of the first in the coalfield to resume work, it was quickly pointed out that the seventy-two men employed there had never belonged to a lodge affiliated to the DMA.[236] When rumours spread in August that attempts were being made by the management of the Pelaw Main Collieries to persuade men to return to work, it was reported that seven men had already been recruited—'all strangers to the district'.[237] However, if pointing the finger at 'strangers' or non-unionists allowed union members to dodge harder questions about the solidarity of their immediate lodge and village, it is clear, as documented above, that not all blacklegging was reducible to this.

231 B. Carr, 'Memories', 18–19.

232 See Chap. 3.

233 BLSA, C900/11043C1.

234 DUL, 1996/7: 4, data gathered by Robert Moore.

235 Notably many of the pits that reopened on pre-strike terms were not members of the owners' organization. It is significant that the Adventure Colliery, which gained a reputation as the most notorious of the 'scab' pits, was owned by Rainton Colliery Company, which did not belong to the DCOA.

236 *WW*, 11 June 1926.

237 *CC*, 13 Aug. 1926.

It may be that blacklegging was simply indicative of a stronger—and perhaps braver—concern with individualism that many within the coalfield shared. For, even if they did not go as far as blacklegging, many proved themselves capable of more ordinary deceptions during the course of the dispute, which were hardly consistent with a collectivist ideology or the spirit of a communitarian struggle. In straightened circumstances, individuals sought to obtain money any way they could and newspapers frequently recorded prosecutions of those fraudulently receiving benefit. In July, for example, Percy Donkin of Seaham Harbour was prosecuted for 'obtaining relief on false pretences, having represented himself as a married man when he was single'.[238] A week later, Thomas Allan of Horden was jailed for not declaring his sickness benefit, which allowed him to qualify for unemployment pay.[239] Petty corruption could even exist within the ranks of union officials and the minutes of Eden lodge record the suspension and subsequent resignation of its treasurer over anomalies in the payment of lockout benefit, with the implication that some had been pocketed.[240]

Others were prepared to further their own personal welfare more directly at the expense of their fellow miners. One woman interviewed fifty years later still remembered how upset she had been when one day her husband arrived home with his relief money, and she had shouted from a neighbour's house to tell him where to leave it. It had vanished by the time she got home.[241] Another man remembered a woman who had gone to talk to a friend, leaving her husband's meal cooking on the stove. When she returned, the pan had gone, with the meal inside it.[242] Such crimes were not always undertaken anonymously, and at Castle Eden in September, four young miners were charged with assaulting and robbing two others and stealing from them a silver watch, a gold ring, and a pin.[243] As the strike neared its end, others took advantage of the season of goodwill. 'An ingenious fraud is being perpetrated at Dawdon Colliery under the guise of carol singing,' reported one local newspaper, 'As the youths make a considerable noise in the yard, singing several carols, others of the gang have been emptying the coal houses.'[244] Indeed, deceit and theft became so widespread that they entered the culture of the coalfield and became the topic of jokes. In

[238] DRO, U/Ea17, Easington Poor Law Union minutes, 22 July 1926.
[239] *DCA*, 30 July 1926.
[240] DRO, D/DMA334/5, Eden lodge minutes, 13 Aug. 1926.
[241] GCLOT, i (Mrs G.). [242] Ibid., iii (R.E.). [243] *DC*, 25 Sept. 1926.
[244] *DCA*, 26 Nov. 1926.

July 1926, a resident of Monkwearmouth submitted his entry to the *Durham Chronicle*'s column of 'local anecdotes':

Geordie, who is on strike, entered the reading room of the Workmen's Club and began boasting to his fellow strikers that he had a splendid load of best coals. A young putter . . . said 'Wey, Geordie, thoo'll hev te be careful whaat thoo's sayin', because thoo'll get ivvory one of them coals stolen one of these neets.' 'Ne fear, lad, aboot that; it'll tak someone very clivvor te get them,' said Geordie . . . 'thoo sees lad, if onybody tries te steal ma coals they'll hev te steal me alang with them, because, thoo sees, hinny, Aa sleep on the top of them in the room upstairs.'[245]

Most startling of all are the allegations of theft from those most potent symbols of coalfield solidarity, the children's feeding centres. In October, 34-year-old William Jacques, a miner of Houghton-le-Spring, was found guilty of the theft of five loaves of bread and two teacakes from the feeding centre in which he had been assisting, his crime only discovered after a policeman noticed that some children were returning home complaining of still being hungry.[246] At a similar court case later in the year, the chairman expressed regret at the many complaints he had heard concerning the pilfering of food and coal intended for children: 'Men were actually stealing coal which was given by the Colliery Company for the purpose of cooking their own children's food, and stealing the coal which was supplied to keep the children warm.'[247]

Of course, even in normal times, reality frequently diverged from the coalfield communities of lore, whose inhabitants were prepared to do anything to help a neighbour in need. 'People are always going on about colliery people being together and getting along with one another,' explained one man, 'You are always seeing it on the television like *When the Boat comes in*. But I don't remember anything like that. The pit was ruled by fear, so families stuck together, that was it.'[248] Even underground, a place of mythic selflessness, at least one boy remembered his workmates as rather less than supportive. Eric Squires began working down the pit in the interwar years as a pony driver and hated it owing to the ruthless bullying he and his pony received from the older men. One took pity on him, but explained, 'There's nothing you can do about it, not when a man's got a wife and hungry mouths

245 *DC*, 10 July 1926.
246 Ibid., 2 Oct. 1926.
247 Ibid., 13 Nov. 1926.
248 Cited in Beynon and Austrin, *Masters and Servants*, 188.

to feed . . . he'll walk all over you for an extra couple of bob.'[249] Nor were dishonest treasurers a phenomenon restricted to the strike months: in August 1926, John Race, treasurer of the Dean and Chapter lodge, was jailed for two months after being found guilty of embezzling funds since June 1923.[250]

And so perhaps it was inevitable that the increased tension of the strike would make things worse. George Hitchin explained: 'Sometimes, when men were weary and irritable, worried and harassed, nerves cracked and anger swelled up in them and fights broke out. I saw more violence in those few weeks towards the end of the strike than ever before.'[251] Ten years later, in the similarly straightened circumstances of the 1930s, a 35-year-old man told John Newsom how poverty had the ability to destroy goodwill, and he spoke with painful candour: 'I'm selfish now, and why? Because I've been so driven that I'm thinking all the time of number one or at any rate of number one's wife and kids. I'd deceive the Means Test man if I was clever enough and I'd steal if there was anything in this village worth stealing.'[252] Presumably it was the same concerns that led some men, in 1926, to defy their union and to blackleg.

## VII

The overlapping and conflict of identities and loyalties is one of the most startling traits thrown up by the evidence. Even in 1926, those who were relatively new to the area might be confused by the colliery culture they joined. Cuthbert Headlam, for example, was nonplussed when he addressed a meeting at Cornsay in September 1925 and found that radical politics did not prohibit an expectation of his support: 'The meeting closed with the singing of the Red Flag and an appeal to me to give a subscription to the Cornsay Colliery F.C. What strange people they are!'[253]

The Durham mining communities included the striker who volunteered as a special constable, the union official who returned to work, and the miner who owned his house and despaired at the relief given to his fellow strikers. Such instances fracture the conventional historical picture of a community united by political outlook, while even at the

[249] E. Squires, *Pit Pony Heroes* (1974), 33. [250] *DCA*, 20 Aug. 1926.
[251] G. Hitchin, *Pit-Yacker* (1962), 84. [252] Newsom, *Out of the Pit*, 21–2.
[253] Diary entry, 10 Sept. 1925, in Ball, *The Headlam Diaries*, 70.

time they broke down the accepted social structure of the pit village, challenging established hierarchies based on respectability. It is also too simplistic to see the coal owners and the Conservative Party engaged in a straightforward battle with the union for ideological control and personal support. Coal owners such as Sir Hugh Bell were bitterly opposed to government reorganization of the industry and had locked the miners out in an attempt to impose heavy wage cuts and lengthened hours. Their paternalism should not be mistaken for pure generosity, for it had strong political advantages. But, at a local level, there was a blurring of boundaries and some kind of mutual understanding, endorsed by the local lodges. There, the colliery hierarchy could be generous with welfare, get on well with the union leaders, and maintain such attitudes during the bitterest dispute that their industry had ever seen. Of course, an attachment to a political ideology does not necessarily have to be consistent. As Martin Bulmer has argued, 'the search for class consciousness presupposes its articulation in a coherent ideological form. But an important feature of images is that they may be fragmentary, ambiguous or uncertain'.[254] Far from being the epitome of the archetypal proletarian with a straightforward class consciousness, in 1926, many Durham miners had not worked out any consistent ideological position and their actions were frequently a reflection of this.

However, for the vast majority of miners, some kind of attachment to their union underpinned all, and other ideologies (including the Labour Party) had to work in conjunction with the union to ensure support. This does not mean that the union and its leaders were immune to criticism; indeed, its centrality could make it more likely. In June 1926, in response to Ellen Wilkinson's denunciation of the barbaric conditions prevalent in the pits of Somerset, Lady Astor wondered: 'If these are the conditions, what on earth have the Miners' Federation been doing?'[255] Her words were somewhat flippant, but they were not entirely inconsistent with the dominant feeling in the coalfield, where the union was seen as a man's insurance and was at least partly blamed if things did go wrong. George Alsop, born in Chopwell in 1911, would later speak revealingly about his time as a putter in the late 1920s. He found himself declared ineligible for free coals and had to search out a 1911 agreement himself, which proved that free coal was an entitlement of every workman at the colliery. 'I took it to the Lodge and showed it

[254] M. Bulmer, 'Introduction', in Bulmer (ed.), *Working-Class Images of Society* (1975), 5.
[255] *HPD*(C), 197, c. 1033.

to them,' he remembered, 'I didn't blame the company; I blamed the union. I blamed the union for not carrying out the agreement that was laid down between the workmen and the coalowners.'[256]

The communitarian ideology espoused by the union was not sufficient to ensure the loyalty of every man. In the face of enormous material suffering, a few chose to put their own self-interest, and that of their family, ahead of any wider loyalty to their workmates or neighbours. However, over the issues for which the strike was fought in 1926, the concerns of the union hierarchy mirrored those of their rank-and-file members. The fact that the lockout was fundamentally about the fight for a decent standard of life was expressed in eloquent terms in the House of Commons again and again by the miners' leaders. It was echoed in the coalfield by those less practised in public speaking but who shared the same determination to fight. John McIlroy and Alan Campbell have described the DMA Secretary W. P. Richardson as 'an essentially moderate man driven beyond moderation by the miners' predicament'.[257] A similar transformation could be seen amongst ordinary union members. One old miner, a child during the strike, later talked poignantly about his father:

> I'm now talking about a very mild-mannered, law-abiding person . . . wouldn't say boo to a goose. That mild-mannered man broke the law during the 1926 strike—obviously in the beginning they went onto the spoil heaps picking out the coal until there was no coal left. So down in Bloemfontein woods . . . [they] cleared the soil away and they started working this seam. So did my father and of course it was against the law—someone was watching for the policeman as well. So that we had a fire during the 1926 strike.[258]

Even when a more radical politics vied with the union for support, it was frequently interpreted in moderate terms. One man, interviewed in the 1970s, recalled the widespread victimization of men in the aftermath of the strike. But, they were not agitators, he insisted, 'they were men that stood up for what they were entitled to'.[259] Perhaps one of the most interesting insights comes from Mary Craddock, a miner's daughter who discovered Communism in the mid-1930s as a teenager. She at once went 'with revolutionary zeal' to convert her father. 'Communists . . . Do you know what I think of them? Riff-raff, that's

[256] G. Alsop, 'A Kind of Socialism' in Austrin et al., *But the World Goes on the Same*, 25.

[257] J. McIlroy and A. Campbell, 'Fighting the Legions of Hell', in McIlroy et al., *Industrial Politics*, 57.

[258] BMOA, 1998/4.

[259] GCLOT, i (N.C.).

what they are,' was his response, 'Corner-enders. Them as wouldn't work if they could. On the look-out for something for nowt. Them's Communists.' However, his defence of his union principles then drew on an analogy with which Communists too would presumably have identified. He gave his daughter a match to snap, and then asked her to do the same with a box full of them. To Mr Craddock such symbolism meant only one thing. 'Unity is strength,' he told her, 'That's why I'm a union man.'[260]

[260] M. Craddock, *North Country Maid* (Maidstone, 1995 edn.; first pub. 1960), 46–7.

# 3

# The Attitudes of Women

> When Martha awoke it was still dark and bitter cold . . . She lay quite still in the kitchen bed, holding herself rigidly away from Robert, whose coughing and restlessness had fitfully broken the night. For a minute she reflected, sternly facing the new day, choking down the bitterness she felt against him. Then with an effort she got up.[1]

In A. J. Cronin's novel, *The Stars Look Down*, Martha Fenwick had a number of reasons for feeling bitter. First was her unexpected pregnancy: 'Him, again, coming home in liquor . . . .' Second was the ongoing strike in the Durham coalfield, of which her husband had been a perpetrator, and into which their savings had disappeared. Yet, though her silent, brooding resentment froze the household, she never faltered in her duty as wife and mother. After her search for food yielded only one loaf, a gift from the baker's son, she divided the bulk between her husband and three sons, keeping the smallest slice to herself. Later, when they returned to work, the quality of the support she gave her working menfolk would contrast with the conduct of her daughter-in-law, the flighty Jenny, who did not come from a pit family herself.[2]

As a novelist, Cronin had to ensure that his characters would be recognizable to his readers. 'His women are as brilliantly painted as his men,' gushed one review quoted on the back cover of the novel's 1965 edition. Certainly, the character of Martha Fenwick resonates with wider representations of the miner's wife, which portray her as a counterweight to the male-based union. Such stereotypes suggest that, as a result of the gendered division of labour within the pit village, the miner's wife was alienated from the collectivist ethos created around pit, pub, and union lodge. Her overwhelming concern with her family made her an opponent of strikes. While some, like Martha, chose to remain rigidly silent if a strike did erupt, others would press (nag) their

[1] A. J. Cronin, *The Stars Look Down* (1935), 9.

[2] Ibid., 9–15, 170–2, 518–19.

husbands to betray the union and return to work.[3] In his study of Durham's Deerness Valley, Robert Moore cited a report of one blackleg who was forced to return to work in 1926 after literally being driven from his house at the end of a carving knife.[4]

4. Cartoon from the *Labour Woman*, August 1926. Despite her sobbing child, the mother strikes a suitably heroic pose as her husband looks away. Reproduced by permission of the People's History Museum.

But, alongside the representation of the wife as strike-breaker, subverting the solidarity forged in the darkness of the mine, runs another

[3] See, particularly, D. H. Lawrence, 'The Miner at Home', in K. Sagar (ed.) *The Mortal Coil and Other Stories* (1971), 109–14.

[4] R. Moore, *Pitmen, Preachers and Politics: The Effects of Methodism in a Durham Mining Community* (Cambridge, 1974), 205.

potent image. Despite the apparent contradiction, this presents women as the real binding force of the community. While men established their social relationships around pit and pub, these were strengthened or even outdone by the informal kinship and neighbourhood networks of their women and the practical and emotional support that accompanied them. Oral and written testimony frequently supports this, with both men and women emphasizing the degree to which women contributed to the creation of a sense of community within the pit villages.[5] During the 1926 dispute, their role in holding the fabric of the village together was emphasized by strike propaganda which was full of heroic testament to the activities of the womenfolk. 'Half my meetings are women,' declared Herbert Smith, addressing a gathering during the strike, 'They are always the toughest half.'[6] Two years later, Ramsay MacDonald paid them a private tribute: 'I really do not understand how the splendid women in the Coal Fields have been managing to keep their houses together during the last awful four or five years,' he wrote to Seaham's Labour councillor, Joseph Blackwell, 'We stand paralysed in front of their heroism.'[7]

During the lockout itself, yet another portrayal of women within the coalfield presented them as the innocent victims of a male struggle. Both sides sought to manipulate this image. Labour leaders spoke of women and children starving in the pit villages while the government stood aside and washed its hands of responsibility;[8] opponents referred to the miners' decision to go on strike as proof of their wanton disregard for the misery of their families.[9] The suffering of women and children could also be detached from the political aims of the miners. Fundraisers could happily provide charity for this group while condemning the policy of the MFGB and the madness of A. J. Cook. Even the Prince of Wales, the later Edward VIII, made a donation to the miners' wives distress fund. He commented that he was unable to take sides, but pointed out

[5] See, for example, Beamish Museum, Oral History Archive (henceforth BMOA), 1993/5; G. Carr, *Pit Women: Coal Communities in Northern England in the Early Twentieth Century* (2001), 108–17.

[6] Cited in P. Foot, *'An Agitator of the Worst Kind': A Portrait of the Miners' Leader A. J. Cook* (1986), 8.

[7] Durham County Record Office (henceforth DRO), D/X1208/10, J. Ramsay MacDonald to J. H. Blackwell, 17 Dec. 1928.

[8] See, for example, speeches made in the House of Commons by two Labour MPs, George Hardie and Arthur Greenwood. *Hansard Parliamentary Debates* (Commons), fifth ser. (henceforth *HPD*(C)), 196, cc. 2521–2 and 198, cc. 1553–6.

[9] Although when Colonel Lane Fox, Parliamentary Secretary to the Mines Department, implied as much in the House of Commons he was furiously interrupted by Labour MPs and forced to apologize for any offence caused. *HPD*(C), 197, cc. 994–7.

that 'it would be an undesirable end to any dispute that one side should have to give in on account of the suffering of their dependants'.[10]

These conflicting images of women, whether as anti-communitarian strike-breakers, coalfield heroines, or simply innocent victims, echoed the ambiguous position that women were accustomed to occupying within the pit villages. For, even during ordinary times, the domestic labour that women undertook within the home was both crucial to and marginalized from the work of the mine around which their lives were organized. Jaclyn Gier-Viskovatoff and Abigail Porter have explored these contrasting female identities, drawing attention to two cartoons of the 1920s published in the MFGB's *The Miner*. One depicted 'the miner's wife as a powerful figure, tossing the aristocratic politician out of her yard'; the other showed 'a beleaguered miner's wife, exhausted, head on the table, still rocking the cradle of her baby with her other foot'.[11] Valerie Gordon Hall made such conflict her theme, and also identified two different types of female identity: 'domestic women' who were subordinate to their husbands and concerned with domesticity and maternal care; and 'political women' who rejected this lifestyle and fought for socialist or feminist causes. Her article suggested that women conformed to one model or the other; rather than that individual women could, perhaps, fit both.[12]

The centrality of the wife and mother has been particularly celebrated in Wales, characterized in the figure of the Welsh mam. It is this region that provided Sue Bruley with the material for two recent essays examining female participation during 1926 itself. She surveyed the varied arenas in which women participated in the struggle, from helping in soup kitchens to taking part in anti-blackleg demonstrations. In the earlier essay, she was careful not to overemphasize their role, having found little evidence 'of women activists in South Wales as being anything other than marginal', nor evidence of a more equitable relationship developing between men and women within households. She took a more positive stance in the later essay, suggesting that, although women were unable to make any lasting challenge to male dominance, there was some evidence of increased gender alignment,

10 *Seaham Weekly News* (henceforth *SWN*), 4 June 1926.

11 J. Gier-Viskovatoff and A. Porter, 'Women of the British Coalfields on Strike in 1926 and 1984: Documenting Lives Using Oral History and Photography', *Frontiers: A Journal of Women's Studies*, 19 (1998), 202–3.

12 V. Gordon Hall, 'Contrasting Female Identities: Women in Coal Mining Communities in Northumberland, England, 1900–1939', *Journal of Women's History*, 13 (2001), 107.

and an indication 'that the status of women had been enhanced'.[13] In neither account, however, did she give any hint that mining women were anything other than supportive of the strike. In this regard, Bruley's essays contrast the work of Alan Griffin in his wider history of mining in the East Midlands. Griffin suggested that one possible reason for the disintegration of solidarity in Nottinghamshire in 1926 was the tendency of women to harass their husbands until 'the unpleasantness of home' outweighed a man's fear of being dubbed a blackleg. He was careful to restrict this possibility to Nottinghamshire, a county in which the increased opportunities for female employment altered the dynamic of the working-class family. In the North East, he argued (and presumably he would also have added South Wales), women tended to support their striking menfolk. His evidence for this was a speech made by Peter Lee, in which Lee boasted of the success of Durham's local councils in providing children with free school meals. Even in Durham, therefore, Griffin implied that a woman would only support of the strike if her children were fed first.[14]

The lives of Durham's women were circumscribed by the pit villages in which they lived. Early marriage, large families, and a punishing domestic routine dictated by the colliery shifts constituted the experience of most.[15] Except in domestic service, opportunities for female employment were few: 807 of every 1,000 Durham women were classed as 'unoccupied or retired' in the 1921 census.[16] During the lockout, the rhetoric of the strikers demanded that every member of the pit village be mobilized behind the trade-union banner, yet women were also required to continue to fulfil their domestic role within a patriarchal structure. This chapter will explore these contradictions and identify the different strands of female involvement during the strike. It seeks to progress beyond stereotypes and establish a more nuanced understanding of the

[13] S. Bruley, 'Women', in J. McIlroy, A. Campbell, and K. Gildart (eds), *Industrial Politics and the 1926 Mining Lockout* (Cardiff, 2004), esp. 244–5; and 'The Politics of Food: Gender, Family, Community and Collective Feeding in South Wales in the General Strike and Miners' Lockout of 1926', *Twentieth Century British History*, 18 (2007), 69, 77.

[14] A. R. Griffin, *Mining in the East Midlands, 1550–1947* (1971), 248–9; MFGB, *Annual Volume of Proceedings for 1926* (1927), 653.

[15] For a more detailed consideration of the lives of Durham mining women throughout the interwar years generally see H. Beynon and T. Austrin, *Masters and Servants: Class and Patronage in the Making of a Labour Organisation* (1994), 154–84. For the high fertility levels of British miners, see S. Szreter, *Fertility, Class and Gender in Britain, 1860–1940* (Cambridge, 1996), 321, 388.

[16] *Census of England and Wales, 1921*, ii, *County of Durham* (1923), p. xxxvi. These figures excluded women's participation in temporary or casual work.

relationship of mining women to such traditionally male institutions as the union and the Labour Party, and their attitudes to strike action itself.

## I

Debates over the political affiliations of women as a social group were prominent in the 1920s, sparked by the enfranchisement of many older women in 1918 and discussions over the further reduction of the qualifying age, to be fulfilled at the end of the decade. Many believed that the female vote was, and would continue to be, of overwhelming benefit to the Conservative Party, especially given the Conservatives' effective use of targeted propaganda. When the Labour Government fell in October 1924, Beatrice Webb commented on the extra two million Conservative votes, cast by 'elderly wives, widows and spinsters trooping into the polling booths, clutching the Conservative polling card as a talisman of safety against the nationalization of women and the confiscation of all property, from millionaires' millions to Post Office savings'.[17] Few Conservatives, however, took the female vote for granted. Cuthbert Headlam feared the prospect of votes for women aged 21, believing the younger female generation to be more radical than their male equivalents:

> The younger women are far less likely to vote Conservative than the young men—and that means an immense increase in the Socialist vote. In our pit villages the women are far wilder than the men—and they are hopeless to argue with—they listen to the sob stuff with open ears—and of course they will be fed nothing else.[18]

As it happened, Headlam lost his seat to Labour's Will Lawther in 1929. More widely, too, the extensions of the vote to women in 1918 and 1928 made no appreciable dent in Labour's electoral performance in the Durham coalfield. The party claimed four seats in 1918 compared with two (plus one Lib–Lab)[19] in 1910; the real breakthrough for Labour in the county was to come in 1922. Reasons

[17] Diary entry, undated, *c.*29 Oct. 1924, in N. MacKenzie and J. MacKenzie (eds), *The Diary of Beatrice Webb*, iv, *1924–1943: The Wheel of Life* (1985), 44.

[18] Diary entry, 1 May 1927, in S. Ball (ed.), *Parliament and Politics in the Age of Baldwin and MacDonald: The Headlam Diaries, 1923–35* (1992), 118.

[19] John Wilson, MP for the Mid-Durham division, was one of only three MFGB-sponsored MPs to not take the Labour whip after the Federation's affiliation to the Labour Party in 1909.

for the increasing success of the Labour Party have been debated many times elsewhere, and the extension of the franchise (to all adult men as well as to the first women voters) is one explanation amongst many.[20] However, local Labour organizations specifically catering for women were flourishing in the North East by the 1920s, and Labour Party conferences frequently singled out both Durham and Northumberland for special praise.[21] These were still developing organizations, far behind their male counterparts, but their growth was constant. The Durham County Labour Women's Advisory Council had been formed in 1920, its main objective being to develop female electoral allegiance to the Labour Party through political education and activity. By 1925, there were 118 Labour women's sections in the county and 25,000 attended the inaugural Labour Women's Gala of that year, when the political speeches and procession of banners through the city mimicked the prestigious annual miners' gala.[22]

In Sidney Webb's constituency of Seaham, the increasing political consciousness of the female population was noticed by Beatrice Webb. In December 1923, she reflected on the campaigning meetings held in advance of the general election. She was pleased that there had been a regular attendance of women, sometimes numbering sixty or seventy, compared with the previous year when 'if there were two or three women one was agreeably surprised'.[23] Webb's own profile in the area may have helped, and her long letters to the Seaham women kept them informed of national political events. But, other women were equally prepared to take on leadership roles. Emmy Lawther had met Andy, one of the notorious Lawther brothers of Chopwell, at Ruskin College, Oxford. She later arrived in Durham as his wife. 'She was sort of more advanced than what we were,' remembered a local woman fifty years later, 'But

[20] See C. Matthew, R. McKibbin, and J. Kay, 'The Franchise Factor in the Rise of the Labour Party', *English Historical Review*, 91 (1976), 723–52. For alternative views see P. Clarke, *Lancashire and the New Liberalism* (Cambridge, 1971); D. Tanner, *Political Change and the Labour Party, 1906–18* (Cambridge, 1990); M. Childs, 'Labour Grows Up: The Electoral System, Political Generations, and British Politics 1890–1929', *Twentieth Century British History*, 6 (1995), 123–44; M. Pugh, 'The Rise of Labour and the Political Culture of Conservatism, 1890–1945', *History*, 87 (2002), 514–37.

[21] See, for example, *Report of the Twenty-fourth Annual Conference of the Labour Party* (1924), 20; *Report of the Twenty-fifth Annual Conference of the Labour Party* (1925), 30.

[22] P. Lynn, 'The Influence of Class and Gender: Female Political Organisations in County Durham during the Inter-War Years', *North East History*, 31 (1997), 46–7; P. Graves, *Labour Women: Women in British Working-Class Politics, 1918–39* (Cambridge, 1994), 162.

[23] Diary entry, 3 Dec. 1923, in N. MacKenzie and J. MacKenzie (eds), *The Diary of Beatrice Webb*, iii, *1905–1924: The Power to Alter Things* (1984), 430.

what a tremendous help she was to us. She got us together on all sorts of things, on picketing, on—during the strike and even before that—when we formed the first welfare in the Wesleyan schoolroom over there.'[24]

Of course, the political allegiances of the men in these women's lives were an important influence. Those women who were most involved with the Labour movement were likely to have had the support of husbands or fathers, and may have owed their political loyalties to them. One woman interviewed in the 1970s remembered how she had attended meeting after meeting during the strike, following the union banner wherever it went. But, her attitude was likely to have been shaped by the fact that her father (with whom she had still been living) was a lodge chairman.[25] Another woman spoke of her mistrust of blacklegs, and directly attributed it to her husband: 'I suppose they thought that was their principle, but it wasn't our principle because my husband was a union man.'[26]

But, while male family members certainly influenced the political outlook of their wives and daughters, this may be simply a reflection of the influence of one family member upon another, rather than evidence of a specifically gender-based authority. Sometimes wives could influence their husbands, and one Seaham resident told Beatrice Webb that it was only when she had got her husband safely into bed (so that he could not possibly leave the house) that he listened with patience to her reading of Webb's letter. Webb looked at the particularly long letter she had just composed, stretching over eleven pages, and wondered, 'What will the poor man do this time?'[27] Political loyalties might be independent of men altogether and could be passed down the female line. One old miner recalled that, just before he married, his wife-to-be went to his mother to ask for any advice.

> [His mother replied] 'What I would do if I were you, I would go along to the [Co-operative] Store there . . . [and] become a member'. And she did that, and we had as much money saved in that Birtley store . . . it carried me and the family right through the strike. That was the way we were. My wife always stuck to the store.[28]

His story has echoes in the opinions voiced by a Welsh miner interviewed by Angela John, whose article has highlighted the importance played by

[24] Gateshead Central Library, Oral History Transcripts (henceforth GCLOT), iv (Mrs S.).

[25] Ibid., ii (Mrs F.).

[26] Ibid., i (Mrs G.).

[27] London School of Economics (henceforth LSE), Passfield 4/15, B. Webb to Seaham women, 18 Aug. 1926.

[28] GCLOT, iii (J.R.).

the Co-operative stores in the coalfield communities of South Wales. In his view, women played 'a remarkable part in political life by sustaining the Co-op'.[29]

Other women challenged the political views of their fathers and husbands. One read 'a bonny lot of books on Communism', because her husband had recommended that she should. But, she was not persuaded and remained solidly Labour, growing confident enough in her political beliefs to argue with Emmy Lawther at a meeting.[30] At least one father encouraged such independent thought and was keen that his daughter should develop her own beliefs (although the daughter believed herself to have been in a lucky minority in this respect). She explained:

> me father . . . he did talk with me about the necessity for women . . . to be interested in politics—because he had no downs on women . . . [He used to say that if] I got interested in anybody, I . . . [should] think about their politics. Because he used to say, they'll stifle you, they'll stifle you . . . It's a long time ago and for your father to talk to you like that is something, isn't it.[31]

Indeed, at least amongst the active Labour women of the county, ensuring that their men remained loyal to the Labour movement was seen as a responsibility. The following year, after the passing of the Trade Disputes Act, the Durham Labour Women's Advisory Council passed an emergency resolution, pledging: 'to do all possible to counteract it by seeing our menfolk, husbands, sons and brothers, contract into the Political Fund. In addition, we will use every possible means to increase individual membership of the Party, to raise funds, and to return Labour majorities at the Local Government elections next spring.'[32]

## II

It is possible to argue that a woman's support for labour organizations, whether a result of her husband's influence or not, was likely to weaken during the times of worst hardship such as a strike, when

[29] A. V. John, '"A Miner Struggle": Women's Protests in Welsh Mining History', *Llafur*, 4 (1984), 85.

[30] GCLOT, ii (D.H.).

[31] British Library Sound Archive (henceforth BLSA), C900/05554C1.

[32] Cited in M. Callcott, 'Labour Women in North East England', *North East Labour History Bulletin*, 17 (1983), 37.

she had to look to her family first and foremost. The *North Mail* noted:

> The men on strike have the excitement of meetings; the simulation of tremendous battles of wits; the consciousness of playing an important part in a vital struggle . . . But, the woman has to sit at home and wait. The niceties of the wordy arguments around her leave her unmoved. She is desperately aware of only one fact. There are hungry children; there is no food . . . The disputes of man threaten more and more to strike at the home . . . Is not the time approaching when women will cry, 'Enough!'?[33]

Despite its political posturing, the *North Mail* had a point, and oral sources suggest that many women found the strike harder to cope with than their husbands. In sharp contrast to men, especially of the younger generation, who often reminisced about a beautifully hot summer and lazy days spent playing endless games of football and cricket, women could not enjoy the stoppage of work so freely.

In normal times, the working-class wife was almost always in charge of managing the household's finances, and the significance of this has long been disputed. Some historians have suggested that it empowered women; others have claimed that it amounted to yet another burden of responsibility.[34] But, in 1926, the women of the pit villages lost any claim to empowerment at all. For seven months they were released from their usual duty of keeping their men supplied around the clock with hot baths, hot food, and clean clothes, but their responsibilities of dealing with the day-to-day feeding and clothing of the family continued essentially as before—only now had to be organized with far less money. During the lockout, one woman was married with a young child, and thinking back years later she felt that the hardest burden had been placed on her own sex: 'The men would have their smoke, but of course the women as usual, they had all the rest of the things to do, such as make do and mend for their children, and see to everything, and help to do cooking . . . And that stuck in my memory as very hard.'[35] It was a divide echoed in coalfields across the country, and in South Wales a Mrs Davies recalled that 'my brothers ended up very, very sunburnt, whereas my mother was worn out'.[36]

33 *North Mail*, 4 May 1926.

34 See the debate summarized in A. Taylor, '"You Never Told Your Mam": Women's Management of Household Finances and Credit during the Interwar Period', *North East Labour History Bulletin*, 30 (1996), 51–4.

35 GCLOT, i (Mrs C.).

36 Cited in Bruley, 'Women', 245.

For many women the lockout further increased their burden of work due to the need to provide alternative earnings. Opportunities for female employment were limited in the pit villages, but the strike prompted younger girls in particular to go into domestic service or take up other tasks. Joe Brennan was one of eight in his family: a couple of his sisters had to go into service during the strike; two others went to work on a farm: 'It helped a lot,' he remembered.[37] For another man, slightly older than Joe, the wages provided by women had an added benefit. He explained that he and his mates would try whenever possible to court 'a lass who was working'. He had two sisters and one brother, and both sisters had taken jobs, one at the store and one at the Birtley tin plate works.[38]

The oral testimony of these men is supported by that of working women themselves. One remembered being the only member of her family bringing in a wage: 'There were three still at school and three on strike. There was nothing else to manage on except my nine and sixpence halfpenny a week.'[39] Perhaps the most striking testimony comes from a woman who had just left school in 1926. In her family, the duty of providing a wage seemed to fall entirely upon mother and daughters. It was her mother who worked at the store, kept pigs, and took washing in, and the family was able to survive the strike because of this. The girl herself also 'used to do a hard day's washing from eight o'clock in the morning till eight at night for one shilling'. Later she went to work in the fields of a nearby farm where her sister had gone into service. 'If it wasn't the field work,' she sighed, 'you were double possing [washing clothes].' It is only at the end of her account that her father makes an appearance, for if she ever did find herself with any spare time then she was enrolled to help him tallow the shoes he was mending, or pick coals at the pit heap.[40]

However, the added hardship suffered by women during the strike may actually have increased their determination to see it through. The awareness that an inglorious return to work would render months of privation meaningless must have acted as a psychological barrier against surrender as the year rolled on, most pronounced amongst those who suffered most. As Jack Lawson remembered with regard to the 1892 strike in Durham:

37 GCLOT, iii (J.B.).

38 Ibid., ii (T.L.).

39 DRO, B1/14, G. Purdon, *It was Grand Toffee: Horner's of Chester-le-Street Remembered*, no page.

40 GCLOT, iv (Anon.).

As the sense of defeat deepened, passions rose until the women and children were as bad as the men. Indeed, the women were the worst, as they always are, on such occasions. The women of the coalfields watch an approaching industrial storm with fear and quaking, for they know what it means more than anyone. Once the pay stops, savings go, debt begins to mount up, then hope dies and there is only one thing left—the will to victory. When this stage is reached, then the world may be against them, death and everlasting damnation come upon them, but they will not retreat an inch. And woe betide the man who would compromise.[41]

An old miner shared similar impressions of the women of 1926: 'But if everybody had been like the women, [well] then. If the men had been as strong as the women we would still have been out on strike. Cos the wives, phew . . . they knew what was suffering.'[42]

It seems plausible to speculate that some women must have broken under the strain and urged their husbands to return to work. As part of the Gateshead oral history project, a few of the dozens of interviewees were asked why they thought their workmates and neighbours might have blacklegged. One man explained that 'some of their wives was nagging them . . . you know'.[43] Another spoke of a friend: 'the poor lad was forced to do it. His mother was a real Amazon . . . His father was forced to do it [too].' He remembered how he had once been at their house and seen his friend's mother throw a pit boot at her husband for not getting the coal in quickly enough. Her husband had immediately jumped to do her bidding.[44]

However, individual men must have returned to work of their own accord too, just as individual women pressured their husbands to do so; what is important is whether or not women as a social group deserve a reputation as the gender most likely to break with the solidarity of the strikers. In fact, although a couple of the Gateshead interviewees did mention the influence of nagging wives, as cited above, the decision was just as often explained without any specific reference to female pressure, and frequently it was geographical rather than gender divisions that were stressed. Only one blackleg was interviewed for the Gateshead project, and he explained that, for him, still a boy living with his parents, it was his *father* who was the determining influence in his return to work: 'Now he was a good father, but he was very militant and he was very bitter because he had lost his job through not being able to get to work.

[41] J. Lawson, *A Man's Life* (1944), 40–1.
[42] BMOA, 1991/82.
[43] GCLOT, iii (J.B.).
[44] Ibid., iii (H.M.).

And fair enough we had to go to work you see, and therefore I was forced into the position I was in.'[45]

Far from being cut off from the moral pressures of the community, the social penalties that prevented many men from returning to work were just as compelling for the women of the pit villages. When A. J. Cook announced that blacklegs should be treated as lepers, he added that their wives and children should be seen in the same light.[46] In most cases, therefore, women risked the same alienation within the community when their husbands or fathers went back to work as those men did themselves, whether or not they had played any determining influence in the decision. Indeed, often they could expect the greatest condemnation to come from the other women of the village, the group amongst whom they would almost exclusively have formed their social relationships. In this respect, it is significant that while many of the men interviewed for the Gateshead project were fairly sympathetic towards the families of blacklegs, remembering that they had a difficult time, the women interviewed were generally far less forgiving: 'We were as bitter against their wives as we were against the men, and that was how the women felt.'[47] These concerns were likely to dampen a woman's enthusiasm for blacklegging. For the miners themselves, such fears also meant that a commitment to wife and family became a reason to support the position of the community, rather than defy it. Three years later, when a strike at Dawdon stopped work at his pit, Londonderry voiced his displeasure in a letter to MacDonald. He told of two men who had visited their colliery manager to explain apologetically that although they wanted to continue working they had been threatened with the break-up of their homes, and 'as they both had wives and families, they could not take the risk'.[48]

The tendency to blame women for a man's decision to return to work may result from inadequate source material. Just as some men may have been influenced by their wives in making a choice to return to work, others may have taken the decision *in opposition* to their wives, but this is likely to have gone unrecorded due to the scarcity of blackleg testimony in the first place, and the improbability of it being

[45] GCLOT, iv (L.H.). The motivation of blacklegs is discussed in some detail in Chap. 2, Sect. VI.

[46] *Stanley News and Consett Chronicle*, 12 Aug. 1926.

[47] GCLOT, iv (Mrs S.).

[48] The National Archives: Public Records Office (henceforth TNA:PRO), PRO 30/69/1714, Londonderry to J. Ramsay MacDonald, 17 June 1929.

mentioned by a blackleg even had it been the case. Furthermore, most testimony comes from men. Those conducting the Gateshead project, for example, put their question about the motivations of blacklegs only to some of the men they interviewed, and not to any of the women. To men who remained loyal to the union, a nagging wife provided an easy explanation for blacklegging, allowing them to see those who defied the union as emasculated and deficient, rather than posing more difficult questions about the solidarity of workers in the industry.

A further problem arises from the fact that class and gender were just two of many identities in the coalfield which overlapped to define their inhabitants. In 1924, Chopwell's new lodge banner was unfurled to display a central picture of Keir Hardie flanked by images of Marx and Lenin (see Illustration 3). Despite the reputation of the village as a centre of militancy, not everyone approved, and upon its appearance the banner was pelted with stones by a group of villagers, mainly women. At face value, this appears to be evidence of an antipathy to political radicalism amongst at least some of the women of the village, and, set out like this, such a conclusion is implied (reasonably enough) by those who have documented it.[49] However, hostility to Communist icons does not necessarily constitute a rejection of class solidarity; furthermore, closer examination reveals that more complex allegiances were at work. One oral history respondent remembered hearing a woman threatening to cut the new banner to pieces, but she suggested that the woman's objections were framed in primarily religious terms: 'She was a Catholic . . . so she didn't want that banner you see.'[50] Even more striking are the recollections of an old man who was a child at the time. He asked another observer to explain to him who Marx was, and was told that he was 'a well known German philanthropist'. It is probable that others, both male and female, shared this rather sketchy awareness of Marx's relevance on the banner. 'But he was a German, and people that . . . had lost their husband or son . . . in the war you know, they were dead against it, this banner.'[51]

## III

If few women actively opposed the strike in 1926, many were actively supportive. Miners' wives and daughters were not excluded from the

[49] See, for example, Beynon and Austrin, *Masters and Servants*, 223.

[50] GCLOT, ii (D.H.).

[51] GCLOT, i (A.W.).

rhetoric of the labour leaders, and most would have had some knowledge of the issues over which the strike was being fought. Bessie Johnson was 17 years old in 1926 and had just moved to Birtley with her family. Her father was a prominent activist and so it is perhaps no surprise that she participated in the strike, helping him to sell newspapers. However, when she was interviewed in her old age she implied that both men and women had enthusiastically taken part in strike-related activity: 'I can remember having meetings at Shadon's Hill', she explained, 'and I remember A. J. Cook coming to speak and . . . *everybody* went to the meetings. Everybody that had any trade union authority or principles, or use, were working [in aid of the strike].'[52] Indeed, the charisma of A. J. Cook rubbed off on men and women alike. In August, the miners' leader spoke at West Rainton to a gathering of hundreds of miners and their wives. The flirtatious banter between Cook and the crowd confirms that the hero's following he commanded amongst the miners extended in at least some cases to the women of the coalfield as well. The *Durham Chronicle* reported:

> Before speaking he removed his coat and hat. 'Do you want a lady's maid?' shouted a man. Cook laughed and assured the women that he had completed his undressing. 'There's a lass wants to kiss you,' cried an enthusiast. 'There's many a lass I would like to kiss here,' replied AJ [*sic*]. ' I have not seen my wife for a fortnight.' Judging from subsequent remarks, Mr Cook has a warm side for the ladies and a mere man called out, 'Don't say too much or the women will get swelled heads.'[53]

Lower wages and lengthened hours had the potential to affect miners' wives just as dramatically—possibly even more so—than it did their husbands. In the event of a reduction of wages, all members of the family would suffer, even assuming that all food was shared out equally. The miner's wife was also the person who would have to deal with the practical reality of a smaller wage bill, with all the anxiety entailed. Nor was a woman's stake in the strike purely a financial one. Much has been written about the punishing routine of the miner's wife, whose never-ending cycle of washing and cooking could be continued around the clock if she had a husband and sons on different shifts.[54] Any increase in hours would therefore mean disruption to her work as well, as the Scottish miners' MP Duncan Graham explained during the

[52] BLSA, C900/05554C1. [53] *Durham Chronicle* (henceforth *DC*), 14 Aug. 1926.
[54] See, for example, S. Webb, *The Story of the Durham Miners, 1662–1921* (1921), 71–2.

Parliamentary debates on the Coal Mines Bill: 'The women are in this thing more than the men . . . Hon. Members seem to think that it is quite a fair and right thing for a miner's wife to be called upon to be up from five o' clock in the morning till twelve and one o' clock the following morning, so that coal can be produced cheaply'.[55]

Another Labour MP, Thomas Cape, argued that the successful passage of such legislation through Parliament would also significantly affect her emotional well-being:

> the reason why the miner's wife stands so firmly beside her husband on this issue is that she . . . wants some alleviation of the suspense and the anxiety she feels when her man is not at home. And there is something more! She thinks about her lad. When she has to send her lad of fourteen years into this dreadful danger zone, it is quite natural that her feelings of mother-love should come out to protect that lad so far as it lies in her power to do so.[56]

Such arguments surely resonated within the coalfield, with its stories of the pit wives who always insisted on getting out of bed to see their husbands and sons off to work in case it proved to be their final goodbye.[57] Indeed, Cape's statement was supported by the Durham miners' MP, Joshua Ritson, who recalled his meeting with a miner's wife in his constituency the previous week. 'I have five of them who go down the pit, and God help me if I have five hours added to my day's anxiety, in addition to my own work,' she had told him.[58]

One way in which women could actively aid the strike effort was through fundraising. At national level, this achieved a high profile through the Relief Committee for the Miners' Wives and Children, the only official fundraising body sanctioned by the MFGB, and led by Marion Phillips, the Labour Party's chief woman officer. However, local women also played a significant part in raising money. Events such as whist drives, dancing, tea parties, and football matches were organized, frequently under the auspices of the women's section of the local Labour Party. Their commitment and ability made a big impression on one 14-year-old boy in 1926. He still spoke of their actions with admiration in his old age, explaining that the activities laid on during the strike had almost always been organized by women:

> They took a very active part . . . .They'd get their washing done and their baking done, and they'd say, 'Howay we'll go and have a . . . ' and they used to

[55] *HPD*(C), 197, c. 455.

[56] Ibid., 197, c. 1259.

[57] Lawson claimed that his mother had always done this. Lawson, *Man's Life*, 33.

[58] *HPD*(C), 197, c. 1240.

organize it. And they were marvellous you know. They'd mebbe get a bit tea party up or something like that. Where in the world they got the stuff from, I do not know, but they managed.[59]

A recent study of Labour's grassroots has argued that constituency party activity became increasingly diverse between the wars. Rather than concentrating on purely electoral matters, social events became more prominent: 'In so doing, parties and associations hoped to integrate themselves within the wider community, so heightening their profile and raising necessary funds and support in the process. As importantly, they served to bolster party morale and sustain the party organization between elections.'[60] As the authors point out, this was a development particularly associated with the local women's sections. It is a trend clearly visible in Durham in 1926, with a constant stream of events hosted by local Labour women's sections to raise funds in support of the lockout.[61] This was as much about providing a social forum for women as it was about political activity for its own sake, and was perhaps particularly important in a mining area where women were excluded from the social and political activities provided by the union.

Other women demonstrated their support of the strike through their attendance on the picket line or at demonstrations. Admittedly, in many instances women played only a supportive role, and men remained the focus of the action. In Birtley, for example, men with slingshots would hide amongst a crowd of women. When the blacklegs arrived the women would suddenly part to allow them to shoot.[62] In Chopwell and its surrounding villages, women helped with the illegal distribution of the locally published strike bulletin, *Northern Light*. Upon the approach of a policeman, one Gateshead interviewee later recalled, the bundle of papers would be handed to the nearest woman. She would hide them under her shawl or apron and, pretending to call on a neighbour, would pass them on until they had travelled through several hands (under several aprons): 'And the police were just unable to catch us, because they really thought it was just women moving from one house to another.'[63]

59 GCLOT, ii (Mr P.).

60 S. Ball, A. Thorpe, and M. Worley, 'Elections, Leaflets and Whist Drives: Constituency Party Members in Britain between the Wars', in M. Worley (ed.), *Labour's Grass Roots: Essays on the Activities of Local Labour Parties and their Members, 1918–45* (Aldershot, 2005), 19.

61 Such as the fancy dress carnival organized by the Seaham women's section in aid of the boot fund, attended by 750 people. *SWN*, 17 Sept. 1926.

62 GCLOT, ii (Mr P.).

63 Ibid., iv (Mrs S.).

However, women acted not only as a cover for men but also in their own right, and reports of women prosecuted for strike-related offences occur with reasonable frequency in the local press. The fact that such cases involved women made it more likely that they would be reported; but they are also an indication only of those women who were actually apprehended and charged. In August, three women were amongst five people charged with intimidating a miner working at Adventure Colliery, for kicking his door, throwing stones, and shouting threats.[64] Three months later, Annie Maughan was summoned before the courts for assaulting a working miner on a bus. She admitted that she had aimed a blow at his face, grazing his forehead and knocking his hat off, but claimed that she had been provoked: the miner had been scoffing her. She was the wife of a striker and had ten children.[65]

Women were frequently involved more anonymously; their presence recorded by the horrified officials who dealt with the disturbances. After an anti-blackleg rally at Mainsforth Colliery involving about 500 people, the police superintendent complained that 'the conduct of some of the women was disgraceful and disgusting, and they were just as bad, if not worse, than the men'.[66] A similar reaction was elicited from a naval officer on duty during the general strike, who was appalled to see 'a large number of women . . . [who] lined the parapet of the railway cutting and bombarded a train as it passed with bottles and stones, while the men looked on and encouraged them'.[67]

Women's involvement in strike activity continued to provoke anxiety throughout 1926. The local press surrendered to clichés and blamed a supposed female tendency towards overexcitement and hysteria. When a deputation of three women visited their local board of guardians to complain about the scales of relief, for example, the *Chester-le-Street Chronicle* described them all as having been in a 'highly excitable frame of mind', almost 'shrieking' at the guardians.[68] Others were similarly reluctant to take female protestors seriously. In November, seven women were charged in the police courts under the Emergency Regulations. The prosecutor remarked, as one might warn a child of the consequences of his actions, that he 'did wish women would keep out of the way in these cases, for they were a positive nuisance. He wished to treat these

64 *DC*, 7 Aug. 1926. 65 Ibid., 27 Nov. 1926. 66 Ibid., 9 Oct. 1926.

67 TNA:PRO, ADM116/2439, diary of Commander H. Tafrell Dorling, 10 May 1926.

68 *Chester-le-Street Chronicle* (henceforth *CC*), 8 Oct. 1926. Dependants of strikers were permitted to apply for poor relief, although strikers themselves were not.

cases seriously. Stones had been thrown which had struck policemen, injuring them.'[69]

Evidence from other regions suggests that women engaged in similar activity across the British coalfield districts in 1926.[70] Their actions were frequently seen by nervous observers as constituting a new development in the conduct of industrial protest. In Wales, the *Western Mail* reported in November that 'a strange feature of the disturbance is the surprisingly prominent part played by the women'.[71] In Nottinghamshire, D. H. Lawrence returned to his childhood home of Eastwood and witnessed a raucous scene as two women were taken to court for insulting and obstructing the police. 'They were two women from decent homes,' he remarked with distaste, 'In the past they would have died of shame, at having to go to court. But, now, not at all.'[72]

Yet, despite such shocked reports, the actions of these women were not historically unique. Historians have documented female participation in mining strikes across the world in the late nineteenth and early twentieth centuries: participation that was often considerable and sometimes violent.[73] Within Britain, too, women's actions in 1926 fitted into a much older history of female protest. Welbourne's history of the Durham and Northumberland miners, published in 1923, described the treatment of blacklegs during the stoppage of 1892. He noted that 'such men as dared work . . . were followed to the pits by crowds of hooting women'.[74] In particular the ritualized protests that met a blackleg as he emerged from the pit were reminiscent of a 'rough

[69] Ibid., 19 Nov. 1926.

[70] For brief snapshots from a variety of different coalfields, see Bruley, 'Women', 230–1.

[71] *Western Mail*, 5 Nov. 1926, cited in D. Beddoe, *Out of the Shadows: A History of Women in Twentieth-Century Wales* (Cardiff, 2000), 91.

[72] D. H. Lawrence, 'Return to Bestwood', in W. Roberts and H. Moore (eds), *Phoenix II* (1968), 259. Such evidence should be contrasted with Griffin's suggestion that Nottinghamshire women opposed the strike. The evidence of Lawrence has also been cited by John McIlroy in his essay on Nottinghamshire in 1926; he similarly questions Griffin's assertion. See J. McIlroy, 'Nottinghamshire', in McIlroy et al., *Industrial Politics*, 212–13, 219–20.

[73] See, for example, D. A. Corbin, *Life, Work and Rebellion in the Coal Fields: The Southern West Virginian Miners 1880–1922* (1981), 92–3; A. DeStefanis, 'Violence and the Colorado National Guard: Masculinity, Race, Class, and Identity in the 1913–1914 Southern Colorado Coal Strike', in J. J. Gier and L. Mercer, *Mining Women: Gender in the Development of a Global Industry, 1670–2005* (New York and Basingstoke, 2006), 195–212; B. Scates, 'Mobilizing Manhood: Gender and the Great Strike in Australia and Aotearoa/New Zealand', *Gender and History*, 9 (1997), 285–309; also, with regard to gold mining, J. Krikler, *White Rising: The 1922 Insurrection and Racial Killing in South Africa* (Manchester, 2005), 78–102.

[74] E. Welbourne, *The Miners' Unions of Northumberland and Durham* (Cambridge, 1923), 275.

music' tradition that had been prevalent in various forms throughout many regions of Britain until the nineteenth century.[75] Certainly, action taken in Durham in 1926 had strong echoes of this, and one woman remembered how the women of Chopwell treated the blacklegs:

> During the strike itself, if a blackleg was coming from the pit . . . it was the women that went to the pithead with their rolling pins, and their tin dishes and their tin pots . . . and we used to meet them coming out of the pit and wherever they lived in this village about fifty or sixty women would sing them home, or we'd get them in the middle of us . . . and we used to be sticking them with our hatpins and making sure that they never came out again.[76]

## IV

If the difference between male and female support for the strike is much less significant than is sometimes portrayed, this is not to suggest that gender divisions were unimportant. Only occasionally is there any hint that the exceptional circumstances of a national strike could break down gender divisions within the home. Mrs Spence, for example, was a young married woman in 1926. Her husband was ineligible for relief, and she had to make a long journey on foot to pick up her own food vouchers: 'I remember one day my hubby, he made Pan Haggerty [a regional dish] out of potatoes and leeks out of the garden, and put pig dripping in . . . He made that three times, before I got home.'[77] Her story has echoes in the anecdotes told in 1984–5 of women who went off to address strike meetings and found to their surprise that they could leave the children with their dad.[78]

On the whole though, the continuation of a specifically female identity was inevitable given the social organization of labour within the pit villages. Even Mrs Spence's husband, as far as can be discerned, cooked for her only three times in the seven months of the lockout. More often, when women did become involved in the strike, the traditional division between male and female roles in the community continued to

75 See E. P. Thompson, 'Rough Music', in his *Customs in Common* (1991), 467–531.

76 GCLOT, iv (Mrs S.). The persistence of earlier forms of female (and male) protest has already been noted in studies of South Wales. See Gier-Viskovatoff and Porter, 'Women', 215; R. Jones, 'Women, Community and Collective Action: The "Ceffyl Pren" Tradition', in A. John (ed.), *Our Mother's Land: Chapters in Welsh Women's History, 1830–1939* (Cardiff, 1991), 35–7.

77 BMOA, 1976/125.

78 J. Stead, *Never the Same Again: Women and the Miners' Strike, 1984–5* (1987), 16.

be reproduced. Even on the picket line, the responsibilities of women were different from those of men. One of the strongest tributes to female support came from one Gateshead interviewee whose wife had been fully involved in the strike effort. He remembered with pride how she had been part of a group of women picketing a chemist's shop that had been selling its merchandise to the wives of blacklegs: 'The women came out in strength. The women picketed the shop [which] . . . had to close . . . and to this very day [1976] . . . Metcalfe's have never been able to open a shop in Chopwell.'[79] The point is that responsibility for picketing a shop fell to women.

In this context it is significant that within the pages of the local press, female activity was most visible in connection with demonstrations regarding poor relief. As with the picketing of shops, this was a question that directly concerned women in their role as homemakers, affecting their ability to care for their families. Thus, when the Chester-le-Street guardians were sacked and replaced by the Ministry of Health appointees, almost 15,000 people poured into Chester-le-Street to protest. The gathering was a mixed one, but included a large number of women, led by women's Labour sections from Ouston, Birtley, and Grange Villa who carried their own banners. It was observed in the press that many pushed prams or carried children.[80] This was followed by a series of mass meetings co-ordinated by the Labour women's organizations. Several thousand women attended one such demonstration at the guardians' offices at the beginning of October, joined by several hundred men. Again, the number carrying babies was noted in the local press.[81]

It is perhaps surprising, therefore, that women did not play a dominant role in the administration of the soup kitchens, though they might be involved. When one old man was asked who had organized the soup kitchens during the strike, he replied that 'it was just the ordinary men in the village'.[82] Amongst the Gateshead interviewees, too, many more mentioned fathers who worked at the soup kitchens than mentioned mothers. Indeed, one explained that her father and the other men of the village had been particularly adept at the cooking of mass meals owing to their recent war service.[83] Such an impression is supported by photographic evidence, and the observations of historians of the lockout in South Wales have pointed in the same

[79] GCLOT, iv (Mrs S.). [80] *CC*, 3 Sept. 1926. [81] Ibid., 15 Oct. 1926.
[82] BMOA, 1997/19. [83] GCLOT, ii (Mrs F.).

direction.[84] One possible reason for the overwhelmingly male presence was the free meal that went with the job. The poor relief received by women during the lockout was frequently inadequate, but it did at least help. Some soup kitchens were set up specifically to serve the single men who were both ineligible for relief and had no family whose relief they could share, but many more were established to provide for the miners' children. By helping with these, many strikers were guaranteed a meal.[85]

The lack of a significant female presence within the soup kitchens may also have been indicative of the strongly male-dominated atmosphere of the union. Used extensively during the lockout of 1921, soup kitchens had already been established as an integral part of the strikers' defence by 1926 and were immediately set up. As such, they were part of a formal, union-sponsored coping strategy and so tended to be organized and administered by its male representatives. Such an interpretation complements the observations of Griselda Carr, who noted that while Northern mining women were often involved in informal protests against blacklegs during 1926, they were not involved in union-organized pickets. She suggested that county and local union officials were always appreciative of women's vital contributions to the struggle, but found no evidence that official advice or instructions were ever issued with regard to involving women in the union's actions.[86] Indeed, male exclusivity permeated the union all the way to the national level, and an attempt by miners' wives to procure affiliation to the MFGB in 1927 was denied. Their granddaughters would meet similar resistance from the NUM when they made the same request after the 1984–5 strike.

Some men remained hostile to any kind of female activity, seeing it as a violation of gender boundaries. In November, Ellen Watson was summoned before the courts charged with throwing a potato at a blackleg, and aroused laughter by describing how she had already got into serious trouble with her husband for going to the demonstration at all.[87] It may be that he objected less to female involvement in the strike than to the specific involvement of his wife. Thus, Mary Craddock, who

[84] Bruley, 'Women', 236; John, 'Miner Struggle', 87. Although Alan Campbell has cited examples which suggest that women played a much more dominant role in the soup kitchens set up in the Scottish coalfield districts. See A. Campbell, *The Scottish Miners, 1874–1939*, i, *Industry, Work and Community* (Aldershot, 2000), 241.

[85] For such suggestions, see BMOA, Peter Talbot, written memoirs; GCLOT, ii (T.L.).

[86] Carr, *Pit Women*, 131, 150.

[87] *DC*, 6 Nov. 1926.

grew up in Rainton in the 1920s, would later remember being taken as a child to see Ellen Wilkinson, then MP for Middlesbrough. She noticed 'the quiet respect with which the miners listened attentively . . . She seemed cross about something, and I wondered why none of the men told her to "howld her gob", as they would have done if she'd been one of their wives carrying on in such a fashion.'[88]

The style as well as the reasons for female protest marked another difference from the actions of their male counterparts. Even the weapons used might be particularly female; the banging of pots and pans to accompany the blacklegs home has already been mentioned. In the autumn of 1925, during a local strike at Chopwell, one official made the mistake of passing through the town on his way home and was so severely bruised and stabbed by hatpins that he required medical attention.[89] Gender itself could act as a weapon in such situations. I have come across no evidence of the white-shirting rituals common in South Wales,[90] but in a culture in which concepts of masculinity centred on physical strength and ability underground, the common portrayal of blacklegs as effeminate was reinforced by the taunts of women. Female protesters might also possess other advantages over their male counterparts. One common tactic employed by the strikers was to break open the bottom of trucks carrying coal, which both temporarily grounded them, and allowed the miners to crawl underneath the truck to steal the coal. One old man remembered that on one occasion a policeman was doing his best to pull people out from under the bottom of the truck:

> and of course he got a hold of Bella's ankle to pull her out you see. When she seed who it was she let a scream out. She said, 'Get away, you dirty bugger. Did you see what he did? He put his hands up my clothes.' You see, well that flattened the sergeant. He didn't know which way to look, you see. He just went away out of the road.[91]

## V

To return to the stereotypes with which this chapter began, it is clear that although all resonate within the coalfield, none satisfactorily or

[88] M. Craddock, *Return to Rainton* (1963), 108.

[89] DRO, C3/8, L. Turnbull, *Chopwell's Story*, no page. The Chopwell miners had come out on strike over a local dispute in June 1925; this merged into the national strike the following year.

[90] See Bruley, 'Women', 239.

[91] GCLOT, iv (L.H.).

fully portrays the experience of the women who inhabited Durham mining communities in 1926. The available sources can never be fully representative: undoubtedly some women would not have looked favourably upon strike action, many others would have been ambivalent or suffered mixed feelings. However, the weight of evidence suggests that many women did support the strike, often actively, and this does not seem to have been restricted to any particular generation. Bessie Johnson was 17 when she eagerly went to see A. J. Cook; Ellen Watson was 38 when she threw the potato at the blackleg; Annie Maughan was 53 when she assaulted the blackleg on the bus. Obviously some women participated more visibly than others. In August, Beatrice Webb noted the strain that the Seaham Labour women were under, given their lack of numbers:

> Our Labour women everywhere seem to be worn out—more than the ordinary folk, as they make it their concern to see to others, and the work in the canteens—in distribution—in trying to raise funds—falls on the few comparatively speaking. Many are on the verge of a nervous breakdown. All look tired and strained.[92]

However, many more women participated outside of an organized structure, and oral interviewees of both sexes tend to refer to female activity in universalized terms when they speak of 'the' women, rather than 'some', 'many', or even 'most'. One old man remembered that when one of his neighbours went back to work, 'the women knocked all his windows out'.[93] Besides, apart from active protest and involvement, there were many other ways in which women could support their men and the union, and the indirect support that a woman provided in her role as house manager was crucial for the survival of the lockout. Bill Williamson has written of his grandmother's contribution to the strike, arguing that her willingness to 'muddle through, to make do with nothing and to scratch resources together as best she could' was central to the family's ability to endure the months that followed, and not only demonstrated her support of the strike but also shaped her consciousness: 'her experience of class conflict, as it were, in the kitchen, was just as sharp as my grandfather's down the pit . . . Her aim, in short, was to protect her family as much as she could; she left the political wrangling to others.'[94] His grandmother was the wife of a

[92] LSE, Passfield 2/4/H, B. Webb to L. Fenn, 7 Aug. 1926. [93] GCLOT, iii (R.E.).

[94] B. Williamson, *Class, Culture and Community: A Biographical Study of Social Change in Mining* (1982), 176.

Northumberland miner, but his words are surely relevant to the miners' wives of Durham.

This is not to say that all women came together with one heart and mind, just as it cannot be said of their menfolk. Quarrels and hostility inevitably broke out on occasion, perhaps particularly so during the tension of the strike. In August 1926, for example, Mary Kennedy was charged with assaulting Mary Myers at a children's feeding centre. A local reporter believed it to be the result of a family argument.[95] Also noticeable is the way in which gender remained a defining stratification within the community, even when men and women did unite in support of their union. In 1926, when the men threw stones, women threw potatoes; when men used pit props as weapons, women used hatpins; when men picketed the colliery offices, women picketed the shops. Opposition to blacklegs was articulated in gendered terms. Just as men dismissed a blackleg as effeminate, so women dismissed a blackleg's wife as lacking the true virtues of a miner's wife. Perhaps the most telling comment in this regard comes from one woman who remembered being taunted by a blackleg making £3 a week, a majestic sum compared to the 22*s*. with which she was managing. But, it did not upset her: 'I thought, look at your house.' She explained: he had an 'awful' house.[96]

In her work on the experience of the women of South Wales during the lockout, Bruley noticed similarities with the miners' strike of 1984–5, but suggested that the major difference was that women in the 1980s were 'much more conscious of gender divisions and the collective power of women', even if the long-term impact of the dispute on gender relations was much more limited.[97] In Durham, the 1984–5 dispute saw an overtly feminist agenda articulated by at least some of the most prominent women involved. Yet, if it is debatable how far this resonated amongst the rank-and-file of the coalfield even in the 1980s, it was certainly not rhetoric that would have been recognized by those involved in the earlier dispute. Of all the women interviewed for the Gateshead project in 1976, only one mentioned hopes of female equality, and even then she did not speak of feminism directly, but presented her hopes within the context of her children (nor did she specify daughters). Her words were also surely influenced by the changing perceptions of women engendered by the passage of fifty years. She explained that the women of the village, especially the young women, 'felt that we should win, and that we wanted something better for our bairns than what

[95] *DC*, 28 Aug. 1926. [96] GCLOT, ii (Mrs R.). [97] Bruley, 'Women', 245.

we'd had . . . And that's why I'm so pleased to be here today, speaking, because this is the year of equality for women'.[98]

It is useful here to compare Elizabeth Roberts' study of working-class women in the North West of England in the early part of the century. Roberts discovered frequent indifference towards the early feminist movement, finding that more women were involved within the Labour Party than within suffrage organizations. She believed that amongst the majority there was little feeling that they or their mothers had been particularly exploited by men. Many were aware of their limited horizons and opportunities, but also knew that these were shared by their husbands and sons, while 'those . . . who perceived their lives in terms of exploitation, saw themselves, and their men, as being oppressed by employers, the rich, the middle classes and the bosses . . . In other words, women who were conscious of exploitation interpreted it in terms of class conflict.'[99] The words of one of the founder members of the Durham Labour Women's Advisory Council suggest that Roberts' conclusions may have relevance to Durham. 'If the Labour movement has to go down in history,' Mrs Jolly explained to her interviewers in 1970, 'one of its finest achievements is how the women stood by their men in 1926, and I really believe that'.[100]

Despite this, the sheer sense of enjoyment resulting from a peculiarly female solidarity, common in normal times but reinforced during the strike, should not be forgotten. Beatrice Webb recorded in her diary her impression of a conference held in October:

> As I looked at the gathering of 400 miners' wives and daughters, in their best dresses and the prettily decorated tea tables, with piles of cake and bread and butter, it might have been a gathering of prosperous lower middle class women . . . Neither were they gloomy—they were in a jolly talkative state of mind; they were enjoying their lives . . . The men and boys were more silent and sullen; some of the elder men were anxious and wistful.[101]

Even those distanced from formal politics could get caught up in the excitement of activity. When Mrs Short reminisced about the treatment meted out in her village to the wives of blacklegs, she touched upon

[98] GCLOT, iv (Mrs S.).

[99] E. Roberts, *A Woman's Place: An Oral History of Working-Class Women, 1890–1940* (Oxford, 1984), 2.

[100] Cited in M. Callcott, '1926: Women Support the Miners', *North East Labour History Bulletin*, 19 (1985), 42.

[101] Diary entry, 24 Oct. 1926, in M. Cole (ed.), *Beatrice Webb's Diaries, 1924–32* (1956), 123.

many of the aspects covered in this chapter: the support of the women of the pit communities for the strike; their antagonism towards the blacklegs' wives who broke the solidarity of the village; and an awareness that the lack of an overt political consciousness did not prohibit an endorsement of the trade union values that their men were fighting for. Overwhelmingly, however, the impression is one of fun and exhilaration:

> Now the same blacklegs' wives, when their men was at the pit used to get their coal, and we didn't get owt [any] coal. So whenever we saw that coal wagon we used to shout at the end of the street, 'The coal wagon, Mrs So-and-So's coal was coming.' She never got a scrap. The women joined together and we took the whole lot. We got it all . . . [Now], my mother was a woman that had no politics about her at all . . . she used to say, 'Give me the bairns! Give me the bairns! There's the coal car, let's go!' And away we used to run, and my sisters and anyone that was in the house . . . she never got a scrap of coal.[102]

Her words are complemented by those of Mary Catleugh, a young girl employed at Horner's sweet factory at Chester-le-Street during the strike. She recalled the trouble there was in Pelton with the blackleg miners, and ended her account of the lockout with a touch of regret: 'We missed the fun, us Horner girls,' she said.[103]

[102] GCLOT, iv (Mrs S.).

[103] DRO, B1/14, G. Purdon, *It was Grand Toffee*, no page.

# 4

# Religious Identities

The extent of religious belief is notoriously difficult to gauge, but in the 1920s the majority of Durham inhabitants were not regular churchgoers. Despite this, the churches and chapels maintained an obvious physical presence within the mining villages. The vicar would have been a recognizable figure, and local clerics often carried out house-to-house visits. Many were defeated by the size of the parish, but most managed to attend at least in times of sickness or death.[1] 'The non-church and chapel going part of the population are with rare exceptions not anti-religious,' observed the vicar of Greenside, 'The clergy are welcome in all homes and often sort [*sic*] in sickness'.[2] Most people participated in religious rituals at times of birth, marriage, and death. Churches and chapels also played an institutional role in village life, and on occasions such as Armistice Sunday the various religious denominations were always heavily represented in both the organization of the day and in the services which formed an integral part of the commemoration.

For children, religious attachments might be formed through their educational experience. Just over a third of Durham's elementary schools were maintained by a religious institution in 1926, the majority of which belonged to the Anglicans.[3] Although almost all non-provided schools also chose to teach non-denominational religion, conformity could be checked in the church schools through regular inspection. Of course, attendance at a church school was no guarantee of absorption of the Christian faith. Castle Eden Colliery School at Monk Hesleden gave its Inspector particular cause for concern, and he wrote a damning report in 1926. Five years later he noted little improvement, admitting with some despair that he had 'never been able to speak

[1] Durham University Library (henceforth DUL), AUC/4/13, Auckland Castle Visitation returns, 1924.
[2] Ibid., Greenside parish.
[3] Durham County Record Office (henceforth DRO), CC/Ed352, school reports, 1926–7.

with much enthusiasm in regard to the religious instruction in this school'.[4]

The significant role played by religious bodies in the coalfield raises questions about the potential conflict between a confessional identity and a class-based one. The impact of Methodism on class consciousness has already been subject to intense historical debate, following Elie Halévy's famous thesis that the lack of revolutionary fervour in Britain in the late eighteenth and early nineteenth centuries was at least partly due to the religious revival of the eighteenth century. Eric Hobsbawm became a celebrated critic of such a claim, pointing to the connections between radical centres of activity and Methodist ones. He argued that amongst the northern miners of the nineteenth century 'Primitive Methodism was so closely identified with trade unions as to become, practically, a labour religion,' although he conceded that Wesleyan Methodism promoted different values.[5] A few years earlier, Robert Wearmouth had sought to demonstrate the connections between the labour movement and Methodism by drawing up long lists of Methodist trade union leaders.[6] The debate was addressed specifically in relation to late nineteenth- and early twentieth-century Durham by Robert Moore in his research on Methodism in the Deerness Valley. He concluded that Methodism inhibited the development of class consciousness and acted to reduce class conflict. His study included a discussion of the specific problems posed by the 1926 lockout, and will be considered below.[7]

However, the impact of Anglicanism on coalfield society has more rarely been explored. Indeed, particularly in studies of the North East and South Wales, it is frequently dismissed as irrelevant. Often the Church of England came late to the pit villages, where nineteenth-century Methodist chapels might have been erected one or even two generations earlier. 'We are not dissenters,' the early Durham miners' leader John Wilson argued, 'There is nothing for us to dissent from.'[8] This was not the case everywhere, and Anglican churches had long been present in some of Durham's mining villages. Yet, it has been

[4] DRO, EP/MH.SJ 11/3, reports of religious instruction for Castle Eden Colliery School, 1926–31.

[5] E. Hobsbawm, 'Methodism and the Threat of Revolution in Britain', in his *Labouring Men* (1964), 26.

[6] R. F. Wearmouth, *The Social and Political Influence of Methodism in the Twentieth Century* (1957).

[7] R. Moore, *Pitmen, Preachers and Politics: The Effects of Methodism in a Durham Mining Community* (Cambridge, 1974), esp. 202–13.

[8] H. Hensley Henson, *Retrospect of an Unimportant Life*, ii, *1920–39* (1943), 80.

represented as the church of the employer class, a church in which the miners, that supposedly most proletarian of occupational groups, failed to find a sympathetic place of worship.[9]

In fact, the Church of England was far from insignificant in the coalfield, evident in the number of its followers. Keenly aware of the 'sociologically significantly different ways of "belonging" to a religious group', Moore attempted to estimate the number of believers in the Deerness Valley based on the recollections of his respondents. He suggested that in 1919 65 per cent of households had at least one member who was 'regarded' as belonging to a church, of which 40 per cent were seen as Methodist households, 10 per cent Catholic, 4 per cent Baptist, and 10 per cent Anglican.[10]

Moore's figure for Anglican adherents, if assumed to be typical of the county as a whole, compares unfavourably to other assessments of Anglican belief. The 1924 diocesan visitation recorded 156,138 people in the county registered as parochial electors in an adult population of 839,765, equating to 18.6 per cent who presumably had a basic sense of themselves as Anglicans.[11] This, of course, was no guarantee of any stronger commitment to their church or to attendance at worship. The number of committed church members was considerably smaller, and in 1924 no more than 8,832 people in the bishopric (just over 1 per cent of the adult population) attended their annual meeting to elect parochial church councillors.[12] However, it also excludes those who may not have been registered but who still felt some kind of identification with the parish church, or, indeed, those who dallied with Anglicanism simply as a change from their usual place of worship.

If Moore underestimated the number of Anglicans, his estimate of the number of Methodists may have been too high. In 1968, Michael Kinnear restricted his calculation of Methodists to the numbers contained in circuit records. He counted a total of 26,279 Nonconformists in the county constituencies of Durham (including 12,722 Primitive Methodists and 9,320 Wesleyan Methodists) and suggested that this figure represented 7 per cent of the county's population. He also revealed

[9] For studies of coalfield communities that emphasize the importance of Methodism see, for example, S. Webb, *The Story of the Durham Miners, 1662–1921* (1921); H. Beynon and T. Austrin, *Masters and Servants: Class and Patronage in the Making of a Labour Organisation* (1994); M. Lieven, *Senghennydd: The Universal Pit Village 1890–1930* (Llandysul, 1994).

[10] Moore, *Pitmen, Preachers and Politics*, 69–70.

[11] H. Hensley Henson, *Quo Tendimus? The Primary Charge Delivered at his Visitation to the Clergy of his Diocese in November 1924* (1925), 159.

[12] Ibid., 37.

wide regional variations: Barnard Castle (which encompassed part of Moore's Deerness Valley) had the highest percentage, at 17.3 per cent of the population. But, Kinnear's percentages should be reduced still further. He did not detail his methodology, but the same figures are reached by measuring the number of Methodists against the number of electors in 1922, a number that is not representative of the entire adult population. If the 26,279 men and women active in Nonconformist churches are compared to an adult population of 839,765 in 1924 (as recorded by the diocesan visitation), the number of practising Nonconformists works out at only 3.1 per cent.[13]

Moore's figures do correlate with contemporary Catholic sources, which claimed that the diocese of Hexham and Newcastle was around 10 per cent Catholic, with 220,471 followers in a population of 2,225,129.[14] Aside from the possibility that Catholic administrators might overestimate the number of their fellow believers, this figure also needs to be treated with caution. County Durham formed only a part of the diocese, and Catholics were not spread evenly within it, tending to inhabit the bigger cities such as Newcastle. However, they were not absent from the mining villages: sixty-three Catholic churches were registered for marriages in County Durham in 1926. Many of these, such as St Cuthbert's at New Seaham, or St Michael's at Houghton-le-Spring, ministered to areas dominated by coal miners.[15]

The relative neglect of the Anglican Church by historians is also surprising given the existence of a rewarding collection of sources. Invaluable for much of this chapter are the visitation returns issued by the Bishop every four years, to be completed by all parish priests in the diocese.[16] In contrast, Methodist sources are problematic. In his study of the British Free Churches in the interwar years, Peter Catterall regretted that 'their minute books cast light on little of value, beyond perhaps the need for new heating systems'.[17] The Durham minute books are no exception; indeed, its northern climate made heating systems all the more important. Moore based much of his research on oral history interviews, an option no longer available, although his decision

[13] M. Kinnear, *The British Voter: An Atlas and Survey since 1885* (1968), 126; the number of electors in 1922 is given in F. W. S. Craig, *British Parliamentary Election Results, 1918–49* (Glasgow, 1969), 338–48.

[14] *The Catholic Directory, 1926*, 216.

[15] Ibid., 202–12.

[16] Although these have recently begun to be tapped. See R. Lee, *The Church of England and the Durham Coalfield, 1810–1926: Clergymen, Capitalists and Colliers* (Woodbridge, 2007).

[17] P. Catterall, 'Morality and Politics: The Free Churches and the Labour Party between the Wars', *Historical Journal*, 36 (1993), 667.

to deposit his findings in Durham University's archives also gives access to his unpublished research materials. Unfortunately, Catholic sources have proved near impossible to find, despite a search of both secular and religious depositories. I have come across only occasional anecdotal evidence, some of it with potentially important implications: the claim of one old man that the Catholics were 'more clannish', the memories of fights between pupils of local Catholic and Anglican schools; the Catholic miners who had money taken off their wages for the upkeep of their local priest.[18]

A discussion of Anglicanism in the coalfield will form the basis of this chapter and the issues raised will then be considered in the light of existing debates about the impact of Methodism. The lack of evidence prohibits any sustained comparison with Catholicism: an exploration of the tensions between a confessional and class identity amongst the small number of Catholics in the coalfield must await another historian.

## I

The popular image of the Anglican Church is of an institution irredeemably bound up with Conservative politics, standing in ideological opposition to the Labour-voting miners. It was with sadness that a Durham vicar noted in 1928 'an underlying suspicion that the Church is a handmaid of the Conservative Party'.[19] Many churches had been built by coal owners as an exercise in paternalism and often with overtones of social control.[20] Such practices were on the wane by the 1920s, but links still existed and employers might continue to provide financial assistance. In April 1926, at a church council meeting at Blackhall, for example, the vicar thanked the local colliery company 'for past favours, kindness and support'. A couple of months later the same council thanked the company more specifically, this time for the gift of paint which had allowed the redecoration of the church.[21] Other contributions were more substantial. At St Andrew's in Beamish, the financial report for 1926 reveals colliery company money again, in this

18 British Library Sound Archive (henceforth BLSA), C707/155/1–3C1; C900/11043; C900/11043C1; W. A. Healy, *Between the Wars: Childhood Memories of Horden* (1996), 75.

19 DUL, AUC/4/14, Auckland Castle Visitation returns, 1928, New Seaham parish.

20 See A. J. Heesom, 'Problems of Church Extension in a Victorian New Town: The Londonderrys and Seaham Harbour', *Northern History*, 15 (1979), 138–55.

21 DRO, EP/Bla6/1, parochial church council minutes, St. Andrew's, Blackhall, 12 Apr. and 5 July 1926.

case a contribution of £75 to the annual stipends of two curates (worth £550 in total).[22]

The practical assistance rendered by the colliery companies to their local churches allowed them to exert a certain amount of moral pressure on their politics. In June, Bishop Henson received a letter from Lord Londonderry complaining about the support being given by James Duncan, vicar of Dawdon, to the strikers at his collieries. 'They [the employers] cannot understand how the Clergy of all men should have nothing but condonation and sympathy for the Strikers,' confided Henson in his diary. He for one thought this rather impolitic, for 'it is the case—although the fact is forgotten and ignored by the "Socialist" clergy—that the chief, indeed almost the only supporters of the Church's work are precisely those same employers'.[23] Londonderry may have felt particularly aggrieved given his personal association with Dawdon's church, St Hild and St Helen. Its font bore a dedication to the four 'Helens' of the Londonderry family, linking 'the founders of much of the region's secular wealth and one of the founders of the region's spiritual heritage'.[24]

As a national institution, the Anglican Church was further linked to an apparently exploitative class through its receipt of coal royalties, from which it made about £400,000 a year.[25] Henson was an obvious local target for criticism: his imposing Auckland Castle residence was a visible sign of such wealth. The hypocrisy of the Church became a common theme in the rhetoric of the miners' leaders, and A. J. Cook railed against the Bishop's opposition to another government subsidy for the industry, pointing out that Henson had 'lived on a subsidy all his life'.[26] Henson privately admitted that his relationship with the miners might have been easier had he lived 'like John the Baptist in the wilderness, clothed not in gaiters but in leather and camel hair, feeding on locusts and wild honey'.[27]

Of course, Henson had neither the intention nor inclination to live as such, and in Durham the reputation of the Anglican Church as a bastion of Conservative politics was given greater weight by the unambiguous political position of its senior clerics. In public, the Bishop's references

[22] DRO, EP/Bea2/76, parochial statistics, St. Andrew's, Beamish, 1926.

[23] Dean and Chapter Library, Durham Cathedral (henceforth DCL), diary of H. Hensley Henson, 2 June 1926.

[24] Lee, *Church of England*, 183.

[25] Report of the Royal Commission on the Coal Industry (1925), ii, Minutes of Evidence (1926), q. 13,813.

[26] *Durham County Advertiser* (henceforth *DCA*), 13 Aug. 1926.

[27] DCL, Henson's diary, 9 May 1926.

to the miners were frequently warm, and at a diocesan conference in November he declared that 'no one can have any personal contact with miners without forming a very high estimate of their potentialities and conceiving a great liking for them'.[28] However, although attached to no political party, his stance was virulently anti-socialist and he despised the class consciousness fostered by the labour movement. One biographer has implied that his 'violent, almost obsessional' dislike of trade unions was near pathological: as vicar of Barking, a younger Henson apparently suffered recurrent premonitions that he would die in a street battle with strikers.[29] He frequently spoke out about the mischief of strike action, both before and during the 1926 dispute, and was revolted by the 'criminal act' of the general strike in particular: 'These hateful economic conflicts fill the whole horizon of life, and nothing any longer matters anything at all. And how squalid the actual issues are! How repulsive in their technicalities are the contentions! How debased, morally and intellectually, are the zealots of "Labour".'[30] His prejudices were shared by his Dean, J. E. C. Welldon, who was also unequivocal in his condemnation of the general strike, suggesting that it was but 'an evil shadow of the treachery of which the Germans were guilty when . . . they invaded the borders of Belgium'.[31] Such opinions were well publicized within the mining community and were often reproduced in the local press. Those who attended the sermons of the two men had an added opportunity to absorb their beliefs, which inevitably crept into their preaching despite Henson's avowed hostility towards the use of the pulpit as a political soapbox. During the general strike, for example, Henson asked that his congregation pray for the strikers, 'that they may be led to a worthier sense of their duty as citizens'.[32]

The pronouncements of Henson and Welldon hardly endeared the two men to the miners or their leaders. Tension had been apparent even before the strike, and at the previous year's miners' gala a phrase had been attached to one of the banners proclaiming, 'To Hell with Bishops and Deans, We Demand A Living Wage'. Later that afternoon, when Welldon was spotted amongst the crowds, men and women surged forward with the cry, 'Here's the Bishop, hoy [throw] him in the river!'

[28] *Durham Chronicle* (henceforth *DC*), 13 Nov. 1926.

[29] M. Grimley, 'Henson, Herbert Hensley', in *Oxford Dictionary of National Biography* (Oxford, 2004), xxvi. 613.

[30] DCL, Henson's diary, 4 June 1926.

[31] *DCA*, 28 May 1926.

[32] O. Chadwick, *Hensley Henson and the Durham Miners, 1920–1939* (Durham, 1983), 14.

He was jostled towards the bank, losing his hat in the process, and a police boat had to be launched to rescue him before he was pushed in.[33] The story has gone down in Durham memory as one of the most memorable gala tales, although in its retelling Welldon usually finds himself getting a drenching. He later shrugged off the incident, suggesting that the miners had mistaken him for Henson, who had recently published his opposition to the minimum wage in a London newspaper, but either man could legitimately have been the target: Welldon's opposition to higher wages was also well-known.[34]

During the strike itself, Henson and Welldon continued to attract hostile comments. In September 1926, Herbert Dunnico, the Baptist preacher and Labour MP for Consett, spoke in particularly vitriolic terms when he addressed a crowd at Stanley. Choosing a rather damning phrase to describe two bishops, he declared that 'two greater representations of anti-Christ he could not imagine'.[35] Henson also received at least one anonymous letter condemning him for his stance. He shrugged it off: 'I suspect that if I had but vented a few canting platitudes about the miners' children etc, etc, the author of this effusion would have exalted me as a very Saint of God.'[36]

In fact, the attitudes of the two most senior churchmen in Durham were at odds with the prevalent mood amongst the Anglican hierarchy in the 1920s. Henson and Welldon were operating within the context of an increasing ecclesiastical involvement in social questions, a radicalism that had grown since the war, with William Temple as its most famous figurehead. This had encompassed issues relevant to the miners even before the upheavals within the coal industry inevitably focused attention. COPEC (Conference on Christian Politics, Economics and Citizenship) was one of the most well-known initiatives of this movement, being an interdenominational group of churchmen driven by a Christian socialist agenda, who came together in a series of meetings between 1920 and 1924 to debate pressing issues of the day. When the coal dispute escalated into a national strike, many religious leaders remained committed to social intervention. The Archbishop of Canterbury made his famous plea for peace during the general strike in May; a group of Anglican bishops and Nonconformist leaders (including Temple) then

[33] O. Chadwick, *Hensley Henson: A Study in the Friction between Church and State* (Oxford, 1983), 167.

[34] See, for example, comments made by Welldon at a meeting of railwaymen in 1924 in J. E. C. Welldon, *Forty Years On* (1935), 207.

[35] *DC*, 18 Sept. 1926.

[36] DCL, Henson's diary, 5 July 1926.

attempted mediation in July. Durham also had a legacy of its own with regard to episcopal intervention in industrial disputes, owing to Bishop Westcott's renowned mediation during the 1892 coal strike. Yet, in 1926, Henson led the opposition to such action, vehemently criticizing his 'feather-headed' colleagues whom he accused of only prolonging the conflict by their actions.[37]

Admittedly, this movement was about peacemaking, and at no time did any bishop express support for the strike: indeed, Matthew Grimley has argued that the experience was decisive in Temple's rejection of socialism as a sectionalist ideology.[38] The Durham legacy of sympathetic clerical involvement could also be exaggerated. Bishop Westcott's mediation in 1892 could be offset, for those who could remember great grandfathers' tales, by the actions of Shute Barrington, Bishop of Durham during the 1810 strike. When Durham Gaol had no more room for striking miners, he had provided his stables as alternative premises for 300 of them.

Nevertheless, in 'this heated "Copec" atmosphere' (as Henson described it), the condemnation of Durham's senior clerics was conspicuous.[39] With a topical description in August 1926, Joshua Ritson declared that he felt inclined to nickname them the 'blackleg bishops', because they were so out of harmony with other bishops around the country.[40] In fact, Henson's opinion was more attuned to the line taken by the head of the Catholic Church, Cardinal Bourne, who unambiguously dismissed the moral claims of the strikers and declared the general strike as nothing less than a sin against God.[41] Henson approved: 'Comparisons are made between Cardinal Bourne's message which was clear, relevant and useful: and that of his Grace and the bishops, which was indefensible, unwise and likely to do much harm in the future.'[42]

## II

Whereas a number of studies have dealt with the attitude of the higher clergy to the strike,[43] little has been written about the attitude of the

[37] Chadwick, *Henson: A Study*, 170.

[38] M. Grimley, *Citizenship, Community and the Church of England: Liberal Anglican Theories of the State between the Wars* (Oxford, 2004), 103.

[39] DCL, Henson's diary, 28 July 1926.

[40] *DC*, 14 Aug. 1926.

[41] *Tablet*, 15 May 1926.

[42] DCL, Henson's diary, 18 May 1926.

[43] For example, Grimley, *Citizenship, Community and the Church of England*; G. I. T. Machin, *Churches and Social Issues in Twentieth Century Britain* (Oxford, 1998).

lower clergy, whose response might be far more relevant to those living in the pit villages. In fact, historians of the twentieth century have rarely followed the lead of their nineteenth century counterparts, a number of whom have explored the ways in which local clergymen in working-class parishes often tried to reconcile the needs and interests of their parishioners with the rather different expectations of their church hierarchy.[44] Of course, a parish priest and a miner still inhabited different social strata. One Northumberland miner's daughter, born in 1918, vocalized memories with which many Durham residents could surely have identified, when she recalled that 'in my youth, the parish vicar was envied somewhat . . . With a car of his own, a chauffeur and domestic help, he was a man apart.'[45] The clergy were further separated from their parishioners by their education. Of the seventy-four priests and curates from the deaneries of Easington, Chester-le-Street, and Houghton-le-Spring in 1926, forty-six held a BA, and thirty of these also an MA. Twenty-one had been educated at Oxford or Cambridge, and thirty-two at Durham University.[46]

Some vicars sought to maintain such distinctions. One admitted in 1924 that he had 'declined several offers from men to read the Lessons in Church, as I prefer doing so myself, and they are not sufficiently educated'.[47] But, more common was the feeling that such factors damaged a priest's credentials within the pit villages. The vicar of Wheatley Hill believed that too much emphasis was placed on the intellectual training of priests: 'The miners do not want "Greek Cube roots".'[48] He was perhaps aware of jokes circulating within the coalfield that stereotyped churchmen as kindly and concerned but basically out of touch. One 'local anecdote' submitted to the *Durham Chronicle* by a resident of Quebec in May 1926 was typical of half a dozen others: 'A clergyman calling upon a woman asked her how it was that she didn't attend church. She replied, "Because I haven't any boots to come in." "Oh, but it's not the boots we look at," replied the clergyman, "It's the souls." "Eh, but," said the woman, "Aa hev ne soles on 'em." '[49]

[44] For example, K. S. Inglis, *Churches and the Working Classes in Victorian England* (1963); H. McLeod, *Religion and the Working Class in Nineteenth-Century Britain* (1984). Lee's recent *Church of England* begins to rectify this, covering the first quarter of the twentieth century, but his period of study remains primarily the nineteenth.

[45] M. Wade, *To the Miner Born* (Northumberland, 1984), 6.

[46] *Crockford's Clerical Directory* (1926 edn.).

[47] DUL, AUC/4/13, Visitation returns, 1924, Eighton Banks parish.

[48] Ibid., Wheatley Hill parish.

[49] *DC*, 1 May 1926.

Some local incumbents further alienated themselves from their parishioners through their political stance. Many shared the political sympathies of Henson; a few could also be as vituperative. Edward Rust, the vicar of Hamsteels, published his opinion of the dispute in his parish magazine in September:

> The trades union leaders have inflicted a far greater degradation in the standard of living among the miners than ever the masters dreamt of. It is pitiful to see the miners and their children . . . grovelling amongst the pit heaps, working as hard or harder to pick up a hundredweight of almost worthless cinders as they would do to get a ton of good coal if they were working for the masters . . . Meanwhile the men and their dependants are living on the dole and on public charity and, most shameful of all, on money extorted by the Soviet tyrants of Russia from their wretched starving victims.[50]

His attitude was consistent with an alleged wider distaste for his working-class parishioners. One later recalled that, 'he used to be late for working-class funerals'.[51]

Other vicars based their opposition to the strike on immediate self-interest. The Revd H. J. Peck condemned the strikers as selfish, claiming that many clergymen had a guaranteed income of no more than £200 a year and so were in a much greater position of poverty than thousands of miners (see Table 2.4 for miners' wage levels).[52] At least one of the local ratepayers' associations formed in 1926 recruited the local vicar as president.[53] But, some had been suspicious of Labour politics even before the strike. In 1924, the vicar of Tudhoe explained that a Labour ideology taught 'people and especially children not to depend upon their own exertions but to look to others for necessities of life'.[54] His worries were shared by his counterparts elsewhere, and a handful of local priests became involved more directly in local politics. The local elections in 1925 saw two Durham clerics stand for positions on the county council, both on anti-socialist platforms.[55]

Such attitudes inevitably affected the standing of these men within the pit villages. In November 1926, one Durham vicar (also a shareholder

[50] *DCA*, 3 Sept. 1926.

[51] DUL, misc. accession 1996/7:4, qualidata reference QD35/Moore, data gathered by Prof. Robert Moore.

[52] *DC*, 20 Nov. 1926.

[53] For example, at Shotton. *DC*, 13 Nov. 1926.

[54] DUL, AUC/4/13, Visitation returns, 1924, Tudhoe parish.

[55] DRO, D/MCF25, record of Durham County Council election contests, 1922–31. At Annfield Plain, Labour held the seat against the challenge of the Revd Fisher Ferguson; at Whickham the Revd Alexander Dunn retained his seat.

in his local colliery company) referred in his Sunday sermon to the scandalous treatment meted out to blacklegs. The next morning he came across a miners' meeting and was booed and jeered.[56] Church-going miners were able to vote with their feet. Rust's anti-socialist tirades led some members of his congregation to abandon his church and attend religious services elsewhere. Disillusioned, the vicar twice appealed (unsuccessfully) to Henson to request a change of benefice. 'I am in the lamentable position of seeing the village going to ruin without being able to help it,' he explained, 'Not being a person of superabundant tact, I have shown my disapproval of their conduct and consequently they treat me as an enemy, forbid their children to come to my church or Sunday School, even insult adults on their way to church.'[57]

However, not every parish priest stood in opposition to the dominant Labour philosophy of the coalfield, despite the popular stereotypes. Some local clerics were actively and openly supportive of the Labour movement and Henson even despaired that far too many had become 'mere parasites of the Labour Party'.[58] In 1922, when Labour became the official opposition in Parliament for the first time, a congratulatory message was sent to Ramsay MacDonald by representatives of the Anglican Church; several vicars of County Durham added their names.[59] One was Harry Watts of Shildon, who articulated the ideals of Christian socialism so vehemently rejected by many of his colleagues. Speaking to the women's section of Easington Labour Party in November 1926 he explained:

> He was not a minister who had gone 'Red', but a Socialist who had gone Minister . . . No one could accuse him of being a traitor to his cloth . . . Christ was the great founder of Socialism, and because the Gospel He had commanded to be preached to the poor was not being preached as it should be, matters were as they were today.[60]

Watts was president of his divisional Labour Party branch and a source of particular aggravation to his Bishop: in September, Henson recorded a surely exaggerated report that, during the general strike, Watts had

[56] N. Emery, 'Pease and Partners in the Deerness Valley: Aspects of the Social and Economic History of Waterhouses, Esh Winning and Ushaw Moor' (Durham University MA thesis, 1984), 141; *DC*, 6 Nov. 1926.

[57] DCL, Henson's diary, 9 Aug. 1926.

[58] Ibid., 6 May 1926.

[59] The National Archives: Public Record Office, PRO 30/69/1346, 'Memorial of the Priests of the Church of England, the Church of Wales and the Episcopal Church of Scotland Presented to James Ramsay MacDonald and the Labour MPs of Scotland'.

[60] *DC*, 27 Nov. 1926.

'led his parishioners at a public meeting to swear with uplifted hand that they would read no other newspaper than the "Daily Herald" '.[61] Harry Watts was not, however, the only parish priest to gain notoriety within the pages of Henson's diary, even if his name occurred most frequently. The incumbent of Easington also drew Henson's wrath as being 'a man of small ideas, Socialist professions, and a reputation which waxes with the distance from his parish!'[62] In 1926, such men lent their support to the strike, took an active part in its political organization, and were prepared to associate themselves with the most notorious of miners' leaders. When A. J. Cook arrived at Ryhope to address a mass meeting in October, for example, the local vicar, Percival Knight, followed him to the platform to lead the crowd in prayer.[63]

That some were so sympathetic to the miners may have been due to the changing social background of the Anglican clergy in the diocese. Recent research by Robert Lee has revealed a conscious attempt by the late nineteenth-century Church hierarchy to extend the appeal of the Church through the ordination of men from a wider social and geographical demographic. By the first two decades of the twentieth century, over half of clerical ordinands in Durham came from lower-middle-class or working-class backgrounds, figures that differed dramatically from those of fifty or a hundred years earlier.[64] It remained the case that very few Anglican clerics came from local mining families, unlike their Methodist counterparts. William Hodgson, vicar of Escomb, was a miner's son and was to stand for and win a Durham County Council seat for Labour in 1928, but Harry Watts was the son of a tutor, and James Duncan, the vicar who had upset Lord Londonderry, was the son of a policeman.[65] Education also continued to distinguish these men: Hodgson and Watts had been undergraduates at Cambridge; Duncan received an MA from Durham. Nevertheless, such men might still retain ties with the class into which they had been born. Henson privately despaired that Hodgson in particular could ever be made into a 'reverently mannered clergyman'. The only chance of this would be 'his removal from the too-familiar environment in which he has been reared. [Otherwise] . . . In constant contact with his family . . . the possibility is remote indeed.'[66]

61 DCL, Henson's diary, 15 Sept. 1926.

62 Ibid., 3 May 1926.

63 *DC*, 23 Oct. 1926.

64 R. Lee, 'Class, Industrialization and the Church of England: The Case of the Durham Diocese in the Nineteenth Century', *Past and Present*, 191 (2006), 174–5.

65 Ibid., 182.

66 DCL, Henson's diary, 4 Apr. 1926.

## III

The majority of local clerics were not so passionate about the strike or the labour movement as either Rust on one hand or Watts and Hodgson on the other, but most remained relatively well-disposed towards the mining population throughout the 1920s. A tentative indication of this can be gained from the 1924 visitation, in which each parish priest, nearly all of whom were still incumbent in 1926, was asked his opinion on the moral standards of his parishioners. Leaving aside the handful who answered only with reference to sexual morals, the replies can be used as a basic indicator of the sympathetic inclination of the priest towards the local population, ranging as they do from those who admired the honesty and kindness of the miners in almost exaggerated terms to those who dismissed the miners as simply 'very self-satisfied'. Concentrating on the deaneries of Chester-le-Street, Easington, and Houghton-le-Spring (which covered the three rural districts with the highest number of miners proportional to the male population: see Table 1.1), an analysis of the responses shows that almost half gave a generally positive answer, and only a quarter gave a generally negative one (see Table 4.1).

A sympathetic parish priest was in a position to make a practical difference to striking miners and their families, even if this was not done in an overtly political way. Church premises were made available to the

**Table 4.1** *'What opinion have you formed as to the moral standards of your parishioners?' Responses of parish priests to the Episcopal Visitation of 1924*

| Response | Chester-le-Street Deanery (16 returns) | Easington Deanery (17 returns) | Houghton-le-Spring Deanery (15 returns) | Total |
|---|---|---|---|---|
| Positive (unqualified) | 1 | 3 | 2 | 22 |
| Positive (qualified) | 7 | 3 | 6 | |
| 'Average' | 1 | 3 | 0 | 13 |
| Very mixed response | 3 | 3 | 3 | |
| Negative (qualified) | 1 | 1 | 2 | 12 |
| Negative (unqualified) | 2 | 4 | 2 | |
| No answer | 1 | 0 | 0 | 1 |

*Source*: DUL, AUC/4/14.

miners' cause at no cost and church halls were frequently converted into feeding centres or given over to fundraising events.[67] Vicars were often personally involved, and at one fundraising concert promoted by Browney lodge, the Revd Harry Hayward acted as chairman.[68] This example is significant as the proceeds were donated to unmarried strikers, whose plight had become particularly contentious because of the battle with the Ministry of Health over their status with regard to relief.

Of course, sympathy for the miners (and particularly their wives and children) was not necessarily synonymous with political support for the strike or its aims. Edward Maish, the vicar of Belmont, not only assisted at his local children's feeding centre but did so with particular dedication: a letter sent to the *Durham Chronicle* by the centre's chairman at the end of the dispute expressed thanks to all its voluntary workers, but specified for the vicar 'an extra word of gratitude . . . for his helpfulness and regular attendance'.[69] Yet, though a willing volunteer, the experience only increased Maish's despair

> that the parents of 210 children in the parish should find it necessary to ask other people to feed their little ones . . . I fear that there are few amongst us who realise how terribly near our country is to far darker times and that most of us are quite ignorant of the battle for existence which our nation is now fighting.[70]

(His response to the question on moral standards had, incidentally, been a relatively positive one.) Charitable offerings could therefore come from those whose political opposition to the strike was well known. In August, Dean Welldon himself presented a trophy cup to the county agricultural show, to be awarded to a Durham miner for the best vegetable tray.[71]

But, these were still gestures that did not have to be made: many would not be so generous. Henson, for example, thought that the focus on the suffering miners' children was misleading. In May, he recorded a conversation with one of his vicars, who had spoken about the strike 'with such fatuity that my patience was strained. He maunders about "the children" as if *they* could be benefited by a total cessation of their parents' earnings.'[72] There were also financial restrictions. Many churches found

[67] For example, at Langley Park and Newbottle. DRO, EP/LP 6/1, maintenance committee minutes, All Saints', Langley Park, 19 May 1926; *DC*, 2 Oct. 1926.

[68] *DC*, 2 Oct. 1926.

[69] Ibid., 11 Dec. 1926.

[70] *DCA*, 3 Sept. 1926.

[71] DCA, 6 Aug. 1926.

[72] DCL, Henson's diary, 3 May 1926. Original emphasis.

themselves forced to cancel or postpone concerts, outings, or other planned events and church coffers consequently suffered. At one church council meeting, a forthcoming dance in aid of the local boot fund was discussed with regard to whether the church should levy a charge for the overheads. At least one member opposed this, arguing that everyone else was providing services free, but after a discussion of the hall's debts it was decided that a charge would have to be made.[73] Nevertheless, most churches reacted sympathetically in the face of declining collections. At Blackhall, the issue was handled sensitively and the usual practice of passing the collection plate around during services was abandoned; instead it was surreptitiously placed by the door.[74]

Most priests therefore offered at least a benign neutrality to the strikers and were able to present their church as an apolitical body rather than a hostile one. The tone of sermons and prayers frequently emphasized hopes for peace rather than any condemnation of events or particular parties. At Esh, the local vicar was careful to stress that both sides had an equal responsibility to work constructively towards a settlement, appealing both to the coal owners to be generous in their offers and to the miners to accept reasonable offers so made.[75] At Blackhall, the vicar opened his parochial church council meetings with a prayer that 'wise counsels would prevail and a lasting peace be secured for the coal industry'.[76] At the very least, vicars stressed that the solution lay in the adoption of religious values. At Trimdon, on 16 May, Alfred Davison recorded in his service book the text that he had used for his Sunday sermon. 'General strike,' he wrote, followed by a reference from Peter's first letter. The Bible citation reveals the vicar's fears about the ongoing industrial crisis, but his message is clearly a religious rather than an overtly political one: 'But, the end of all things is at hand,' Peter warned, 'Be ye therefore sober, and watch unto prayer. And above all things have fervent charity among yourselves: for charity shall cover the multitude of sins' (1 Peter 4: 7–8).[77]

The success of local priests in presenting their churches in this way allowed many miners to avoid a fundamental conflict of interest between their religion and their politics, even in the politicizing atmosphere of the strike. On the whole it appears that most priests were successful

[73] DRO, EP/Pen77, church council minutes, All Saints', Penshaw, 10 Aug. 1926.
[74] DRO, EP/Bla6/1, church council minutes, St Andrew's, Blackhall, 5 July 1926.
[75] *DC*, 16 Oct. 1926.
[76] DRO, EP/Bla6/1, church council minutes, St Andrew's, Blackhall, 26 Apr. 1926.
[77] DRO, EP/Tr26, register of services, St Mary's, Trimdon.

in maintaining their body of worshippers and did not see the conflict between a Labour and a Christian identity as a serious one. When the Bishop's visitation of 1928 asked ministers how far spiritual life had been affected by political agitation, the answers were remarkably upbeat. Of fifty-four returns analysed, twenty-one ministers felt that political agitation affected the spiritual life of the parish either very little or not at all, and only five saw it as a serious problem (see Table 4.2).

Admittedly, a priest's perception of conflict within his parish might be very different from that of his parishioners, but many none the less testified to an overlapping of political radicalism and church attendance, as at Wingate, where the parish priest recorded that 'many of our people belong to the Labour Party but remain quite faithful members of the Church'.[78] Most striking was Thomas Fenton Fyffe's opinion of his Cornforth parishioners, or at least his opinion of his church as a place immune to the tensions of worldly political life: 'The Secretary of the Women's Conservative Association, who is also our Church Secretary, is on the best of terms with her fellow-Communicant, the Secretary of the Women's Labour Association. One Church Warden is a Conservative and the other Labour and they are the most sincere of friends.'[79]

**Table 4.2** *'How far is spiritual life in the parish affected by political agitation?' Responses of parish priests to the Episcopal Visitation of 1928*

| Response | Chester-le-Street (17 parishes) | Houghton-le-Spring (16 parishes) | Easington (21 parishes) | Total |
|---|---|---|---|---|
| Not at all | 4 | 4 | 9 | 21 |
| Very little | 2 | 7 | 6 | |
| Some | 3 | 2 | 0 | 5 |
| Dwindled since strike | 1 | 1 | 3 | 5 |
| Legacy of strike continues | 1 | 0 | 0 | 5 |
| Very much | 2 | 0 | 2 | |
| Don't know/no answer given | 4 | 2 | 1 | 7 |

*Source*: DUL, AUC/4/14.

78 DUL, AUC/4/14, Visitation returns, 1928, Wingate parish.

79 Ibid., Cornforth parish.

## IV

It is worth making a comparison here with one of the conclusions of Robert Moore's research. Moore argued that one of the main characteristics of Methodism in this period was a sense of fellowship, which allowed religious identities to shroud political ones. He suggested that the language of the chapel, the activities it organized, and the cementing of communal relations by, for example, high rates of intermarriage, all contributed to a specifically Methodist outlook which was able to erode a class or occupational identity.[80] Amongst a core of believers, the communal nature of chapel life helps to explain how a religious attachment could be maintained through the strike. Thus, social activities continued throughout the hungry months of 1926, with Sunday school excursions, choir trips, and other activities presumably helping to alleviate boredom and lift morale.[81]

However, such phenomena were not restricted to the Methodist chapels but were characteristic of any established religious community, whether Methodist, Anglican, or Catholic. In 1924, therefore, the vicar of St Cuthbert's, Benfieldside, described the cementing force provided by the varied social activities of his church, ranging from Shakespearean plays and needlework, to football and Morris dancing.[82] One man, born the son of a Durham miner in 1918, remembered that as a child, 'the [parish] Church was the centre of our social life'.[83] During the lockout itself, Anglican miners threw themselves into constructive activities with the enthusiasm of their Methodist counterparts and, in July, a group of Pittington miners undertook a major cleaning of their church. The work occupied them for eight days, and was preceded each morning by prayer.[84] Similar renovation and cleaning work went on at parish churches in Sacriston, Seaham, and Crawcrook.[85]

Church-based sports teams were another way of reinforcing a religious communal identity. Writing about the first half of the century, Jack Williams has suggested that Northern England had a greater concentration of church-based football and cricket teams than any other part of

[80] Moore, *Pitmen, Preachers and Politics*, 124–32.

[81] See, for example, events chronicled in *DCA*, 30 July 1926 and *DC*, 18 Sept. 1926.

[82] DUL, AUC/4/13, Visitation returns, 1924.

[83] BLSA, C900/11043C1.

[84] *DCA*, 30 July 1926.

[85] *DC*, 7 Aug., 11 Sept. 1926; *Blaydon Courier*, 30 Oct. 1926.

the country.[86] Certainly, in Durham, such teams figured prominently in local league tables, and local vicars (of whom the youngest often themselves played) took advantage of this to appeal to the young. In September 1926, for example, a service exclusive to sportsmen was held at Seaham's parish church, where the vicar was also the captain of the cricket club. The proceeds were donated to a local cricketers' fund.[87] Sometimes such activity was a direct response to the lockout, and in June the members of the Thornley Catholic Young Men's Society held a sports gathering to relieve 'the tedium of idleness'.[88]

Admittedly, an involvement with social activities did not necessarily foster a real engagement with a particular church, and certainly not with its ideology. One woman born in 1909 told of how she and her brother never went to church, though they went on the organized walks and loved to attend the field days, when tea and sports were provided.[89] However, a concern with a specifically religious community helps to explain why many Anglican priests took the apolitical line that they did. Their philosophy professed the openness of the church community to the whole population, and could not give those involved in the mining industry preferential attention, despite their numerical superiority. Not every member of every Durham village belonged to a mining family and the baptism register of St Nicholas' in Hetton-le-Hole gives some sense of the church as a non-occupational focus for the community. As well as a number of miners who brought their children there to be baptized in 1926, so also did an innkeeper, school teacher, housekeeper, butcher, fireman, blacksmith, motor driver, joiner, and waggonwright.[90]

At a time when almost every event chronicled in the local press was donating its proceeds to the children's boot fund, therefore, churches remained faithful to other charitable causes. Thus, when Harvest Festival was celebrated at Houghton-le-Spring, offerings of fruit and flowers were in aid of a local hospital; in Sacriston they were donated to charities for the blind, deaf, and dumb.[91] Priests who were overtly sympathetic to the strike could also find themselves boycotted in turn. 'Since the strike when the Vicar led the TUC in Shildon, and sermons were unbearable, I have either gone to South Church or stayed at home on Sunday

[86] J. Williams, 'Churches, Sport and Identities in the North, 1900–1939', in J. Hill and J. Williams (eds), *Sport and Identity in the North of England* (Keele, 1996), 113.

[87] *Seaham Weekly News* (henceforth *SWN*), 10 Sept. 1926.

[88] *DCA*, 25 June 1926.

[89] BLSA, C900/05554C1.

[90] DRO, M42/93, baptisms of St Nicholas', Hetton-le-Hole.

[91] *DC*, 23 and 30 Oct. 1926.

evening', complained one woman about (predictably) Harry Watts.[92] At times, this alternative emphasis could lead to tension, and the friction is almost tangible in the transcript of a meeting of the Houghton District Education Committee. The discussion was about the supply of coal to schools, and whether to accept coal from Adventure Colliery. If they did, then they would be using blackleg coal. If they did not, then the schools would lack heating and may have to send the children home:

| | |
|---|---|
| Committee member: | I am not in favour of buying coal from a colliery where they are working against our cause. |
| W. Law, Vicar of Lyons: | What do you mean by 'our cause'? |
| Committee member: | The coal miners' cause. |
| W. Law, Vicar of Lyons: | But this is an Education Committee meeting.[93] |

However, it is less clear whether such a communal attachment necessarily supports Moore's main contention, that 'the effect of Methodism on a working-class community was to inhibit the development of class consciousness and reduce class conflict'. Although he stopped short of describing Methodism as a 'blackleg religion', admitting that no particular evidence of Methodist disloyalty to the strikers could be found, he argued that those union leaders who were Methodists tended to favour conciliation over more aggressive forms of industrial conflict, and often remained on good terms with the colliery hierarchy. Three of the collieries in his area of study, for example, were owned by the Pease dynasty: strong Quakers who operated a vigorous paternalistic control and were rewarded by a compliant workforce and good relations with a moderate union leadership.[94]

In fact, although Moore admits that this ethos was changing (1926 itself, he suggests, would prove its death blow), the fact that the Pease family shared a Nonconformist religion with many of their employees did not necessarily mean that they were sympathetic to collective politics. J. A. Pease and A. F. Pease both served under Asquith in the early part of the century: Arthur resigned from the Liberal Party in protest over the 1909 budget; his cousin Joseph achieved higher office and held several Cabinet positions before 1915. The public statements of both men echoed the policy of the MAGB, by which body Joseph (by then Baron

[92] DCL, Henson's diary, 28 July 1926. [93] *DC*, 23 Oct. 1926.
[94] Moore, *Pitmen, Preachers and Politics*, esp. 26, 81–4, 167, 207.

Gainford of Headlam) was chosen as a representative to give evidence to the Sankey Commission in 1919. There, he first damned nationalization as something that would be 'nothing less than a disaster to the nation', then dismissed the temperatures of over seventy degrees in which some miners had to work: 'I do not mind working in heat.'[95] Four years later, Arthur showed a similar incomprehension of the demands of their workforce: 'I cannot see why they object so much to working an extra half-hour when they have nothing particular to do with their time.'[96] Later, as President of the Federation of British Industries, 1927–8, J. A. Pease would also become one of the principle opponents of Mondism.[97]

Moore also contends that although Methodism did produce political leaders amongst working men, it did not produce leaders who were willing to articulate and pursue class interests as such. This is perhaps true of the earlier period of which he writes, when William Crawford and John Wilson embodied the tradition of Lib–Lab politics and Methodist outlook; it may even be true of the interwar Deerness Valley. It is not true, however, of the wider outlook of the miners' leaders across the Durham coalfield in 1926. Men such as Jack Lawson and Peter Lee had fully cast aside the Lib–Labism of their elders to embrace the new politics of Labour, but the religious and associational life of the chapel remained central to their identity. Indeed, as Stuart Howard has pointed out, for Jack Lawson the rejection of the old pit culture and his adoption of Methodist ideals of self-improvement were closely linked not only with self-advancement but the advancement of his class: his ultimate aim was Parliament and the pursuit of working-class interests.[98]

In 1921, Sidney Webb had placed the influence of the Methodists as central to his history of the DMA:

> Is it too much to say that it is very largely upon the foundation laid in Durham by the humble 'ranters' of 1821–1840, and by the efforts of the families which have passed under their influence that, in the various parts of the county, Co-operation and Friendly Societies, Trade Unionism and the Labour Party have been in the subsequent generations developed?[99]

Indeed, a survey of the most senior miners' representatives in both the political and industrial life of the county reveals that, of eleven men,

[95] *Parl. Papers, 1919*, xii (1), Coal Industry Commission (1919), ii, Reports and Minutes, Apr. 1919, qq. 19670, 19977.

[96] Cited in Moore, *Pitmen, Preachers and Politics*, 91.

[97] M. W. Kirby, *Men of Business and Politics: The Rise and Fall of the Quaker Pease Dynasty of North East England, 1700–1943* (1984), 118–19.

[98] S. Howard, 'Leisure in the Pit Villages: Meaning and Change', *North East Labour History Bulletin*, 27 (1993), 19.

[99] Webb, *Story of the Durham Miners*, 126.

**Table 4.3** *Religious affiliations of Durham Miners' Association officials and miners' MPs*

| Name | Position/Constituency | Religious affiliation |
|---|---|---|
| DURHAM MINERS' ASSOCIATION OFFICIALS | | |
| James Robson | President | Methodist New Connexion |
| W. P. Richardson | General Secretary | Primitive Methodist (choirmaster) |
| Peter Lee | Executive Committee Secretary | Primitive Methodist (lay preacher) |
| Tom Trotter | Treasurer | Primitive Methodist |
| J. E. Swan | Compensation Secretary | None |
| James Gilliland | Durham Miners' Association Agent | Primitive Methodist (lay preacher) |
| MINERS' MPS | | |
| Joe Batey | Spennymoor | Primitive Methodist |
| Jack Lawson | Chester-le-Street | Wesleyan Methodist (lay preacher) |
| R. Richardson | Houghton-le-Spring | Anglican |
| Joshua Ritson | Durham | Wesleyan Methodist |
| W. Whiteley | Blaydon | Methodist New Connexion (choirmaster) |

*Source*: J. M. Bellamy and J. Saville (eds), *Dictionary of Labour Biography*, i–iv (London, 1972–7); Wearmouth, *Social and Political Influence*, 109–43; *Durham Chronicle*, 10 Apr.–3 July 1926.

nine were Methodists in some form, of whom five played an active part in their chapels. Primitive Methodists predominate over those of the Wesleyan and Methodist New Connexion faiths, but the sample is probably too small to be relevant in this respect (see Table 4.3).

Amongst lesser officials too, Methodism was an important reference point. From September to December 1926, the *Durham Chronicle* ran a series of columns featuring the chairmen, secretaries, and treasurers of local lodges after an appeal to such men to send in biographies. The information of the fifty-one men included testified both to the continuing importance in their lives of religion generally and Methodism specifically. In features that ran to only a few sentences, over one-third thought it important to mention their faith. Thirteen were Methodists (again, often local preachers), two belonged to other Nonconformist denominations, and one was Catholic.[100]

[100] *DC*, 4 Sept. – 25 Dec. 1926. The tally was as follows: Primitive Methodist, 9; Wesleyan Methodist, 7; United Methodist, 1; Presbyterian, 1; Salvation Army, 1; Catholic,

The influence of Methodism, and particularly Primitive Methodism, upon the leaders of the labour movement in the county is therefore undeniable, although of course the predominance of Methodists amongst the union and Labour Party hierarchy does not necessarily mean that they also predominated amongst the most class-conscious of the rank-and-file. The emphasis on education within Methodist families and the opportunities it provided for men to polish their public speaking skills at the pulpit made it easier for a Methodist to accede to positions of office. Indeed, one of Moore's respondents pointed out that many miners' leaders left the chapel once they had made a name for themselves within the union: 'They began to put politics before religion—used religion as a stepping stone.'[101]

Nevertheless, the demonstration of such influence does somewhat dent Moore's claim that the Methodists of the Deerness Valley in 1926 sought to separate out their religion from their politics. He argued that Methodists as a collective group virtually avoided taking sides at all by volunteering to work in soup kitchens and other forms of relief work. He quoted one respondent: 'We often wondered which side people took in 1926, but we tried to be Christians and we agreed to differ. 1926 didn't make much difference in the chapel. "Things said" didn't lead to any falling off. At times, the chapel cemented things together.'[102] One of Moore's most interesting observations in this regard was the quiet yet solid response to the lockout in the village of Waterhouses. He argued that, elsewhere, an attempt to unite the men under a socialist leadership and appeal to more overtly socialist principles was divisive, and led to higher rates of blacklegging and more trouble with the police. In contrast, Waterhouses remained united owing to the leadership of the Methodist lodge secretary Matt White, a man who retained his early Liberal ideology in all but name, and made appeals to traditional community values. Moore claimed that 'this is the best example of the unifying effect of the social and political implications of the Methodist view of the world'.[103]

1. No mention of a specific church was given by thirty-two others, although a reference to temperance campaigning by one suggests that he may also have been associated with Nonconformity. See also Wearmouth, *Social and Political Influence*, 109–43, for descriptions of twenty-six prominent leaders of the Durham miners in the late nineteenth and early twentieth centuries who were Methodists.

[101] DUL, 1996/7:4, data gathered by Robert Moore.

[102] Cited in Moore, *Pitmen, Preachers and Politics*, 211.

[103] Ibid., 212.

Yet, men and women across the county involved themselves in soup kitchens, and many later spoke of their work without acknowledging either a religious motivation or even a religious affiliation. Although chapels and churches frequently provided the premises, the workers in the soup kitchens were of all religions and none; moreover, those Methodist lodge leaders who organized and helped out at their local kitchen were hardly doing so in an attempt to retain an apolitical stance. Soup kitchens were not neutral: for many, they symbolized the determination of the strikers. Nor was Matt White himself afraid to criticize religious leaders if they adopted a position at odds with the needs of the miners. In Moore's unpublished research materials, an exchange is recorded in which one of the local Methodist ministers wondered in 1926 if his salary for the following quarter could be secured by raising a loan on the Manse. According to the respondent, Matt White was furious: 'I move he gangs [goes]; putting the Manse in pawn for his own salary when everyone around is hungry.'[104]

Perhaps one reason for Moore's conclusions is his reluctance to distinguish between the different Methodist sects. He admitted that he had expected to find significant differences between the Primitive, Wesleyan, and Methodist New Connexion traditions, but that this had not proved to be the case: 'While in one village it was said that the WMs were a bit staid and snobbish, the same was said of the PMs in the next village.'[105] In fact, this reveals very clearly that Wesleyan and Primitive Methodists *did* view each other differently; and these divisions are even more apparent after an examination of Moore's own unpublished research materials: staid and snobbish, perhaps, but more importantly there is evidence of an attachment to a particular sect according to lines of social class. 'The Wesleyans were anti-socialist right through,' one of his informants told Moore, while another explained that the division between Wesleyan and Primitive frequently reflected that of private and colliery houses.[106]

Such statements are echoed in other sources, in other geographical areas. One of the Gateshead interviewees in 1976 told his interviewer that of the two Methodist chapels in Birtley, 'One was the Wesleyan Methodists which was more the shop-keeper, respectable wing of

[104] DUL, 1996/7:4, data gathered by Robert Moore.
[105] Moore, *Pitmen, Preachers and Politics*, 231.
[106] DUL, 1996/7:4, data gathered by Robert Moore.

the Nonconformist movement, and then there was the Ravensworth Road Primitive Methodist chapel . . . .'[107] Another man, interviewed for Beamish Museum, explained:

> the deputies, the overman, the keekers, the token cabin man and of course all the clerks in the colliery offices, they were a strata above the ordinary working man. As they went to the Wesleyan chapel, if you were a working man . . . you went to the Primitive Methodists . . . Now there was quite a difference and it was different policies as well. There was the Liberal colour. Wesleyan would be yellow whilst Primitive Methodist of course would be red . . . if he was a deputy he would be a Wesleyan lay preacher of course.[108]

Those miners who belonged to the Wesleyan church may therefore have been exposed to a more conservative rhetoric and belonged to a congregation less exclusively made up of the manual working classes. This might indeed have rubbed the edge off a nascent class consciousness; although the Wesleyan lay-preacher Jack Lawson may have objected to any assertion that he was losing touch with the class into which he had been born. But, for the Primitive Methodists, their choice of sect in itself could be interpreted as a mark of class identity.

Rather than focusing on variations between Primitive and Wesleyan Methodism, recent work on the impact of Nonconformity on working-class communities has stressed the difference between the beliefs of the chapel hierarchy and the beliefs of its adherents. In his study of the Rhondda, part of that other great Nonconformist stronghold, Chris Williams pointed out that until the First World War, the labour movement continued to be regarded with considerable hostility by the chapel establishment, with ministers often speaking out against local Labour political candidates and occasionally even expelling Labour Party or trade union activists from its chapels.[109] After the war, such a clear relationship with political Liberalism broke down, but Robert Pope has suggested that this simply led to a loss of direction amongst the Nonconformist leadership. While the chapels were often at the forefront of relief activities in times of hardship such as 1926, he argued that they were unable to translate this into practical advice for social improvement.[110] His conclusions are supported by Peter

[107] Gateshead Central Library, Oral History Transcripts (henceforth GCLOT), i (C.H.).

[108] Beamish Museum, Oral History Archive (henceforth BMOA), 1998/4.

[109] C. Williams, *Democratic Rhondda: Politics and Society, 1885–1951* (Cardiff, 1996), 114.

[110] R. Pope, *Building Jerusalem: Nonconformity, Labour and the Social Question in Wales, 1906–1939* (Cardiff, 1998), 247.

Catterall, who has claimed that although the number of individual Nonconformists active and influential in the labour movement in the period was considerable, the free churches as institutions did not become supporters of the Labour Party, nor greatly influence it during the interwar years: 'Labour candidates did not mean Labour chapels. The number of Free Church Labour MPs in Durham did not mean that the chapels had, like the miners, converted to Labour politics.'[111]

Just as some of the Anglican clergy might fulminate against the strike, so too did some of the Nonconformist leaders. On 30 April 1926, the Revd F. Weekes preached at Chester-le-Street and warned of the dangers of the impending crisis: 'During the course of industrial dispute many people forget that there are two sides to the question—three sides. The third side is the patient public, which suffers most, and which in the end has to pay.'[112] Some chapel hierarchies continued to be hostile once the dispute had begun. At least one (Wesleyan) Sunday school made it clear that children could not expect special treatment just because their fathers were on strike. At this school, stars were given to children for good attendance, but absences were excused if there was a reasonable explanation. On 1 August the teachers decided that 'we do not give a star to children for absence on account of having no boots'. Clearly, poverty due to the strike was not a satisfactory excuse.[113]

But, Methodism was also conditioned by the ethos of the pit villages in which it existed. Like the parish churches, life and worship in the chapels was inevitably influenced by the rhythm of the coal industry: in April 1926, for example, a Sunday school in Chester-le-Street planned its field day deliberately to coincide with a pay Saturday, and then had to postpone it when the strike broke out.[114] Many Methodist leaders therefore acted similarly to their Anglican counterparts in the adoption of a more sympathetic stance. The overall attitude of the Primitive Methodist church was demonstrated at its district synod held at Chester-le-Street at the beginning of May. Representing thirty-five

[111] Catterall, 'Morality and Politics', 676–9. [112] *CC*, 30 Apr. 1926.

[113] DRO, M/CS195, Sunday school teachers' minutes, Perkinsville Wesleyan Methodist Church, 1 Aug. 1926. I have been able to find only very limited evidence regarding Sunday schools, whether Anglican or Methodist, and this has prohibited anything other than brief references. Their role in cementing religious and social attachments would benefit from further study and would perhaps prove more fruitful over a less geographically restricted area.

[114] Ibid., 18 Apr. and 30 May 1926.

circuits and over 18,000 members, eighty-five delegates met and carried a unanimous resolution:

We believe that the provision of a living wage for the worker is the first call upon industry and that industrial reorganization should keep that end in view. Recognising however, that all reorganization is essentially slow, we call upon all involved in our industries to cultivate the spirit of goodwill for the sake of the community and to work unitedly for settlements just and permanent.[115]

Like the Anglican churches, the Methodist chapels also took a leading role in fundraising during the lockout: in June, the Stanley Primitive Methodist Circuit forwarded £30 to the DMA to be used for miners and their families; in August, Handel's *Messiah* was performed under the auspices of the Seaham Free Churches in aid of the children's boot fund; in September, a concert was given in Sherburn Wesleyan Methodist Church and proceeds donated to the distress fund.[116] They were often led by men who were rooted in the community. In 1926, for example, Ralph Cummings of Silksworth celebrated his jubilee as a Primitive Methodist local preacher. Born at New Seaham Colliery seventy years previously, he had moved to Houghton-le-Spring as a child and begun work in the colliery blacksmith's at the age of 11. Later, he had moved to Silksworth Colliery and there worked as a pick sharpener. He had also been a member of Haswell Co-operative Society for over twenty years.[117] Meanwhile, some individual ministers were vocal in their political beliefs. In January 1926, the Revd E. B. Storr gave a lecture at New Seaham Primitive Methodist church entitled 'Keir Hardie: The Worker's Friend'.[118] Another minister active in the Waterhouses Primitive Methodist circuit took part in a question and answer session at the Young People's Institute in 1926 and was asked whether or not Christ was a Socialist, and, if so, how he differed from Marx. The minister replied: 'The direct answer to the former part of this question is Yes! . . . For myself, I think that the Christian church would have stood better today if in the past we had sung fewer hymns of the joy in heaven when we get there, and meditated more upon the common joys which ought to be our united heritage here.' As Moore points out, he then went

[115] *CC*, 7 May 1926.

[116] *DCA*, 11 June 1926; *SWN*, 13 Aug. 1926; *DCA*, 1 Oct. 1926.

[117] *DC*, 26 June 1926.

[118] DRO, M/Sea 23, minutes of trustees' meeting of New Seaham Primitive Methodist Church, 23 Jan. 1926.

on to tie such rhetoric to a Methodist belief in personal regeneration:

Jesus was a socialist I have said. He contemplated a transformed order of society, but that transformed society is *to be affected by personal centres of renewal*. The outward is to be transformed by the inward . . . [whereas Marx] affirmed that the hope of the future, like the existing present, depends upon the inward being transformed by the outward.[119]

## V

Despite the dominance of church and chapel in the communal and trade union life of the pit villages, Durham's secularist heritage remained an important reference point for many. As David Howell has suggested with regard to an earlier coalfield:

The contradictory pressures are important, but the dominant cultural legacy was most probably that enshrined in the ballad 'A Pitman Gan te Parliament' celebrating Tommy Burt's election as the first miner's MP for Morpeth in 1874. Here the enemy was no coalowner but the Bishops 'guslin' away on five thousand a year.' It proved to be a formidable cultural legacy.[120]

In the late nineteenth century, the North East had more National Secular Society branches than any other region in the country, although Edward Royle has suggested that this owed more to Charles Bradlaugh's appeal as a Liberal than to a more direct anti-religious appeal.[121] These branches had waned by the interwar years, as had the national movement, but Durham retained a strong anti-clerical strain. One ex-miner, compiling a history of the Brandon district in which he grew up, included a series of jokes:

The vicar was walking down the street when he met one of his parishioners, Geordie, who told him that he'd had a dream.
What were you dreaming about?
Wey, Aa dreamt Aa was in Hivin [Heaven].
And did you have a good time?
Aye, 'As met an aad [old] sweetheart of mine and we travelled about.

[119] DUL, 1996/7:4, data gathered by Robert Moore. Original emphasis.

[120] D. Howell, *British Workers and the Independent Labour Party, 1888–1906* (Manchester, 1983), 43.

[121] E. Royle, *Radicals, Secularists and Republicans: Popular Freethought in Britain, 1866–1915* (Manchester, 1980), 64.

And did you not think of getting married?
Aye, but there was nee parsons there.[122]

The division between those who told such jokes and those who did not, between those who attended the pub and those who attended the chapel, was recorded by Lawson, who converted to Methodism as a young man:

> Silently, subtly, almost unconsciously I had been building a barrier between myself and my old gambling habits and rough and ready life—an intellectual and moral barrier . . . And looking back now, I see that it was inevitable that I should ultimately seek the company of the serious-minded people who gravitated together and formed the 'Society'. This, of course, has no reference to the 'Society' column, but as anyone in a colliery knows, it is the Methodist Society.[123]

Such concerns were not the monopoly of the Methodists, and one Anglican vicar recorded in 1928 a 'marked cleavage between children who play street games and those who come to Sunday School'.[124]

Just as members of a church might cling to a religious identity, non-believers might also articulate a secular one, setting themselves against the sober respectability of the church and chapel-goers. Jack Parks, who had been a Chopwell militant in the 1920s, spoke as an old man of his disdain for 'the church and chapel people [who] never interested me at all. I've always found that what they did was take hold of the soft jobs at the Co-op Society and man the committees.'[125] Similar prejudices were held by some of the non-Methodists interviewed by Moore. The Methodists were snobs, believed one. He explained that the local preachers were a race apart, who never swore and were less corruptible, but spent too much time in 'glory hallelujah', and pretended to be more widely read than they were: they 'out-Wesleyed Wesley'.[126]

Inevitably, the trials of poverty and hardship made some question the religious traditions that they had absorbed as a child. One old man, reminiscing about the strike, remarked bitterly that 'this chap we learned about at school, our Maker, our Saviour and the only one who helped you when you were in trouble, must have gone on strike the day I was born or else he was a Tory'.[127] Many more were indifferent, and, across the denominations, religious bodies had to compete for followers in the face of a rapid multiplication of secular interests. 'People read

122 L. Moran, *The History of Brandon Colliery, 1856–1960* (Houghton-le-Spring, 1988), 261.
123 J. Lawson, *A Man's Life* (1944), 67–8.
124 DUL, AUC/4/14, Visitation returns, 1928.
125 GCLOT, iv (J.P.).
126 DUL, 1996/7:4, data gathered by Robert Moore.
127 BMOA, 1991/82.

little except newspapers and light novels,' complained one vicar, while another summed up the problem as simply 'sheer indifference and nine public houses'.[128] In 1932, George Lansbury, leader of the Labour Party and himself a devout Anglican, declared that the miners hated Henson. Privately, the Bishop denied that this was the case, but suggested that it would almost be preferable, 'for their hatred would at least indicate a consciousness of existence'.[129]

However, if the miners rejected those such as Henson, Welldon, and Edward Rust, this did not necessarily mean a rejection of a religious framework. In fact, many expressed their disapproval in religious terms, arguing that the unsympathetic stance of such men compromised true Christian values. 'We are told that unless we believe in Mr Cook we cannot be Christians!', Rust commented incredulously.[130] Even the most senior clerics found their godly credentials questioned: at a meeting at Easington Colliery, Emanuel Shinwell described Henson as, 'at best, only a humble apology for a Christian'.[131] Perhaps the most striking incident in this regard was the treatment of the son of the vicar of Chopwell, who had broken a local strike the previous year and begun work as an apprentice at the pit. After waiting for him to return to the pithead, his hecklers used religious imagery to mock him:

> We walked him up to the church gates and we sang 'Lead Kindly Light' and 'Rock of Ages' . . . When we got to the church gates he thought we'd take him straight to the vicarage but instead we took him through the churchyard and some of the women as they passed the graves picked up wreaths and put them round his neck. He had a nervous breakdown after it.[132]

Even those who spoke out most vehemently against the paid representatives of the church did not necessarily reject a religious framework. Of the eleven Durham leaders listed in Table 4.3 earlier, only one did not belong to any particular religious denomination (or, at least, no religious affiliation was so important to him as to be recorded in any biography). Yet, John Swan left some clues to his beliefs in the novel he wrote in the 1930s. In one scene, a group of miners called at a friend's house to find him reading the Bible:

> 'You read the Bible yet you do not go to Church,' said Nichol.
> 'The Bible is one thing, the Church is another,' replied Old Jake.

[128] DUL, AUC/4/13, Visitation returns, 1924, Boldon and Swalwell parishes.
[129] Chadwick, *Henson: A Study*, 179–80.
[130] *DCA*, 3 Sept. 1926.
[131] *DC*, 7 Aug. 1926.
[132] DRO, C3/8, L. Turnbull, *Chopwell's Story*, no page.

'You are bitter about the Church.'
'Not at all. They are too anaemic to bother with. If there is a crisis in the nation or with the nations, you can always depend upon them being on the wrong side, anti-Christian, but there are good and sturdy men in the Church, only theirs is a small voice. The Church as a body has no more fight in it than a bag of feathers.'[133]

## VI

National figures of church attendance calculated by Currie, Gilbert, and Horsley in the 1970s reveal considerable consistency during the interwar years. Across the country, the number of Church of England communicants on Easter day hovered around 2.3 million throughout the 1920s. The country's Methodist membership also remained stable, with around 440,000 Wesleyans and 200,000 Primitives.[134] There is little evidence that events in 1926 had any impact on this. Church records show that in Durham, throughout the period of May to December, the lockout made little difference to the numbers attending communion every Sunday. Numbers might fluctuate dramatically from week to week and from church to church but suggest no decisive overall pattern of decline. Even at that year's harvest festival, at which the display of food might have antagonized a hungry congregation, the vicar of Esh noted that the donations were largely in excess of those of the previous year.[135] The Methodist chapels also seem to have maintained their followings. When Willington Primitive Methodist Sunday School celebrated its anniversary services in June, it was recorded that 'notwithstanding the prolonged stoppage of work at the local collieries, the offertories were equal to those of last year'.[136] Indeed, if devotion to a church can be measured in financial terms, the Chester Moor Wesleyan Methodist collection journal is informative. From the beginning of May to the end of November 1925 the amount received in collections was £10 17*s*. 1½*d*. The corresponding period in 1926 shows a reduction of shillings, rather than pounds, and the sum came to £10 0*s*. 2*d*.[137]

[133] J. E. Swan, *The Mad Miner: A Saga of the North* (1933), 59.
[134] R. Currie, A. Gilbert, and L. Horsley, *Churches and Churchgoers: Patterns of Church Growth in the British Isles since 1700* (Oxford, 1977), 128–9, 143.
[135] *DCA*, 8 Oct. 1926.
[136] Ibid., 11 June 1926.
[137] DRO, M/CS56, Chester Moor Wesleyan Methodist Church, collection journal.

As Moore rightly suggested, a moderate approach allowed many miners and their families to avoid a choice between their religion and their politics; instead, they could continue to be faithful to both. However, he also commented that the most salient feature of institutional Methodism was its unwillingness to take sides in the dispute, 'unlike the Church of England which unequivocally sided with the owners'.[138] The evidence collected above has shown that this criticism is far from justified. At the very least, most Anglican clergymen trod a wary path through a range of opposing political opinions. In 1928, the rector at Burnmoor recognized that the faithful parish priest was liable to fall between two stools, 'of being accused by the snobs of being too friendly with the working people, and by the working people of being too friendly with the snobs'.[139] Ultimately, however, most adopted very similar strategies to those that Moore argued were pursued by the Methodists: Anglican vicars, with certain exceptions, usually maintained an apolitical stance and were able to retain the affection of their parishioners. Some went further and demonstrated active support. 'Which will you be? The trade unionist or the pastor?' demanded Henson to a group of trainee teachers in 1927, but the experiences of his own clergymen proved that some hoped to be both.[140]

Where clergymen did take up a more hostile position, they were shunned by the mining community. But, although the unpopular stance of men such as Henson and Welldon encouraged a great deal of personal animosity amongst many within the pit villages, this did little to change the population's attitude to the Church itself, and even less to religious belief. The bishops tended to be seen as representatives of the attitudes of their class rather than their Church and, amongst miners' propagandists, they were resented as a throwback to a more deferential age: 'The day has passed when we had to take our hats off to the squire and bow to the bishop', declared Ritson.[141] Indeed, religious imagery continued to permeate the coalfield. Religious symbols appeared again and again on lodge banners, whether casting the values of the union in moral terms through images of the Good Samaritan, or through more direct representations of religious buildings: Durham Cathedral was a particularly popular image. On gala day, the most important day in the calendar of the DMA, not only were these images carried

138 Moore, *Pitmen, Preachers and Politics*, 212.

139 DUL, AUC/4/14, Visitation returns, 1928, Burnmoor parish.

140 Lee, *Church of England*, 178.

141 Chadwick, *Henson: A Study*, 166.

5. Cartoon from *The Miner*, 23 July 1926
Image supplied courtesy of Leeds University Library.

through the streets, but the timetable of organized events culminated in mid-afternoon with a special miners' service in the cathedral.

Rather than Moore's claim that this dampened class consciousness, it seems to have been the case that it was precisely because of the occupational solidarity of the miners that many religious leaders felt they needed to adopt such a conciliatory attitude. Those church leaders who preached a more conservative politics were criticized, while those who were supportive of a more radical agenda were the men who were accepted as the true Christians. Methodism (as Anglicanism) therefore adapted to the traditions and culture of the pit villages in which it operated, and many of its local leaders might go much further than the formal Methodist hierarchy in their support of the labour movement; its lay preachers and rank-and-file adherents would often go further still. Issues of class identity were therefore as important in influencing the religious life of the village as religious identities were in affecting class consciousness. As E. P. Thompson suggested, this was a process of mutual accommodation:

Where there is an established working-class or plebeian culture, Methodism does not simply move in and 'adopt' this . . . [rather] some process of selective adaptation . . . can often be seen to be going on. There is a transformation of certain components . . . of the culture, but the perpetuation and adoption of others: in this sense there is an adoption *by* the pre-existent community *of* Methodism.[142]

While both religious and secular identities remained important within the coalfield, they never threatened to override a more fundamental loyalty to the strike or a wider occupational consciousness; rather they might be appropriated for such ends. In the national publication *The Miner*, several cartoons published during the strike cast the miners in the role of Jesus: whether carrying a cross on the way to Calvary; surrounded by jeering businessmen; or through a representation of Baldwin as Pontious Pilate, turning away from the crowd and washing his hands.[143] Another portrayed Baldwin and a coal owner advancing threateningly towards a miner, wielding clubs labelled 'Eight Hour Day' and 'Wage Cuts' (see Illustration 5). In front of the miner is a ghostly figure of Jesus, his arms outstretched in a gesture of peace. 'Who's this fellow butting in?' asks the coal owner, 'Don't know—never saw him before. Anyhow, he's got no business here,' is Baldwin's reply.

[142] E. P. Thompson, 'On History, Sociology and Historical Relevance', *British Journal of Sociology*, 27 (1976), 398–9.

[143] *Miner*, 25 Sept. 1926; 20 Nov. 1926.

# 5

# The Influence of Education

In 1926, Hilda Ashby was a 12-year-old miner's daughter. Years later, she remembered travelling to school during the lockout:

> I can always remember sitting in the [train] carriage full of girls—we were all miners' daughters . . . we were all talking about the strike and one of the girls said 'I hardly dare tell you this . . . ' And we were just getting to know where babies came from and we said 'Are you going to have a baby?' And she said 'No—my father's going to be a blackleg.'[1]

There is no way of knowing whether, when Hilda Ashby and her friends chatted about the strike, they were any clearer as to its details than they were about the mysteries of childbirth. They were, however, evidently aware of the shame associated with blacklegging. For it was impossible to shelter children from the lockout in villages in which most of the inhabitants were on strike and whose dramas were played out publicly. The grandparents of 6-year-old Bill Pears lived next door to a blackleg. They would not let him go outside to watch when the blackleg was escorted back from work, but they could not prevent him peeping from an upstairs window and observing 'the whole street, I think probably the whole village, out booing and hissing at him'.[2] Children were occasionally used in a more calculated way for propaganda purposes, as they had been used before and would be again. At the unofficial gala organized at Burnhope in July, for example, one banner was given to a group of children to carry. 'We may be hungry, Daddy,' it read, 'but never give in to longer hours. For remember—that we who are children today are the Miners of the future.'[3]

As part of a family unit, children more usually found themselves incorporated into the survival strategies of their parents. One remembered

[1] H. Ashby, 'Wait Till the Banner Comes Home!', in K. Armstrong and H. Beynon (eds), *Hello Are You Working? Memories of the 'Thirties in the North-East of England* (Whitley Bay, 1977), 38.

[2] Gateshead Central Library, Oral History Transcripts (henceforth GCLOT), i (B.P).

[3] *Miner*, 23 July 1926.

that 'we tried to help in the struggle as best we could. As little boys . . . we couldn't do much. But we raided that big mountain of coal lying on the surface at the Dolly Pit [Newbottle Colliery]. We weren't alone. It was *our* coal. *Our* fathers had dug it . . . We felt no guilt.'[4] This little boy would go on to represent Labour as a Scottish MP; he also spoke with the hindsight given by nearly seventy years. Even if he did hold such fierce political conviction at the time, however, many more children must have engaged in such pursuits simply because they were anxious to help their parents, or because they enjoyed the adventure. It seems unlikely that they would not have known or asked why such actions were necessary. After all, even in more normal times children grew up surrounded by the mining culture of their fathers. Later, Jack Lawson would remember how as a child he would hear 'the men . . . talk of the day's work. All of us, boys and girls alike, knew the technical pit terms as well as if we worked there.'[5] For William Collins, born in 1925, the colliery held more straightforward delights. As an old man he enthused about the pit yard as 'one of the best playgrounds ever invented'.[6]

Away from home and family, the other major influence in almost every child's life was his or her experience of school. In 1920, 151,282 children lived in County Durham, of whom the vast majority (148,580) attended public elementary schools.[7] These could provide a dramatically different frame of reference to that imparted at home. In the mid-1920s, a teacher at Bearpark Council School noted that 'unfortunately our children have two vocabularies, home and school, with the result that talk in school in the form of oral composition or discussion has a great number of faults'.[8] It is an observation that anticipates later stories of children who grew up in the 1950s in the aftermath of the 1944 Education Act, who told of adopting different identities if their attendance at either a grammar school or a secondary modern clashed with the dominant street culture in which they had grown up.[9] The first half of this chapter will consider how far her words could be taken as being more widely

[4] W. Hamilton, *Blood on the Walls: Memoirs of an Anti-Royalist* (1992), 15. Original emphasis.

[5] J. Lawson, *A Man's Life* (1944), 31.

[6] British Library Sound Archive, C900/11043.

[7] The National Archives: Public Record Office (henceforth TNA:PRO), ED120/20, proposed scheme of education, adopted by Durham County education committee, Oct. 1920.

[8] Durham County Record Office (henceforth DRO), E/C/G6, syllabus book, Bearpark Council School, n.d., *c.*1927?

[9] The issue also has modern resonances, with different dialects based on ethnic group. See, for example, S. A. Anderson and S. Butler, 'Language and Power in the Classroom: An Interview with Harold Rosen', *The English Journal*, 71 (1982), 24–8.

true of the division between home and school life: whether the schools of County Durham were seen as alternative centres of community, in competition with a specifically mining identity; or whether they were absorbed into the dominant culture shaped by pit and union lodge.

As well as the specific values that schools promoted, the wider implications of education might also affect the way in which both children and adults viewed their surroundings, through the aspirations created and the social mobility engendered. A recognizable character in literature is that of the clever miner's son, alienated from his peers in his struggle for academic success but eventually able to escape the pit and his mining background. Paul Morel in D. H. Lawrence's *Sons and Lovers* is one of the best-known examples, echoing, of course, the experience of his author.[10] Perhaps even more common is the character of the clever miner's son who ultimately finds that he *cannot* escape the pit. A. J. Cronin's *The Stars Look Down* closes with its hero, Davy Fenwick, descending down the mineshaft once again, his hopes of a parliamentary career dashed by a smirking Joe Gowlan.[11] For different reasons, Huw Morgan in Richard Llewellyn's *How Green Was My Valley* also remains a pitman: despite his academic potential, he turns down the possibility of grammar school because he wants to be a collier like his father.[12] How far these were recognizable stereotypes in the mining villages of Durham, and whether or not the strike made any difference to the way in which education was seen will be discussed in the second half of this chapter.

## I

There were various ways in which the values promoted in elementary schools might clash with the radical coalfield identity to be found elsewhere. In the 1920s, some schools had only recently cut their ties with the colliery owners: Durham County Council completed negotiations with Lord Londonderry over the purchase of his Seaham schools as late as 1912.[13] But, the colliery hierarchy could continue to exert a subtle influence over school life. In September 1926, 11-year-old

[10] D. H. Lawrence, *Sons and Lovers* (1913). For details of Lawrence's schooling and his sense of isolation at his board school and then as a scholarship boy at Nottingham High School, see J. Worthen, *D. H. Lawrence: The Early Years, 1885–1912* (Cambridge, 1991), 75–94.

[11] A. J. Cronin, *The Stars Look Down* (1935).

[12] R. Llewellyn, *How Green was My Valley* (1939).

[13] TNA:PRO, ED120/20, A. J. Dawson, 'Report on School Accommodation', May 1924.

Thomas Bickle won a national essay competition run by the Royal National Lifeboat Institution. Malcolm Dillon, chairman of the Seaham Harbour branch, was asked to present the award, and he congratulated Bickle in front of a crowded school hall, speaking of the proud naval tradition that had created a glorious British Empire. But, Dillon also happened to be chief agent of the Londonderry collieries, and he could not resist the opportunity to remind the assembled children that 'the boys and girls of today possessed advantages which were not dreamed of by the children of half a century ago. He remembered the days when there was no free education unless it was provided by private benevolence, such as the schools . . . erected and maintained by the Londonderry family.'[14]

The majority of Durham's elementary schools were run by the county council, and in 1926 only 35 per cent were maintained by a religious institution, compared with 56 per cent of elementary schools throughout England and Wales.[15] The impact of religious identities within the mining communities has already been explored,[16] but even those managed by the local authority remained vulnerable to political manoeuvrings. In 1922, for example, the Labour Party lost its dominance of the county council, and its opponents took the opportunity the following year to declare it compulsory for all schools in the area to mark Empire Day 'by specifically impressing upon their pupils . . . the sentiments of loyalty and patriotism for which the occasion stands'.[17] At New Seaham's girls' school in the early 1920s there were special lessons on the subject and the singing of 'National Songs', while a neighbouring Anglican school was visited by the aptly named Revd H. Churchyard, who 'addressed the children on their duty as citizens of a Great Empire'.[18] The glories of Empire were complemented by the glories of the monarchy and a day's holiday was proclaimed in several schools when the Duke of York married Elizabeth Bowes-Lyon in 1923.[19] Again, the Londonderry family were prominent in their involvement. In 1924, Londonderry paid for several children to attend the British Empire

[14] *Seaham Weekly News* (henceforth *SWN*), 24 Sept. 1926.

[15] DRO, CC/Ed352, school reports, 1926–7; *Parl. Papers, 1928*, ix (39), Report of the Board of Education, 1926–7, 120.

[16] See Chap. 4.

[17] *Durham Chronicle* (henceforth *DC*), 1 May 1926.

[18] DRO, E/NE55, logbook, New Seaham Girls' School, 30 May 1922; E/NE63, logbook, Seaham Harbour Church of England Girls' School, 24 May 1922.

[19] For example, DRO, E/NE55, logbook, New Seaham Council Girls' School; DRO, E/NE31, logbook, Seaham Harbour Roman Catholic School, 26 Apr. 1923.

Exhibition at Wembley. In Seaham Harbour's National girls' school, the five pupils chosen had all lost their fathers during the Great War.[20]

The significance of this is arguable, for love of monarch and Empire was not the sole preserve of the Right in the interwar years. However, when Labour regained its majority on the county council in 1925, its councillors thought the issue important enough to rescind the 1923 resolution, permitting headteachers to exercise their own discretion regarding Empire Day from May 1926 onwards.[21] Unfortunately, it is difficult to gauge the extent to which headteachers took advantage of their freedom when Empire Day fell in the early days of the 1926 strike, not least because that year it coincided with Whitsuntide when children were on holiday anyway. But, at least one school accommodated this by celebrating a week early, gathering its children together to sing the National Anthem, after which each child made a Union Flag to take home.[22]

As well as the overall ethos of the school, the content of the education provided could be used to propagate certain messages. In 1917, the Durham Director of Education, A. J. Dawson, argued against the inclusion of foreign languages in the curriculum: 'French—or indeed any language other than their own, and a very small amount of that—is not for the masses; the inclusion of such subjects in their school curriculum will only make them conceited and dissatisfied with their station in life.'[23] Four years later, he drew up a list of recommended books for school libraries.[24] Though dominated by the usual boarding school stories, some sought to promote particular values, such as Mabel Quiller-Couch's *Little Book on Temperance* and *Little Book on Thrift*.[25] Most notable are the books that were connected to the coal industry, presumably included because of their assumed relevance to the reader. W. J. Claxton's *Journeys in Industrial England* gave an account of a trip down a coalmine. After describing the darkness and the punishing work, the author reassured his reader: 'We think that the miners must long for the sunlight and the blue sky above them. But they are perfectly contented with their lot.'[26] Another stands out for the vehemence of its anti-union message. In G. A. Henty's *Facing Death: A Tale of the*

[20] DRO, E/NE63, logbook, Seaham Harbour National School (girls' dept.), 25 Aug. 1924.

[21] DRO, CC/E143/6, Durham County Council education committee circulars.

[22] DRO, E/C8, logbook, Brandon Council Infants' School, 21 May 1926.

[23] TNA:PRO, ED19/48, A. J. Dawson, 'Higher Tops: A Paper Prepared for the Consideration of the County Education Committee' (Newcastle-upon-Tyne, 1917), 14.

[24] TNA:PRO, ED19/48, A. J. Dawson, 'Library Catalogue', Sept. 1921.

[25] M. Quiller-Couch, *A Little Book on Temperance* (1914); *A Little Book on Thrift* (Oxford, 1912).

[26] W. J. Claxton, *Journeys in Industrial England* (1914), 32.

*Coal Mines*, recommended by Dawson for senior classes, the hero was a young Black Country miner who chose to leave the pit when his workmates struck over pay. His mother was distressed:

They've promised to give some sort of allowance to non-unionists, Jack.

Yes, Mother, but I'd rather earn it honestly. I'm too young to join the union yet but I have made up my mind long ago never to do it. I mean to be my own master, and I ain't going to be told by a pack of fellows at Stafford or Birmingham whether I am to work or not, and how many tubs I am to fill. No, Mother, I wasn't born a slave that I know of, and I certainly don't mean to become one voluntarily.[27]

Henty, best known for his adventure stories, was enormously popular in the interwar years, but the inclusion of such books on a recommended reading list does not mean that they found their way into every library of every school, still less that every child read them; Dawson himself may not have read them and known exactly what he was recommending. It is also likely that the sons and daughters of miners read about their public-school counterparts and their negotiation of fagging, prep, and Varsity matches with enthusiasm, losing themselves in a world for which familiarity was no prerequisite for enjoyment. Mary Wade, the daughter of a Northumberland miner, won a novel as a Sunday school prize in 1923. It told the tale of Molly, an only child whose cousin arrived from France for the holidays. Years later, she acknowledged that 'like a lot of stories from that period very little of the content seemed true to life, and there were many characters whose background and lifestyle were away beyond our wildest dreams'. Nevertheless, she remembered how as a child she had 'cherished' the book.[28] It is ironic that, with so many such stories beginning with a young boy scared about his first day at boarding school, in some cases its reader might be a young boy soon to embark upon his first descent into the pit.

## II

If schools often conformed to traditional values, they were also fundamentally influenced by the colliery culture that surrounded them.

[27] G. A. Henty, *Facing Death: A Tale of the Coal Mines* (1908), 84.

[28] M. Wade, *To the Miner Born* (Northumberland, 1984), 28.

A substantial majority of their pupils were the children of miners, and would leave school to become miners or the wives of miners themselves; it was therefore inevitable that the practical working of the coal industry would affect the mechanics of the education system. Some schools in the county were used to closing on 'cavilling Monday', for example—the day's holiday from the pit when the new working places for the next three months were allotted. This may have initially been introduced for pragmatic reasons, as a de facto response to children being taken out of school for their father's holiday. However, in June 1926, at least one school closed for cavilling Monday as usual, despite the fact that no cavils were being drawn and no men were working down the pit.[29]

The domination of the mining industry meant that a seven-month strike had the potential to make a considerable impact on a child's experience of school, and this was most apparent with regard to school attendance. The worst problem was the state of children's boots, especially when the schools restarted after the summer break, to more wintry weather and even more stretched expenditure at home. Several school logbooks attributed low attendance to such a cause, one headteacher blaming the combination of inadequate footwear and wet weather for the fact that some children were absent for almost the whole of the Christmas term.[30] Children might also be enlisted at home to help the family through the crisis. By 6 May, a school in Houghton-le-Spring was missing several boys who were reported to be gathering coal.[31] The same school was still having problems in November, when in one week alone, fifty-seven boys were absent for such a purpose.[32] Even once children arrived at school, the strike might continue to compromise their learning. On 29 October, one headteacher regretted that, having endured two weeks without fuel, the school was now very cold and attendance was down heavily as a consequence. Heating was only restored on 9 November.[33] Another recorded emotively that 'it is so extremely cold today that the little ones have wept all day. For three weeks the children have been so extremely cold that they could not write in their books, their hands refusing to grip the lead pencils'.[34]

[29] DRO, E/E90, logbook, Wheatley Hill Boys' Council School, 28 June 1926.

[30] DRO, E/NE81, logbook, Seaham Harbour Church of England Infants' School.

[31] DRO, EP/Ho685, logbook, Houghton-le-Spring Council School (boys' dept.), 6 May 1926.

[32] Ibid., 5 Nov. 1926.

[33] DRO, E/NE36, logbook, Seaham Harbour Council Girls' School, 29 Oct. – 9 Nov. 1926.

[34] DRO, E/C62, logbook, Sherburn Hill Council Infants' School, 25 Oct. 1926.

However, despite the frequency of such complaints in school logbooks, statistics reveal no discernible overall drop in attendance due to the strike. Average school attendance across the county stood at 90 per cent in 1925, 90.5 per cent in 1926, and 90.7 per cent in 1927. The figures show little difference even when restricted to the most concentrated mining districts: average attendance figures for the Houghton-le-Spring, Chester-le-Street, and Easington education divisions hovered around the 90 per cent mark throughout these years.[35] This can be partly attributed to the fact that issues such as the quality of children's boots were not unique to the months of the lockout but were a familiar complaint throughout the 1920s.[36] The strike may have made things worse, but in the deprivation of the interwar period as a whole, it did not create the problem.

Absences were also offset by the fact that certain measures taken during the strike actually became an incentive to attend school. For example, schools became the usual place in which repaired or second-hand footwear was reallocated to the most necessitous children. In November, one school logbook made the familiar complaint that the combination of poor boots and wet weather had taken its usual toll on attendance. A couple of weeks later it noted that attendance had since risen, partly attributed to better weather but also due to the fact that it had hosted another distribution of footwear.[37]

Most importantly, schools played a key role in the feeding of miners' children. The 1906 Education (Provision of Meals) Act was already in effect in Durham in the early 1920s, but the strike led to an enormous extension of its application throughout the county. At the end of April 1926, just over 3,000 children were being fed at eighteen different centres across the county. A month later, over 50,000 were being fed in 226 centres, and attendance remained substantially higher than this until mid-December. At the peak of school feeding in the week ending 23 July, 351,281 dinners and almost as many breakfasts were served to 64,509 school children in over 300 different feeding centres, while a further 30,000 dinners and over 22,000 breakfasts were served to 5,680 younger children.[38] The beneficial results were celebrated by a variety of

[35] DRO, CC/Ed283, Durham County Education Committee, summary of school attendance.

[36] For example, DRO, E/NE55, logbook, New Seaham Council Girls' School, 17 Feb. 1922.

[37] DRO, E/E41, logbook, Blackhall Colliery Council School, 5 and 26 Nov. 1926.

[38] DRO, CC/ED398, elementary education committee reports.

commentators. Two years later, Dr Eustace Hill, the Medical Officer of Health for County Durham, suggested that, despite the depressed conditions, the lack of any marked sign of deterioration amongst school children might be attributed to 'the good results of the feeding of these children during 1926, during which their condition had shown marked improvement, [which] had not yet worn off'.[39]

However, depending on the Poor Law authority, the provision of school meals could, in some cases, act as a deterrent to school attendance. At the beginning of June, seven of the twelve boards of guardians operating in the county were (in line with Ministry of Health guidelines) deducting the price of school meals from the overall amount of relief a mother received, and at the end of the year the Board of Education complained that rather than school meals being an added incentive to send children to school, some parents had preferred to keep their children at home in order to get full payment from the guardians.[40] At one school it was noted that when the guardians changed their policy and decided to subtract two shillings and sixpence from the allowance of those receiving school meals, the number of children attending the feeding centre dropped immediately from one hundred to fifty-five.[41] The provision of meals also distracted attendance officers from their usual work, and the logbook for a New Seaham school explained that 'some families [are] taking advantage of the fact that no attendance work is being done by the attendance officer who is busy with feeding arrangements elsewhere'.[42]

## III

The widespread provision of school meals was important for more than simply its effect on school attendance. It reinforced the central position of the elementary schools within the pit villages and tied them tightly to the concerns of the mining community. Even those pupils who were not the children of miners would have found it hard to forget that

39 TNA:PRO, ED50/77, 'Investigation into the Physique of School Children in Durham Coal Mining Areas', Apr. 1928. The feeding programme was wound down very quickly at the end of the strike and virtually ceased 1927–8.

40 DRO, U/CS305, emergency committee correspondence; TNA:PRO, ED50/77, Board of Education minutes, 7 Dec. 1921.

41 DRO, E/E24, logbook, Wheatley Hill Council Senior Girls' School, 5 Nov. 1926.

42 DRO, E/NE55, logbook, New Seaham Council Girls' School, 10 Sept. 1926. A similar observation was made at Ushaw Moor. See DRO, E/EC128, logbook, Ushaw Moor Council Infants' School, 14 June 1926.

a significant proportion of their classmates had fathers who were on strike. It often meant a reorganization of the school day, for example, as children had to leave lessons early in order to get to church halls and other distribution centres; if the meals were being served in the schools themselves, it meant a physical reorganization of school buildings as classrooms turned into kitchens and dining halls. In contemporary commentary the feeding centre was frequently represented as the village in miniature and, even in 1926, it was already being portrayed as the symbol of coalfield solidarity that in popular memory it would later epitomize. When the Sherburn Hill canteen was opened, a local newspaper reported that 'all classes of the community are bounded together in this effort of feeding the children, and the committee is representative of all the public institutions in the village'. Members of trade unions, the miners' lodge, religious bodies, and the Labour Party were involved in that particular canteen, but perhaps the most important presence, given the nature of the recipients, was that of the teachers, and the secretary was J. G. Huntley, headmaster of the boys' school.[43]

Through the use of their premises, schools could also become a focus for other community activities. School buildings were used for a variety of functions and, during the seven months of the strike, the Easington district education committee granted the use of several of its school halls for a meeting of the Women's Institute, a social event organized by Shotton Primitive Methodist Cricketers, and for the weekly meetings of the Blackhall branch of the Communist Party. The arrangement with the last organization is telling, for it was not the case that every organization was permitted use of the space and, for whatever reason, an application by the Yorkshire Penny Bank to hold weekly meetings was refused.[44] Many of the events organized in the schools revolved around the needs of a community specifically defined by its mining constituents. The Chester-le-Street education committee resolved in September that all applications for the use of school halls for dances and whist drives in aid of the children's boot funds should be accepted during the stoppage.[45] Chopwell's council school was used exclusively for such a cause: in October the district education committee decided that it could be used for a weekly whist

[43] *Durham County Advertiser*, 2 July 1926.

[44] DRO, E/NE/A5, Easington district education subcommittee minutes, Sept.–Dec. 1926.

[45] DRO, E/NC/A6, Chester-le-Street district education subcommittee minutes, 21 Sept. 1926.

drive in aid of the children's boot fund, but that no other applications should be granted.[46]

The fact that the schools became such centres of support for the miners also sheds light on the attitudes of individual teachers, whose enthusiastic involvement can be taken as an indication of at least some level of sympathy. The large-scale feeding of school children could not have been supported without the cooperation of the schools, but many teachers took their involvement beyond any obligatory role. Huntley was not the only headteacher to take a leading position in the canteen committees. The headmaster of Dawdon's council school became the chairman of the Seaham committee, while the headmaster of Seaham's upper standard school joined him as its secretary.[47] At the neighbouring committee established in New Seaham, the secretary was the headmaster of Byron Terrace School.[48] It is also notable that school meals continued to be provided throughout the summer holidays (the number of centres actually increased),[49] made possible only by the willingness of teachers to supervise. In July, the staff of a school in Houghton-le-Spring 'mutually arranged to do the clerical work and issuing of tickets for school feeding during the holidays'.[50] At Shotton it was noted that during the summer holidays, the arrangements for the feeding of the children had continued as usual, 'the whole of the staff having volunteered for duty'.[51]

In addition, teachers and headteachers often sat on boards of guardians and were therefore in a position further to influence the distribution of relief: two Labour guardians on the notorious Chester-le-Street board were headmasters of local schools.[52] It is revealing that, five years later, when a teacher's name was put forward in 1931 as a possible election agent for Ramsay MacDonald in Seaham, the Prime Minister expressed concern that this might break some regulation. No regulation would be broken, he was assured, but 'there is a history behind this subject in Durham. The Authority has been criticized in the past on the ground that it was freely giving its teachers leave of absence in order to attend boards of guardians and other similar bodies (of course almost always

[46] DRO, E/N/A7, Blaydon district education subcommittee minutes, 20 Oct. 1926.

[47] *DC*, 22 May 1926.

[48] *SWN*, 21 May 1926.

[49] DRO, CC/Ed398, elementary education committee reports, 8 Sept. 1926.

[50] DRO, EP/Ho685, logbook, Houghton-le-Spring Council School (boys' dept.), 22 July 1926.

[51] DRO, E/E43, logbook, Shotton Colliery Council Boys' School, 8 Oct. 1926.

[52] DRO, CC/Ed398, elementary education committee reports, 8 Sept. 1926.

as Labour members).'[53] As well as their time, teachers might also be generous with their money. By the beginning of October 1926, the boot fund of the West Stanley Teachers' Association, in operation since May, had collected more than £85 from teachers in the Stanley area. The local press was fulsome in its admiration: this figure did not include the special individual efforts of most of the schools in the area, one newspaper reported, so could not be seen as a full and complete account of what had been done and continued to be done by teachers all over the area.[54]

Of course, as discussed earlier with regard to local religious leaders, a teacher's desire to ensure the well-being of the miners' children was no guarantee of political sympathy with the miners' cause. One school logbook suggests disagreement existed even over the policy of school meal provision. A July entry hints at conflict, noting that 'there has been some trouble at school between the assistants and the Head Assistant in connection with the feeding of the ch[ildre]n'.[55] However, some teachers were explicit about their political allegiances. Figures do not exist for 1926, but Durham County Teachers' Labour Group had 177 members by 1934.[56] Those who did not share such political beliefs feared that a Labour county council was involved in some kind of conspiracy, and in November 1926 the *Blaydon Courier* cited a letter published in the *Newcastle Daily Journal* from an 'Assistant Teacher', which complained of 'headmasters . . . appointed not for their ability as masters, but for Labour platform speaking'.[57] Cuthbert Headlam was similarly put out. Speaking to the Durham Municipal and County Federation in 1932, he wondered, 'Is it surprising that we have so many Socialist school teachers in this County when it is generally known that Conservative and Liberal teachers have but little chance compared with Socialists of being promoted to headships of County schools?'[58]

Whatever the reasons for their appointment, at least some of these teachers were actively prepared to promote their principles within an educational context. The headmaster of Hedley Hill School, John Towers, was already well known in Labour circles as secretary of Deerness Valley

[53] TNA:PRO, PRO 30/69/1716, correspondence between J. Ramsay MacDonald and H. B. Lees-Smith, Mar. 1931. William Coxon, a local school teacher, was put forward to head MacDonald's campaign. Later that year, after the formation of the National Government, Coxon stood against MacDonald as the Labour candidate.

[54] *Stanley News and Consett Chronicle*, 7 Oct. 1926.

[55] DRO, E/E82, logbook, Wheatley Hill Council Infants' School, 12 July 1926.

[56] DRO, D/X1099/4, Durham County Teachers' Labour Group, members' register, 1934.

[57] *Blaydon Courier*, 6 Nov. 1926.

[58] DRO, D/MCF1, speech of Cuthbert Headlam to the DMCF, 1932.

Labour Party when he was called before magistrates in December 1926, charged with assaulting two of his pupils. The two boys, aged 11 and 14, had attended the school feeding centre regularly throughout the lockout. On 1 November, their father broke the strike and returned to work, but for a fortnight did not receive pay. His children remained entitled to be fed during this period, but Towers forbade them to return to the feeding centre and, when they did so, he caned them—particularly savagely, the bench was told. Towers was found guilty, fined forty shillings and stripped of his teacher's certificate.[59] This was an exceptional case and one that would be raised several times in Parliament.[60] Not all teachers would be so discriminatory and other anecdotes give more ambiguous impressions. One old woman later remembered how, during the strike, her father kept her from school one day to take her to the circus. The next day, she, too, was reprimanded and caned by the headmistress: 'She told the class the whole tale, that the strike was on, and I was going to the soup kitchen and getting help, and my dad was wasting money by taking me to the circus.' In fact she had been ineligible for school meals owing to her mother's savings, and, after her furious parents visited the school, she was given an apology.[61] But, the political allegiances of this teacher are less clear than those of Towers: a steadfast unionist was as likely to deplore those who selfishly took advantage of communal help as an angry ratepayer who resented the fact that she was funding it.

However, the integration of teachers within the mining community makes it likely that most had at least some understanding of the miners' concerns. A lack of adequate sources prohibits any detailed analysis of teachers' backgrounds, but anecdotal evidence suggests that some came from the mining villages themselves. One woman, a child during the strike, was asked years later whether her teachers had been sympathetic. She answered: 'Well, most teachers had fathers who were miners. My teacher's father was a miner.'[62] Even if they were not from a mining background themselves, younger teachers often found themselves intimately involved with the mining community, whether they liked it or not. When a Board of Education official lamented a serious staff shortage in Durham in March 1921, he observed that:

> the position of lodgings and 'life' in the colliery villages is one which militates to a tremendous extent against staffing. In many cases I have found that young

[59] *DC*, 4 Dec. 1926.

[60] *Hansard Parliamentary Debates* (Commons), fifth ser. (henceforth *HPD*(C)), 199, c. 1937; 202, cc. 1086–7, 1257–67; 204, cc. 661–7, 674–5.

[61] GCLOT, i (Mrs P.).

[62] Telephone conversation with Elizabeth Stoves, 13 June 2005.

girls straight from a training college have had to lodge in a miner's cottage—the only place—have meals with the family and either retire to a friendless bedroom or take a walk outside while the miner had his wash in front of the fire where he stripped entirely as a matter of ordinary procedure.[63]

Three years later, Hensley Henson was also concerned. In a letter in which he regretted the poor state of housing in the county, the Bishop gave an example of a three-roomed miner's cottage, in which seven adults were living, including two women who were teachers in council schools.[64]

A sympathetic approach on the part of teachers might influence what was taught in the classrooms. One 1920s' syllabus book for Browney Council School began with the rather charming instruction that, for the youngest children, history was to be about 'stories . . . to excite wonder and imagination'. It then outlined a programme for older children which took them through British history beginning in 1066 and ending in 1815. Apart from outlining the dates to be studied by each year group, much was left to the individual teacher to decide, but, as the syllabus instructions repeatedly stressed, social life and social conditions were always to be emphasized, whether teaching Norman, Tudor, or Georgian history. In the final 'standard eight' (12- and 13-year-olds), the chronological narrative was to be abandoned and 'the industrial history of England' was to be studied.[65]

This focus on a local, social history is striking when looking through the few syllabus books that have survived, and makes a surprising contrast to the expected concentration on kings, queens, and battles. It was a syllabus that rooted children within the specific history of their class and their region: at a school in Shotton, teachers were instructed that, for every age group, local and county connections were to be emphasized in history lessons, while, in geography lessons, the landscape and environment of County Durham were to be studied using local maps.[66] The most detailed surviving instructions are found in the syllabus book of Bearpark Council Mixed School. For its 'standard eight' classes, a programme of nineteenth-century history was to be followed. Alongside the teaching of the 1832 and 1867 Reform Acts, German and Italian unification, and other important British and European historical events,

[63] TNA:PRO, ED88/17, Board of Education minutes, Mar. 1921.

[64] Henson to Londonderry, 4 Mar. 1924, in E. F. Braley (ed.), *More Letters of Herbert Hensley Henson* (1954), 29.

[65] DRO, E/C/G155, syllabus book, Browney Council School, n.d., *c.*1924–6.

[66] DRO, E/St/G20, syllabus book, East Stanley Council School, n.d., *c.*1926–31.

much of the syllabus was devoted to industrial domestic history. The topics were arranged under headings, one of which was entitled 'How the State Helped the People' and looked at 'the Factory Laws, the Reform of the Poor Law, Social Insurance, the fixing of wages by the State and Public Health'. Next came 'What people did for themselves,' which examined trade unionism, the Co-operative movement and friendly societies. The final section was entitled 'Industrial Harmony' and considered welfare work, and conciliation and arbitration.[67] Nevertheless, some were keen to give their young pupils an awareness of something wider, too. At Browney Council School, whether in ignorance or conscious opposition to the views that Dawson had outlined a decade earlier, students were to be taught French.[68]

## IV

If the elementary schools were frequently integrated into the dominant mining culture of the village, the education they provided and the aspirations they might create could encourage pupils to look away from the coal industry. In practice, the opportunities for social mobility through education were severely limited. During the year ending December 1919, ninety-seven ex-pupils from maintained or aided schools in the county entered a course of further education, including six who gained places at Oxford or Cambridge.[69] Twenty-one were ex-servicemen in receipt of army grants and twenty more held some kind of local exhibition or scholarship. It is likely that most of these were the children of working-class parents, even if the number of those who came from mining backgrounds is impossible to deduce. However, they were the exception. Far more representative were the 4,553 male elementary school leavers who joined the coal industry; or the 3,654 female leavers who returned to their family homes as domestic workers (a further 950 entered domestic service).[70]

The problem was partly a financial one: the Geddes Axe had taken its inevitable toll on Durham County Council's earlier hopes to make all secondary places free by 1923 and the target was only finally reached

67 DRO, E/C/G6, syllabus book, Bearpark Council Mixed School, n.d., *c.*1927.

68 DRO, E/C/G155, syllabus book, Browney Council Mixed School, n.d., *c.*1924–6.

69 TNA:PRO, ED120/20, A. J. Dawson, 'Educational Problems in the County of Durham', n.d. *c.*1920?, 80.

70 Ibid., 53. Figures are for the year ended July 1919 in the County Part III Area, i.e., Durham County, excluding the City of Durham, Felling, Hebburn, Hartlepool, Jarrow, and Stockton.

in September 1926.[71] In 1924, 325 of 462 children offered fee-paying places at County Durham secondary schools declined them, compared with only nine out of 686 who turned down free places in the same year.[72] Once at secondary school, miners' children remained vastly outnumbered by the children of other social groups. In 1921, a survey of Durham secondary schools counted pupils according to their father's occupation. In the table of results, alongside the number of artisans' children (38 per cent), retail proprietors' children (14 per cent), and labourers' children (4 per cent), the survey did not even list 'miners' children' as a separate category.[73]

Fees were not the only financial obstacle to secondary education. An extended schooling delayed a child's ability to bring in wages for the family and in 1927 the headmaster of one upper standard school regretted that 'several promising boys have been removed from school on attaining the age of fourteen, as their parents state they cannot afford to keep them from work when it is available'.[74] In this respect the strike offered a few extra months of grace, as in the absence of any available employment such considerations were put on hold. 'Owing to strike several children decided to stay at school into next term,' reported one headteacher in June 1926.[75] For most parents and their children, however, the strike only exacerbated financial problems. As general secretary of the Kent miners in the 1970s, Jack Dunn still resented passing the examination for the local grammar school only to be forced to turn down the place because 'poverty and the 1926 lockout' meant that his parents could not afford the uniform.[76] There were surely Durham children in the same situation.

The cost of secondary schooling aside, the success of the clever miner's child was further inhibited by prejudice. One old man who left school in the early 1920s later recounted how, buoyed by his success at elementary level, he had enthusiastically applied to secondary school: 'And my oral [examination] composed of this: "Oh, and where does your father work?" "Ouston E Pit, Birtley, sir." That's all I was asked. I failed my oral.'[77] Parents could also hinder a child's educational advance. In 1922, Mary Crossling of Crook wrote to Lloyd George to request that her 12-year-old daughter be exempted from school until after the summer

[71] TNA:PRO, ED53/547, miscellaneous Board of Education documents. [72] Ibid.

[73] Ibid. The terms 'artisan' and 'labourer' were not further defined in the document.

[74] DRO, E/NE69, logbook, Seaham Harbour Upper Standard Council School (boys' dept.) [75] DRO, E/C68, logbook, Bearpark Council School, 30 June 1926.

[76] M. Pitt, *The World on Our Backs* (1979), 90. [77] GCLOT, iii (H.M.).

holidays, 'as I have to become a mother again in May'. It is clear where, in her opinion, a daughter's place lay: 'I think mother's that are like my self at present we should be granted the oldest girl I think for girls going to school to such a age its wants abolishing.'[78]

Yet, even an extended education was no guarantee of social mobility. In 1920–1, of 459 boys over the age of 12 who left Durham secondary schools as full-time pupils, 172 entered professional, commercial, or clerical occupations, and twenty-two went on to further study at university, but 125 left to join industrial or manual occupations. The position was similar for girls: of 460 secondary school leavers, 191 entered teaching, and seventeen went to university, but almost a hundred returned to home life.[79]

However, men and women could also take advantage of educational opportunities as adults, and by 1926 the North East had established itself as a leading district for the Workers' Educational Association (WEA). Its annual report for 1925–6 described an exceptionally good year in the district, its thirty-two branches having involved 2,300 students in fifty-four tutorial classes, over 100 one-year classes, and three weekend schools.[80] For adult learners, the problem was not so much one of finance as one of time. In 1926, as Labour MPs fought against the introduction of legislation permitting an eight-hour day, Nottinghamshire's George Spencer appealed to the government benches:

> We are not all Bolsheviks. Some miners love literature, some love art and some love science, and they seek to improve their minds in the best way that they can. How can they do it unless they have reasonable time at their disposal at the end of the day's work? It has been my experience to fall asleep over my books when my day's work was done.[81]

In this respect, the enforced leisure time afforded by the lockout proved a boon to those wishing to take advantage of the educational opportunities available. The 1926–7 WEA annual report for the district was enthusiastic:

> In spite of the ever-present financial and other difficulties we have made some progress. We began the year in the midst of industrial crises, when many of our members and students were compelled to think of other things apart from

[78] TNA:PRO, ED18/48, M. Crossling to D. Lloyd George, n.d., *c.*Apr. 1922. Grammar as in original.

[79] TNA:PRO, ED53/547, Board of Education documents.

[80] Trades Union Congress Library Collections, London Metropolitan University (henceforth TUCLC), WEA/Central 4/2/1/1, WEA annual report, 1925–6.

[81] *HPD*(C), 197, c. 1197.

culture . . . Yet, it is pleasing to record that some are taking greater interest in our work than they have done previously . . . We were quite concerned at the early part of the session as to whether we would be able to hold as many classes as in the previous session, and we were agreeably surprised to find that number exceeded.[82]

The number of three-year tutorial classes organized by the Durham University Joint Tutorial Committee and subsidized by Durham's local education authority supports this assertion and suggests that, particularly for men, the lockout offered the freedom to take advantage of such opportunities. In the academic year beginning Easter 1926, twenty-nine of these classes ran at various centres in County Durham, attended by 406 men and 166 women, with an average attendance of 19.7 members per class. This marked an increase both in absolute and average terms when compared to 1925–6. In that year, only twenty-four classes had been available and had been attended by 277 men and 134 women, giving an average attendance of 17.1 members per class.[83]

Others took advantage of their enforced leisure time to pursue alternative means of self-improvement. George Hitchin began learning the violin in 1926, taught by another striking miner.[84] At least one of his counterparts in South Wales, Bert Coombes, would do the same, although Coombes' teacher was a friendly young doctor.[85] Public libraries were also busier. At the end of May, the report for Stanley library noted that the strike had arrested the decline which usually characterized the spring quarter: 'There has been a very large issue [of books] of 10,425 . . . This is the highest Spring issue ever recorded being 27 per cent above the last and 14 per cent above the previous highest, that of 1921 which was also a strike quarter.' The reports relating to the summer and autumn periods also noted an increase, and in December it was recorded that 'the continuance of industrial trouble has turned the attention of very many strangers to the Library'. Most of the extra use was by adults.[86]

## V

Only a generation before the lockout, as a young man in the 1900s, Lawson had felt that his yearning for self-improvement alienated him from

[82] TUCLC, WEA/Central 4/2/1/1, WEA annual report, 1926–7.

[83] TUCLC, WEA/CJAC 4/2, Central Joint Advisory Committee on tutorial classes, annual reports, 1925–6 and 1926–7.

[84] G. Hitchin, *Pit-Yacker* (1962), 85.

[85] B. L. Coombes, *These Poor Hands* (1939), 179.

[86] DRO, UD/Sta54, reports to Stanley Library committee.

many of his peers. Indeed, the bookish culture of the chapel was one reason why Methodism had so appealed to him: there, 'no longer was I "queer" or "alone."'[87] By the time he came to write his autobiography in the 1940s, Lawson suggested that the miner's relationship to learning had been transformed, mainly due to the raising of the school-leaving age and the increased availability of working-class education.[88] But, even if attitudes were changing in the 1920s, it is clear that in a county with a population of nearly one and a half million, the few thousand men and women who took part in some form of adult education remained a tiny minority, as did the few hundred children who progressed to secondary school, or the few dozen who then went on to university. During 1926, for every George Hitchin with his violin, there were many more miners who simply basked in the unusual experience of a summer spent outdoors, filling their days with football or card-playing. Even the Stanley library report noted that the bulk of the increased borrowing was from the fiction department, 'and reflects a primary desire for recreation'.[89]

Robert Moore's research on social mobility in the Deerness Valley reflected his primary interest in Methodism, but he found that many in the interwar years still harboured ongoing prejudices about those who sought to escape the pit. His attempts to quantify the degree of social mobility in the valley proved ambiguous and inconclusive, as he was the first to acknowledge, but he did uncover a degree of contempt amongst many of the non-Methodists he interviewed with regard to the social ambitions of their Methodist counterparts. One man was particularly scathing:

> X's sons (X is a terrible snob); one is in administration, another in insurance . . . affected, regard themselves as a cut above the others . . . Getting out of mining was regarded as 'Fruits of the spirit'—this was said from the pulpit . . . The boy who stayed in the pit . . . was a spiritual failure [in the eyes of such people].[90]

Perhaps such sentiments stemmed from envy, for the reluctance to send a son to follow his father down the colliery shaft was widespread amongst many pit families, of any religion or none. Tales of the parental opposition faced when a boy first announced that he wanted

[87] Lawson, *Man's Life*, 71. [88] Ibid., 67.

[89] DRO, UD/Sta54, reports to Stanley Library committee.

[90] R. Moore, *Pitmen, Preachers and Politics: The Effects of Methodism in a Durham Mining Community* (Cambridge, 1974), 149. See pp. 246–8 for Moore's attempts to tabulate social mobility.

to start working at the colliery occur again and again in miners' autobiographies.[91] It is also possible that, by the late 1920s, a greater worry over the state of the industry, whether engendered by the strike itself or the subsequent depression, had further increased parents' hopes that education could give their children an alternative chance in life. Inspectors were impressed at a meeting of mothers at a school in Bishop Auckland in 1928:

> The courage of these mothers, their cheerfulness, and their interest in these times in educational welfare pure and simple is something to note. They expressed the main difficulty of living on the dole was the provision of boots and of meat—further that any sacrifice on their own part was worthwhile for the sake of the children, who as one mother explained in her own words were 'the people of the future'.[92]

More often than not, school logbooks of the 1920s record the success of open days, at which parents were invited to visit the school, talk to the teachers, and see their children's work, and they note with satisfaction the numbers of parents who attended. In 1923, for example, Seaham Harbour's Catholic school, with around 420 pupils, recorded the attendance of nearly a hundred mothers at an open day.[93]

For many, aspirations revolved fundamentally around effecting an escape from mining altogether rather than social mobility for its own sake, as evidenced by the relative lack of interest in achieving a higher status within the coal industry. This was partly due to the lack of opportunity, and a Board of Education memorandum of March 1925 noted the deficiencies of mining instruction in Durham, promising that the authority was preparing to 'tackle the question in earnest'.[94] Yet, three years later, an inspection report remained pessimistic, blaming the structure of the industry. It pointed out that although the miners were

> anxious to give their children the chance of clean work . . . owing to the small proportion of officers to rank-and-file in coal mines there was little hope of advanced education benefiting mineworkers. It was for this reason that the County was behindhand in respect of Technical Education. Chances of promotion were so few that it did not seem worthwhile.[95]

[91] See, for example, M. Craddock, *A North Country Maid*, 26; Hitchin, *Pit-Yacker*, 57–8; Also F. Zweig, *Men in the Pits* (1948), 28.

[92] TNA:PRO, ED50/77, 'Investigation into the Physique of School Children'.

[93] DRO, E/NE31, logbook, Seaham Harbour Roman Catholic School, 21 Mar. 1923.

[94] TNA:PRO, ED35/4061, Board of Education memo, 3 June 1925.

[95] TNA:PRO, ED50/77, 'Investigation into the Physique of School Children'.

But, it was also because demand remained limited. During the strike, George Hitchin was tempted by evening classes at his local school, but decided against it, 'as most of the subjects taught, such as mathematics, technical drawing and science, were, so I thought, designed for one to get on in the mining industry. I did not want to get on; I wanted to get out.'[96] In the end, he decided to complete a correspondence course in bookkeeping and commerce, to enable him to apply for a job as a clerk:

> When I had garnered a theoretical knowledge of commercial subjects, I scanned the local press for vacancies . . . I got no replies. Clerks were two-a-penny. Sadly I put away my books and contemplated my future with frustration and some fury. I was suffering from social and economic claustrophobia; unable to break out of the barrier of prejudice and ignominy that surrounded me.[97]

He returned to the shaft-bottom after the strike, although would eventually leave in the 1930s to become a teacher.

Further evidence of such preferences is provided by returning to the tutorial classes held in 1926–7 (see Table 5.1). It is significant that classes in humanities and the arts could equal or exceed the popularity of economics and science-based subjects: amongst both men and women, literature and drama classes were better attended than any other. Indeed, the subject titles are remarkably eclectic. At Chilton, twenty-five students studied 'the History and Literature of the Hebrews'; while one-year classes at both Chopwell and Greenside taught their pupils Esperanto.[98]

A generational difference can also be detected in subject preferences. In 1927, a national competition was announced by the Miners' Welfare

**Table 5.1** *Workers' Educational Association tutorial classes in Durham, 1926–7*

| Subject | Number of classes | Male students | Female students | Total students | Average number of students per class |
|---|---|---|---|---|---|
| Literature/Drama | 13 | 187 | 93 | 280 | 21.5 |
| Economics | 10 | 143 | 37 | 180 | 18 |
| History | 3 | 36 | 13 | 49 | 16 |
| Sciences | 2 | 29 | 11 | 40 | 20 |
| Art | 1 | 11 | 12 | 23 | 11.5 |
| TOTAL | 29 | 406 | 166 | 572 | 19.7 |

*Source*: TUCLC, WEA/CJAC 4/2.

96 Hitchin, *Pit-Yacker*, 88.
97 Ibid., 88–9.
98 TUCLC, WEA Central 4/2/1/1, WEA annual report, 1926–7.

Fund, offering several university scholarships. Across the country it attracted a couple of thousand applicants, 313 of whom came from the Durham coalfield. The competition ran in two streams: one for those working as miners; the other for the sons and daughters of miners. Of the 188 Durham miners who applied, seventy-nine declared that, if successful, they would like to study mining engineering, thirty-two put down some other branch of engineering, and twenty-one put science (of whom many probably had mining in mind, commented the judges). Only thirteen were tempted by music or art, and one by theology. The same trend was evident when looking at candidates from all the districts. Even the judges were surprised that despite the great amount of unemployment there was clearly 'still a very large number of workers who have faith in the future of coal mining in this country'.

The choices of the second category told a different story. Of 116 young people who applied from Durham, only two wanted to study mining engineering, and the balance between science and arts subjects was much more even: twenty-eight specified music or art. Over half hoped eventually to become teachers. Perhaps those of the younger generation were more idealistic than the working miners; perhaps they still fought against their place in the pit where their fathers had accepted it. Even so, the non-mining aspirations of both groups make poignant reading. The breakdown is not available for Durham alone, but amongst those miners across the country who hoped to become archaeologists, dentists, gardeners, plumbers, missionaries, and politicians, two candidates aspired to be captains of ships. They were both working miners and it is hard to think of anything that could have provided a greater contrast to their existing employment.[99]

Education was not the only means through which some hoped to thwart the inevitability of pitwork, and other ways out were sought. Professional sport was one such route. Jimmy Seed was born at Blackhill in the late nineteenth century and began work as a miner, as expected, at the age of 14. He later remembered being rejected after a trial with Sunderland Football Club: 'I was low in spirits because I had come to loathe working in the pits, and success at football seemed my only escape.'[100] His break came after his second trial for the club, and he would go on to enjoy a successful career as a player and then as one of Charlton Athletic's most renowned managers. For another miner, Jack

[99] TNA:PRO, ED54/23, First Report of the Selection Committee Appointed by the Trustees, Oct. 1927.

[100] J. Seed, *The Jimmy Seed Story* (1957), 65.

Potts, the lockout itself would prove his ability. He won several of the numerous running and walking competitions organized for the strikers and would go on in the 1930s to represent England several times in the international cross-country championships.[101]

However, as celebrated as the success story was, an alternative stereotype in popular culture was that of the promising young sportsman who was ultimately denied his chance to excel by the capricious cruelty of the pit. In an article written during the Second World War, Jack Lawson lamented the accident that had maimed and eventually killed a young Durham miner: 'He was so strong and agile . . . First-class [football] teams had their eyes on him . . . Then—a broken back.'[102] It is no coincidence that of the three Fenwick boys in *The Stars Look Down*, it is Hughie who dies in the pit disaster, his football boots shiny in readiness for the trial he had been due to have with Newcastle United that weekend.[103]

## VI

The value of education was central to the philosophy of the DMA and several of its lodge banners carried images or phrases that emphasized its importance. It was a message echoed by individual leaders, and Jack Lawson would later note in his autobiography: 'I can hardly remember when I could not read. And that marked a great difference between my generation and that of my parents, for in their day only the fortunate ones were able to read. It was the acquisition of that simple art of reading which began the battle against unnecessary poverty.'[104] A celebration of learning could be witnessed at a national level, too. In June 1926, the former Scottish miner John Wheatley warned the House of Commons that the miners were not a servile class who would willingly submit to longer hours and lower wages:

> the day when you could get that class of people in this country has gone forever . . . Your miners in future will be recruited from the mining villages. They will be the sons of miners; they will be the sons of people who are as well educated in political economy and industrial history, taking them in the mass, as are probably any other million of people in any other walk of life in this country.[105]

[101] E. Farbridge, *Recollections of Stanley* (DRO typescript, 1973), no page.
[102] J. Lawson, *Who Goes Home? Broadcasts and Sketches* (1945), 65.
[103] Cronin, *Stars Look Down*, 198–206.
[104] Lawson, *Man's Life*, 19.
[105] *HPD*(C), 197, c. 1013.

Yet, it was inevitable that the pursuit of education at the highest levels, whether for its own sake or to fulfil social aspirations, would remain elitist, continuing to isolate men and women from the culture in which they had been born. Jennie Lee was at Edinburgh University in May 1926 and helped out with her local strike committee. She was devastated when the TUC told its members to return to work and was torn between pride and pity when she heard that the miners in her Fife home were determined to hold out. Lee was not only from a mining family, but from one which was strongly active in a Labour movement that she had been brought up passionately to believe in. She would be an MP for a mining seat by 1929 and spend the rest of her life in politics; in Aneurin Bevan she would marry an ex-miner. She would never lose sight of her roots; yet even she was distanced from the communities that she wanted to help by the education which enabled her to do so. 'I longed to go home,' she remembered of May 1926, 'But, I dared not leave my books. In June, the following month, I had to sit my finals for my M.A. degree and my teacher's certificate.'[106]

The value bestowed on education by men and women such as Jack Lawson and Jennie Lee was not shared by all. Many of the miners' leaders were proud of an anti-intellectual tradition common to many within the labour movement. Even the ennobling prose of Jack Lawson could not disguise Herbert Smith, a man who was happier on the football terraces than with his nose in a book: 'He was a miner naked of book learning, and unashamed. You don't get coal by the book, and the chief thing about his job was to know pits and their ways; and pits he did know.'[107] Others objected to the content and purpose of formal education, harbouring suspicions that it was being used as a means of disseminating conservative values. In 1928, the Labour and Co-operative MP for Sheffield Hillsborough, A. V. Alexander, had expressed his disgust to the House of Commons that he had 'been unable to find in any elementary school class, any boy who knew who Robert Owen was or when the Co-operative Movement began, although in the matter of Henry VIII's wives they were well-informed'.[108] A few years later, J. P. M. Millar, the general secretary of the National Council of Labour Colleges, cited Alexander's words as yet further proof that 'education at every period is an instrument moulded to serve the interests of the dominant class in that period, whether that class is one of slave-owners, feudal lords or capitalists'.[109]

[106] J. Lee, *This Great Journey: A Volume of Autobiography, 1904–45* (1963), 71.
[107] J. Lawson, *The Man in the Cap: The Life of Herbert Smith* (1941), 85.
[108] Cited in J. P. M. Millar, *Bias in the Schools* (1936), 9. [109] Ibid., 1.

Robert Colls has argued that colliery education in County Durham in the period 1831–70 became a key battleground in the struggle for social hegemony between owners and pitmen. While owners' education primarily acted as a strategy of social control, the miners' self-education was seen as a means by which such control might be wrested back.[110] However, at least by the 1920s, the distinction was not so clear cut. With regard to adult education, the numbers of those involved in the coalfield was laudable, but for many it was simply a form of recreation, while it was pursued by others for purely personal gain. Such a conclusion ties in with Chris Baggs' work on South Wales, which sought to deglamourize and depoliticize the Miners' Institute libraries. He suggested that, unlike institutions such as the CLC and the Plebs League, the libraries were not part of the autodidactic, aspirational tradition of the coalfield and that, as in Durham, fiction was preferred to politics. He suggested that, none the less, 'their lasting achievement was to provide hundreds of thousands of books and newspapers to tens of thousands of readers. It sounds simple enough, but it was no mean feat.'[111]

Meanwhile, the formal education of children could never be wholly divorced from the concerns of the local community. One old miner, born in the early 1920s, recalled his school days seventy years later with an indignation born of hindsight: 'every morning we'd stand to attention in the classroom and sing, "God save the King". There was a map of the world. India was red. Africa was red. Canada was red. Australia was red. The British Empire had taken over the world. There were fifty-six children in the class, and none of them had nay byuts [boots]!'[112]

Much as an education encouraged children to see a bigger picture, many of Durham's elementary schools were situated in colliery districts, and were populated by miners' children. They thus remained particularly sensitive to local issues. Indeed, often it was the children who carried the political beliefs of their parents into their schools. One teacher began work in an elementary school at Deaf Hill in 1926. It was her first job, and she was a colliery boilerminder's daughter herself: 'I remember being at the boys' school then and they all jumped up on their desks and shouted "Blacklegs" as these men went past on the way to work. You

110 R. Colls, '"Oh Happy English Children!": Coal, Class and Education in the North-East', *Past and Present*, 73 (1976), 75–99. See also the subsequent debate, ibid., 90 (1981), 136–65.

111 C. Baggs, 'The Miners' Institute Libraries', *Planet*, 118 (1996), 52.

112 Cited in M. Hudson, *Coming Back Brockens* (1994), 135.

**6.** 'Committee Women Distributing Boots to Children after Holidays', by Annie Hillary, aged 12. *Labour Woman*, October 1926.
Reproduced by permission of the People's History Museum.

can imagine children doing that today, but they did that then. Jumped on their desks . . . "Blacklegs".'[113]

During 1926, therefore, in villages in which the vast majority of the population were on strike, the schools themselves became an integral part of the strikers' defence. In August 1926, the *Labour Woman* ran a national drawing competition on its children's page, and asked for contributions under the title of 'School after the Holidays' (see Illustration 6). A month later, one prize went to Annie Hillary, a 12-year-old from County Durham. She knew what 'school' meant to her.

[113] Miss Parkin, 'The Three R's', in T. Austrin et al. (eds), *But the World Goes on the Same: Changing Times in Durham Pit Villages* (Whitley Bay, 1979), 50.

# 6

# Memory and Experience

In 1972, a woman sat down to write her memoirs. Both the wife of a Durham miner and the daughter of one, she was prompted by the renewal of industrial conflict. 'The miners are on strike,' she explained, 'And that has made me stop to think. This is 1972 and the last time I heard those same words were in 1926 when I was nine years old. But what of the years in between? I was inclined to dismiss them as commonplace . . . .' She then went on to consider those years, particularly her time as a nurse during the Second World War, and ended her memoir with the conclusion that they were perhaps not so commonplace after all.[1] But, it seems remarkable that her instinctive reaction was to see the mining strikes as marking the primary reference points of her life, especially as, being a woman, she was not directly involved in the industry itself.

The miner's fascination with his past was a staple of twentieth-century sociological studies of the mining community. It was noted by Ferdynand Zweig in the 1940s, then again a decade later, in Dennis, Henriques, and Slaughter's famous study of a Yorkshire mining town.[2] In 1975, Martin Bulmer suggested that a 'shared history of living and working in one place over a long period of time' was one of the characteristics of a mining community ideal type.[3] More recently, James Fentress and Chris Wickham's study of social memory included a discussion of collective memory amongst the working classes. It is no coincidence that they chose to use the mining community as their first illustrative example.[4]

A strong collective memory is associated with the subjugation of self, and here again miners have been represented as the archetypal

[1] Durham County Record Office (henceforth DRO), D/MRP87/2, L. Wild, *The Years Between*, no page.

[2] F. Zweig, *Men in the Pits* (1948), 10; N. Dennis, F. Henriques, and C. Slaughter, *Coal is Our Life: An Analysis of a Yorkshire Mining Community* (1969 edn.; first pub. 1956), 56.

[3] M. Bulmer, 'Sociological Models of the Mining Community', *Sociological Review*, 23 (1975), 88.

[4] J. Fentress and C. Wickham, *Social Memory* (Oxford, 1992), 116.

proponents. When Stuart Howard listed the characteristics of miners' autobiographies, in first place he put 'the (paradoxical) practice of writing a plural autobiographical account—a history of selves rather than self'. Howard explained that rather than the 'celebration, admonishment or defence of the self', which is the conventional purpose of such writing, miners' autobiographies tend to fulfil a social function and adopt a collectivist rhetoric.[5] Although he did not use any examples from Durham to illustrate this particular point (he did use them elsewhere), they would have been easy enough to find. John Wilson, one of the most important figures in the early history of the DMA, published his autobiography in 1910. He was keen to impress upon his readers that 'the writing of this life's story has necessitated the use of the first person more than is at all palatable to me, for it was impossible to write it in any other form . . . it [is] obnoxious to me'.[6]

Amongst wider society, the miners have claimed a tragic-heroic status elusive to other working-class groups. One example is in the memory of clashes between strikers and the authorities early in the century at Tonypandy, where troops were sent in 1910 to maintain order during the Cambrian Combine dispute. The mistaken belief that Churchill, as Home Secretary, ordered them to fire on striking miners has achieved more notoriety than the actual deaths of two railwaymen at neighbouring Llanelli, shot and killed by troops during a national rail strike less than a year later. Even in the twenty-first century, it is Tonypandy and not Llanelli which remains more important to a working-class history.[7]

The mythologized events at Tonypandy join other strikes, lockouts, and pit disasters in the creation of a heroic national and regional chronicle of coal and its people. The dates of the national strikes in 1893, 1912, 1920–1, 1926, 1972–4, and 1984–5 are ones that echo loudly within the collective memory of the mining community. At a local level the dates of the most terrible Durham pit disasters also resonate: 1880, when 164 men and boys were killed at Seaham; 1909,

[5] W. S. Howard, 'Miners' Autobiography: Text and Context', *Labour History Review*, 60 (1995), 90–2.

[6] J. Wilson, *Memories of a Labour Leader* (Firle, 1980 edn.; first pub. 1910), 315.

[7] Troops were sent, but Churchill was initially criticized for not sending them fast enough. Once there, they only briefly came into contact with strikers. See M. Gilbert, *Churchill: A Life* (1991), 220–1. For the potency of events at Tonypandy (in contrast to Llanelli) in Welsh memory, see also D. Smith, 'A Place in the South of Wales', in his *Wales! Wales?* (1984), 55–64.

when 168 were killed at Stanley; 1951, when eighty-one were killed at Easington. The development of a social memory of mining became infused with a public narrative that stressed the suffering, resilience, and cohesiveness of the mining community, and care must be taken in the analysis of oral and written memoirs, which could absorb and reflect the romantic framework. Indeed, in their uncritical use of such material, academics have occasionally been guilty of the same. One historian recently claimed that, in 1926, 'family and neighbourly solidarity was very evident from the oral history respondents. When I asked "Jean" (born 1913, Pontycymer) how people managed during the Lockout she said, "everybody clung together and shared" '.[8]

Sometimes the gap between memory and reality was stark. Mark Hudson considered himself a Londoner, but his father had grown up in Horden. As a child, Hudson was constantly told tales about his mining heritage: 'the dignity with which the miners bore the cruelty of their labours, the richness of its [the village's] communal life'.[9] He soaked up the heroic image without questioning. Then, in the early 1990s, he visited his father's birthplace. He found that his family had not in fact lived in Horden since time immemorial; his grandfather had not even been a miner. He spoke to the current manager at the pit, and was told that the narrowest seam at Horden was two and a half to three feet—hardly comfortable, but, as far as the manager knew, Horden had never been a colliery in which a miner lay on his side to hew coal. Hudson felt almost cheated: 'another image from my ancestral mythology was casually shattered'.[10]

This is not to negate the importance of oral accounts, the value of which is now widely accepted. After all the usual caveats that the historian must apply to any source, they can provide perhaps the most effective means of accessing processes of historical memory and the creation of narratives. Luisa Passerini, in her famous study of Turin's working class, defended her use of subjective oral evidence:

> This subjective dimension does not allow a direct reconstruction of the past, but it links past and present in a combination which is laden with symbolic significance. While these oral sources have to be placed in a proper framework, they are highly relevant to historical analysis. These testimonies are, first and

[8] S. Bruley, 'The Politics of Food: Gender, Family, Community and Collective Feeding in South Wales in the General Strike and Miners' Lockout of 1926', *Twentieth Century British History*, 18 (2007), 73.

[9] M. Hudson, *Coming Back Brockens* (1994), 2.

[10] Ibid., 145.

foremost, statements of cultural identity in which memory continuously adapts received traditions to present circumstances.[11]

Nor should a scholarly exploration of a mining past be about shattering myths. As Raphael Samuel and Paul Thompson wrote:

> When we do encounter myth, our first instinct, it seems, is to devalue it, to rob it of its mysteries, to bring it down to earth. Recently spurred on by the revelations of Eric Hobsbawm and Terence Ranger's *The Invention of Tradition* (1983), Anglo-Saxon historians seem happiest at work puncturing legends, proving the modernity of much of what passes for old, showing the artificiality of myth and its manipulable, plastic character . . . Yet, myth is a fundamental component of human thought.[12]

Myths were important to the mining communities of the 1920s, just as they would remain important to future generations. Even if they could not be grounded in any empirical 'truth', the emotional validity that could be generated by a mining mythology could powerfully affect social relationships and social actions. This chapter will look first at the way in which the 1926 strike is remembered by those who lived through it, both in the romanticized story passed down in collective memory, and in the more diverse memories of individuals. It will consider the way in which a certain view of the past became important during the strike, when memory was able to build solidarity as well as commemorate it. In turn, '1926' became a part of this history, and would go on to dominate the memory and shape the actions of future generations.

## I

It is rare to read contemporary commentary on the 1984–5 miners' strike in which the ghosts of the 1926 dispute are not invoked at some point. Comparisons were frequently made then and have frequently been made since: certainly, both the sequence of events and the positions taken by those involved display an astonishing number of historical parallels. However, in 1984, the awareness of some kind of historical repetition meant far more to those involved than merely an entertaining

[11] L. Passerini, *Fascism in Popular Memory: The Cultural Experience of the Turin Working Class*, trans. R. Lumley and J. Bloomfield (Cambridge, 1987), 17.

[12] R. Samuel and P. Thompson, 'Introduction', in Samuel and Thompson (eds), *The Myths We Live By* (1990), 4.

search for symmetry, and the dispute took on a personal dimension as many consciously acted out the roles of their grandparents. As Andrew Richards has noted, in 1984–5, 'the impetus for collective action amongst miners, and the form that such action assumed, was *historically* driven', as collective means of survival such as soup kitchens were set up in conscious imitation of those that had fed the strikers sixty years earlier.[13] In fact, a preoccupation with the heroic tales of earlier times meant that the miners risked losing a sense of perspective on their own era. In February 1985, as it became obvious that the denouement to the 1980s' dispute would echo that of the earlier one, Kenneth O. Morgan begged the miners to shed their obsession with their past, which served only to reinforce their historical role as doomed heroes in a futile struggle:

> The analogies with a heroic past have been inaccurate and fatally damaging to the miners' cause. The evocation of history has been a snare and a delusion . . . the miners would be best served in future if they simply forgot their history, however inspiring . . . the miners, throughout Britain, [should] turn their gaze away from the seductive appeal of the grandeur and tragedies of pre-war years.[14]

Morgan was making a formidable request, for even before the events of the early 1980s encouraged systematic parallels to be drawn, the image conjured by '1926' had become the keystone to a militant, heroic, and tragic past. This was encouraged by the union, and when the second volume of the official history of the MFGB was published in 1953, even its title, *Years of Struggle*, summed up a certain image of the miners, which would then be exemplified in its pages.[15] But, it was also part of a traditional history passed down from generation to generation. Dave Douglass has commented that 'when you're a kid in a pit village you don't get Goldilocks and the Three Bears or Little Red Riding Hood as a bedtime story. You get Churchill and the '26 strike.'[16] Meanwhile, the remorseless demonization of those, and the children of those, who had blacklegged in the interwar years demonstrated the continued importance of these events in the memory of local communities. Examples could be given from the Durham area,

[13] A. Richards, *Miners on Strike: Class Solidarity and Division in Britain* (Oxford, 1996), 32. Original emphasis.

[14] K. O. Morgan, 'A Time for Miners to Forget History', *New Society*, 71 (1985), 283–5.

[15] R. Page Arnot, *The Miners; Years of Struggle: A History of the Miners' Federation of Great Britain (from 1910 onwards)* (1953).

[16] D. Douglass, 'Worms of the Earth: The Miners' Own Story', in R. Samuel (ed.), *People's History and Socialist Theory* (1981), 61.

but there are few quite so terrifying as the words of a Welsh miner, speaking in 1978 of those who had joined the company union in the late 1920s and early 1930s:

> Dick Clarke—hated . . . Edmund—branded. Jim Challinger branded . . . these men have been branded in these localities. *BRANDED.* If you talk about Seary . . . the first name that springs to mind is that he was a scab. It's a name that's with them forever. *BRANDED.* It'll be with their children. *BRANDED.*[17]

A powerful currency resides in '1926' which could shape the way in which people constructed their own stories. After the close of the 1984–5 dispute, for example, Roy Ottey wrote his autobiography. He had resigned his membership of the NUM Executive in October 1984 in protest at Arthur Scargill's leadership, and his autobiography is a justification of his position. It begins with one of his earliest memories: his father sitting outside under a blazing sun, playing cards while on strike in 1926. Ottey was born in November 1924, and perhaps he genuinely could (or thought he could) remember the summer of 1926. In any case, it is an important element of his defence that he should first set out his credentials as a genuine member of the historical mining community.[18]

By 1984–5, certain images of 1926 had long constituted a coherent collective memory of the earlier strike. They emphasized the solidarity of the villages, the charisma of the mining leadership (another 40-something Arthur), the treachery of blacklegs in Nottinghamshire and elsewhere, the apathy of the Labour Party, and the heartlessness of the government. Even those who blacklegged in 1926 might begin to 'remember' the lockout within this received tradition, and were able to reconstruct a sanitized version of the strike. One old man in Durham was too young to remember the 1926 strike for himself but had absorbed stories of the dispute from his father, who had been a miner at Adventure Colliery in East Rainton. During the lockout, Adventure's workforce gained a reputation as the most disloyal in the county. Most men had resumed work by July and the colliery became the scene of numerous protests and demonstrations until the dispute ended. Its lodge ultimately split from the DMA as a result.[19] But, the story told by the Adventure miner's son carefully negotiated its troubled

[17] Cited in A. Burge, 'In Search of Harry Blount: Scabbing between the Wars in one South Wales Community', *Llafur*, 6 (1994), 58. Original emphasis.

[18] R. Ottey, *The Strike: An Insider's Story* (1985), 4.

[19] *Durham County Advertiser* (henceforth *DCA*), 13 Aug. 1927.

history. Although he admitted to the moderate tradition of the colliery and the return of some of its members to work, its reputation as 'the "scab" pit' is glossed over:

> The Adventure being a small, family like colliery—was hardly a hotbed of political or industrial unrest but, like the rest of the Durham coalfield, the Adventure men supported . . . [the MFGB]. Some of the Rainton men were in the Durham Miners' Association, others were not, but it made little difference. The whole of the coalfield—indeed the whole industry—was in dispute, and that included the Adventure. Some of the non-union men and safety men worked during the strike . . . But, my father was one of the strikers . . . By August, forty men had restarted at the Adventure. The drift back to work had begun, but more noticeably in other parts of the country than in the Durham coalfield. By October there was a marked weakening in the strike nationally. On 29 November 1926, the miners were balloted . . . The strike was over.[20]

The adaptation of the past has not been limited to those with family reputations to defend. Aside from issues of solidarity, one of the most consistent images of 1926 involves the weather. Dozens of accounts of the strike, whether oral history interview or written memoir, mention the long, beautiful summer of that year. One man, born in 1910 and interviewed more than sixty years later, described 'a marvellous summer, you know, it was glorious all day long. I think it only rained once in a month. It became very oppressive and sort of thundery, and then it just downpoured for about an hour and that was it, the sum total of the rain for that time.'[21] Indeed, a folk memory of the hot, dry summer days would be yet another feature of that dispute to resurface during 1984–5, and at least one account later described the 'strike weather' of 1984 as 'so reminiscent of 1926 . . . '.[22] However, figures taken from Durham Observatory show that temperatures for May 1926 were lower than the monthly mean for the period 1919–39; June almost exactly matched the interwar average; and only in July, August, and September did temperatures exceed the usual for the interwar years. Even then, they were not exceptional. July 1926 saw an average maximum temperature of 69.3°F; this was exceeded half a dozen times in the interwar years when average July temperatures rose into the seventies. Temperatures then dropped in October and November, falling below the interwar

20 J. H. Fenton, *Going Back a Bit* (Southport, 1994), 27.

21 Gateshead Central Library, Oral History Transcripts (henceforth GCLOT), iii (Mr B.).

22 H. Francis and G. Rees, 'No Surrender in the Valleys', *Llafur*, 5 (1989), 58.

average; October 1926 was the coldest October of the interwar years.[23] The psychological effect of colder weather in the later months of the strike might be expected to have been more important than the heat of the earlier months, as belts tightened, the strike dragged on, and coal shortages hindered attempts to keep warm; but it is a fact that few refer to.[24]

Yet, despite the meteorological record, the weather remains important to a romanticized image of the strike. No doubt miners, used to working long hours underground, were likely to remember and enjoy the summer as a consequence of being out of work, particularly because most of those who have recorded their memories are those who were young in 1926 and therefore most likely to have taken advantage of good weather to be outside, playing sport and other activities. It is also no doubt partly due an inevitable element of nostalgia. It is notable that it falls to Bert Coombes, locked out in South Wales in 1926, to provide an alternative opinion, writing as he was in the 1930s, and not as an old man reminiscing about his youth. He recalled the hot summer days that allowed miners to escape on long walks in the countryside, but his account is a bitter one: 'We walked up the mountains because the grass was soft and because we could not endure the sight of those prosperous-looking cars flashing along,' he explained.[25] Perhaps most importantly, however, blue skies and a blazing sun remain central to the popular memory of 1926 because they complement the prevalent imagery of the strike. An ordinary summer would lack the same romantic currency.

## II

Despite such a coherent communal narrative, the most striking impression gained from the various written and oral memoirs of those involved in the 1926 lockout is of their variety. Although recollections frequently include such similar features as the demonization of the blackleg or the benevolence of the weather, there is no sense of a unified regional experience. The memories of those who lived through the struggle are

[23] G. Manley, 'The Durham Meteorological Record, 1847–1940', *Quarterly Journal of the Royal Meteorological Society*, 67 (1941), 374–5.

[24] One exception is the testimony of the South Wales miner Edwin Greening, '1926 in Aberdare', *Llafur*, 2 (1977), 36.

[25] B. L. Coombes, *These Poor Hands: The Autobiography of a Miner Working in South Wales* (1939), 177.

fractured and diverse, differentiated by an individual's age, gender, and personal circumstances.

A disproportionate number of recollections come from those who were children or young adults in 1926, men and women given voice by the rush of oral history projects conducted from the 1970s onwards, particularly around the fiftieth anniversary of the strike in 1976. For many who were children, the most significant difference that the strike made to their lives was the provision of school meals. These are recalled with varying levels of enthusiasm, dependent presumably both on the quality of meals at their specific feeding centre and on the usual standard of their mother's cooking. Laurie Moran was particularly impressed, describing 'breakfasts of hot cocoa and thick slices of bread and jam . . . [and] dinners of which my favourite was a generous helping of mixed beef and boiled potatoes'.[26] Yet, for others, the strike played no memorable part in their childhood. Some were sent away to stay with relatives,[27] but, for many more, the events of the lockout paled into insignificance—at least in their later recollections—amidst descriptions of their school days, their first encounters with the opposite sex, and, for men, the important ritual of their own first day at work in the pit. Mary Wade grew up in Northumberland, but her account of her childhood is similar to many from Durham. She mentioned the strike only twice, and then only in passing: it was of little consequence to her 8-year-old self, and the overall tone of her childhood memories is a happy one.[28] Another old miner's memory of the strike was eclipsed by his vivid recollection of another event, albeit one similarly entrenched in the mythology of the coalfield. Asked directly about the lockout by an interviewer, he replied:

> I can remember the strike, but there is one thing that stands even more clearly in my mind. It was before the strike, I was about ten or eleven year old, and they were fetching a man, dead, from the pit . . . If I shut my eyes I can see myself standing at the top of Mersey Street there and the men pulling this cart along with the corpse in.[29]

For the strike to be remembered in more explicit detail, children had to be witness to some of its more exceptional aspects. One old woman's

[26] L. Moran, *Seven Decades of Destiny* (Durham, 1996), 7.

[27] As noted in school logbooks. See, for example, DRO, E/C62, logbook, Sherburn Hill Council Infants' School, 11 June and 10 Sept. 1926.

[28] M. Wade, *To The Miner Born* (Northumberland, 1984).

[29] G. Alsop, 'A Kind of Socialism', in T. Austrin et al. (eds), *But the World Goes on the Same: Changing Times in Durham Pit Villages* (Whitley Bay, 1977), 24.

only memory of the strike was of herself aged 6, breaking her mother's strict instructions and following her 13-year-old brother down to the coal heap to help him search for coal. But, that day the heap collapsed and one of her brother's friends was killed. Her brother fled the scene, and she still remembered the jolting feeling as she clung to his back. She had to hold on tight, because he could only secure her with one hand—with the other he managed to save his coal pickings.[30]

Whereas children may not have understood what was going on, for those who were young adults in 1926 an awareness of the strike could be less easily evaded. Even so, for such people the strike is rarely remembered as heroic tragedy but often as a time of happiness and amusement. This feeling was particularly pronounced in men, for whom a seven-month strike meant a seven-month escape from the awfulness of pit work, and who, if they had no family to support, often saw the lack of money as a reasonable exchange for a free summer. Tommy Lawton was born in 1906:

> What were these things [spending restrictions] when you compared them with the possibility of having long lies in bed on a morning, of all the sunshine that we could get, and all the green grasses and all the lovely lasses that were there just to go out with, providing they had a job and a few coppers to spare. But, just think, we wouldn't have to get up at one o' clock in the morning on a cold wet winter's morning. We wouldn't have to go down in the pit, wet cold pits, we wouldn't have to be there for eight hours slugging and pushing, sweating and swearing, and wishing to hell we were back out of the pits oh no all that would be finished with and we could go on from day to day enjoying ourselves in our own way, and ninety per cent of young men saw it that way.[31]

Even when such young men became actively involved in the strike effort, some felt that collecting for charity remained far preferable to filling coal tubs. For one, the strike 'meant that as a youngster of fifteen I was liberated from the pit. All my pals of fifteen and sixteen—all of us—we were really liberated. It was a magnificent glorious summer so that we spent the time more or less tramping and getting into Newcastle with collecting boxes.'[32] For another who was slightly younger, an old man when he was interviewed, the summer months of 1926 were associated in his memory with his last summer of freedom before he started work

[30] Telephone conversation with Elizabeth Stoves, 13 June 2005.

[31] GCLOT, ii (T.L.).

[32] M. Ridley, 'Making a Contribution', in Austrin et al., *But the World Goes on the Same*, 58.

himself: '1926 strike, that's when I left school. It was the best summer we ever had, 1926.'[33] It is through such memories that the strike is perhaps most strongly represented as a festival of the oppressed. In July 1926, a local speaker addressed a crowd of 4,000 at West Rainton: 'We are not dying of starvation . . . a pit is a horrible place and a man will live longer at bank,' he declared, 'This is the best holiday we have ever had.'[34]

Personal circumstances therefore determined the impact that the strike had on the life of an individual, affected by age, gender, and geography.[35] Frequently it was when public events interacted with personal landmarks that the most powerful memories were created. Bessie Johnson shared the sense of liberation of her male peers. The strike coincided with her coming of age, and her dominant memory of 1926 was of the laying down of wooden boards in the fields to allow dancing in the evenings: 'When I hear the tune Valencia, I always think of the general strike. I was 14ish, 15ish, I was just getting the idea of . . . dancing and things like that . . . it was terrific in that strike.'[36] For others, the personal experiences of May to December 1926 were so important that they almost eclipsed the lockout altogether. Mrs Churcher's first son was born on 14 April 1926 and was christened during the general strike. Interviewed fifty years later, the events of the strike provided only the background to her first experience of motherhood: she remembered that the pickets allowed the women to go home from the christening in a taxi (the men had to walk) but they still had stones hurled at them in places. She continued to date her experiences of the strike in terms of her newborn baby: 'Well, when he was about six weeks old, it was when they started bringing the blacklegs home . . . .'[37]

For both men and women, old and young, 1926 did not therefore have to be remembered as tragedy. The 1926–7 football season saw Newcastle United Football Club crowned as champions, captained by Scottish signing Hugh Gallacher who secured a club record and his place as Tyneside hero with thirty-nine goals over the season. Presumably, a considerable number of those who cheered on their team in the first few months of the season (whether inside or outside

[33] K. Armstrong (ed.), *Horden Miners: The Lives of Two Horden Miners in their Own Words* (Peterlee, 1984).

[34] *Durham Chronicle* (henceforth *DC*), 31 July 1926.

[35] For geographical differences within County Durham see Chap. 1, Sect. VI; for ways in which gender affected memories of the strike see Chap. 3, Sect. II.

[36] British Library Sound Archive (henceforth BLSA), C900/05554C1.

[37] GCLOT, iii (Mrs C.).

St James' Park) were miners on strike. It was also a triumphant season for Sunderland Association Football Club, who finished in third place. Home attendances at Roker Park were considerably lower than usual, but 18,000 watched from the terraces in September 1926 to see Sunderland beat Bury 3–0 in the first home game of the season.[38] Other activities might also create happy memories and distract from grimmer realities of depression and lockout. By September, over 225,000 passengers had taken advantage of the cheap half-day excursions offered during the summer months by the LNER rail company.[39] Many miners spent their days tending their allotments (by 1920 there were 50,000 registered allotment holders in the county),[40] and in September, Easington Colliery's workmen's club attracted a record 209 entries in its annual vegetable and flower show.[41] Such events were held all over the county: fifteen different leek shows were reported in the *Durham Chronicle* in just one week in October.[42]

Indeed, 1926 as a year of suffering at all was challenged by contemporaries, just as it has been debated by historians.[43] In October, Beatrice Webb noted in her diary:

> The surface facts show no exceptional distress: indeed, the pit villages look clean and prosperous and the inhabitants healthy . . . Various people told us that the men and boys had benefited by the rest, sun and open air and abstinence from alcohol and tobacco. And the women were freed from coal-dust and enjoying regular hours; whilst the school children, through the ample supply of first-class food . . . were certainly improved in health and happiness. The one want was clothing and boots.[44]

In fact, a village that looked 'clean' was not necessarily an affluent one in a community that valued respectability; nor was Webb taking into consideration the mental well-being of her husband's constituents. The typicality of Seaham can also be questioned, as before the strike its collieries had been relatively prosperous compared to elsewhere in the coalfield.

[38] B. Graham, *The History of Sunderland AFC, 1879–1995* (Houghton-le-Spring, 1995), 66.

[39] *Blaydon Courier* (henceforth *BC*), 18 Sept. 1926.

[40] H. Beynon and T. Austrin, *Masters and Servants: Class and Patronage in the Making of a Labour Organisation* (1994), 118.

[41] *Seaham Weekly News*, 17 Sept. 1926.

[42] *DC*, 16 Oct. 1926.

[43] See, for example, S. Thompson, ' "That Beautiful Summer of Severe Austerity": Health, Diet and the Working-Class Domestic Economy in South Wales in 1926', *Welsh History Review*, 21 (2003), 552–74. The same applies to the wider years of the depression. See C. Webster, 'Healthy or Hungry Thirties?', *History Workshop Journal*, 13 (1982), 110–29.

[44] Diary entry, 24 Oct. 1926, in N. MacKenzie and J. MacKenzie (eds), *The Diary of Beatrice Webb*, iv, *1924–43: The Wheel of Life* (1985), 105.

However, her impressions were echoed by another (middle-class) observer, Durham's Medical Officer of Health, who reported at the close of the year that, 'despite the industrial dispute . . . the mortality statistics are the most satisfactory in my experience . . . I have not the slightest doubt myself that the steps which have been taken to safeguard the health of our young population by providing suitable meals in necessitous cases have been fully justified from the health standpoint.'[45]

Such interpretations can be contested. In 1926, County Durham endured its first and only serious smallpox epidemic of the twentieth century. Eleven people died and a total of 5,791 cases were reported.[46] In Parliament, Joe Batey blamed the Ministry of Health for cutting the scales of poor relief, declaring that 'through the impoverishment of our people by the action of this Tory Government we have had an epidemic of smallpox in Durham the like of which we have never seen before'.[47] But, suggestions that the health of the miners had improved during the strike were not limited to those outside the mining community who might either misunderstand the nature of the suffering or misrepresent it for political ends. The benefits of the lockout were also frequently used as propaganda by the miners' leaders, who contrasted the current situation with the detrimental effects of several hours spent sweating underground each day. One Durham delegate told the MFGB conference in October:

> Conference after conference we come here as delegates and we hear pitiful tales of people starving and going back to work. There is no man in this Conference can prove to me that ever a man has died of starvation since the lockout . . . I want to know if we are improving in health why we should talk about defeat . . . I say we will never collapse as long as we are improving in health . . . The man who does not go back to work will live fifteen or thirty years longer. He is getting parish relief. We have played our parts in the mines . . . We work our souls out day in and day out. It gets on my nerves.[48]

It is certainly likely that the distress in Durham was less severe than that of many other coalfields, due to the Labour-dominated local councils. Even in October, when the most flagrant 'abuses' of the guardians had been curbed, Churchill despaired at the extent of outdoor relief:

[45] The National Archives: Public Record Office (henceforth TNA:PRO), MH57/94, 'Effect on Poor Law System of the General Strike and Coal Dispute', n.d., *c.*Dec. 1926?

[46] DRO, CC/H14, Annual Report of the Medical Officer of Health (County Durham), 1926.

[47] *Hansard Parliamentary Debates* (Commons), fifth ser. (henceforth *HPD*(*C*)), 199, c. 2005.

[48] MFGB, *Annual Volume of Proceedings for 1926* (1927), 891–2.

'No wonder the struggle continues! It is difficult to see why it should not continue indefinitely, the miners adhering to their principles while subsisting on the outdoor relief of their families and becoming gradually habituated to an indigent idleness.'[49]

However, the ability of different families to survive seven months without a miner's wage varied enormously. In some families younger children were sent out to earn money; some had savings or a war pension to turn to, or their allotments. Some, like the family of the young George Hitchin, had a lodger who was not a miner, and whose money provided a valuable income.[50] Others did not, and despite Churchill's claims, the rates of poor relief were not high for those who were, or who came to be, entirely dependent upon them. The difficulty of ascertaining an 'average' wage in the coal industry has already been discussed,[51] but even a miner earning the minimum of 7*s*. 6½*d*. per shift would take home a little over £2 a week (assuming a working week of five and a half days). In contrast, the Ministry of Health's notorious circular 703, issued at the beginning of the dispute, advised a weekly payment of 12*s*. for a miner's wife and 4*s*. for every child, up to a maximum of £1 12*s*. a week.[52] The generosity of the guardians varied, but even the Chester-le-Street board was adhering to this in June 1926: they incurred the Ministry's wrath because they imposed no maximum; did not deduct money if children received free school meals; and, most importantly, refused to deny relief to strikers themselves.[53] At least one Durham poor law union provided relief entirely in kind in order to prevent women from sharing money with their husbands, providing a stark contrast to the traditional, romanticized image of the soup kitchen in popular memory as a symbol of communal help and solidarity.[54]

The DMA could do little to supplement this. It began 1926 with funds of £141,357, and paid out (with help from the MFGB) a total of £352,364 in dispute benefit over the year, amounting to just over 45*s*. per member.[55] There were many, therefore, who did remember the

[49] TNA:PRO, T172/1558, draft memorandum written by the Chancellor, Oct. 1926.

[50] G. Hitchin, *Pit-Yacker* (1962), 81.

[51] See Chap. 2, Sect. I.

[52] Trades Union Congress Library Collections, London Metropolitan University, HV4546, Ministry of Health to Boards of Guardians (Circular 703), 5 May 1926. This compared to the normal rate of unemployment benefit of 18*s*. for a man, 5*s*. for his wife, and 2*s*. for each child.

[53] DRO, U/CS305, emergency committee correspondence.

[54] TNA:PRO, MH57/94, 'Effect on Poor Law'.

[55] *Parl. Papers, 1928*, ix (781), Return showing details of Membership, Income, Expenditure and Funds of Registered Trade Unions with 10,000 or more Members in the Years 1925 and 1926, 4–5.

strike as a time of hardship and suffering, and this was particularly the case for those who had the added responsibility of a family. This was appreciated decades later by those who had been children at the time, perhaps in the light of since having been parents themselves. One man explained: 'It was a lovely summer as far as I was concerned . . . But, I also know that my father died during the strike and left my mother with six children.'[56] Even those who were young and single might be expected to take responsibility for the well-being of others in the community. One 19-year-old lived at Ouston and helped at the soup kitchens there. He later recalled that '1926 was a terrible year . . . I used to be peeling potatoes till I was sick.'[57] Yet, if parents worried for their children, those children could at least benefit from school feeding centres and were appealing recipients of public charity. For young, single men, there was no relief except whatever the local lodge and charity could provide. When Harry McCormick told an interviewer in the 1970s that three of his teenage friends had died during the course of the strike, he put it down to malnutrition.[58]

Small luxuries became prized, and one man, then married with a wife and baby, still remembered fifty years later how he used to look forward to the weekly distribution of relief on a Wednesday when he would go 'straight into the little shop at Pelton Fell, there, and get a big tin of tomatoes'.[59] With regard to Northumberland, Bill Williamson suggested that there was 'a massive shift in the pattern of purchasing', having looked at Co-operative sales figures which revealed that butchers' and drapers' departments were the most heavily hit.[60] In Durham, perhaps one indication of a significant reduction in the amount of money available are the figures for drunkenness. In the three months ending 31 November 1926, coinciding with the final months of the lockout, only 245 people were convicted for drunkenness in County Durham, the lowest total of any quarter since 1919.[61]

Some broke under the strain, and the most tragic cases did not live to record their tales of suffering: there were seventy recorded suicides in County Durham in 1926, compared to forty-seven in 1925. One was 51-year-old Michael Welch of Stanley. His wife told the coroner that he

56 GCLOT, iii (H.M.). 57 Ibid., iii (Mr F.). 58 Ibid., iii (H.M.).

59 Ibid., iii (J.D.).

60 B. Williamson, *Class, Culture and Community: A Biographical Study of Social Change in Mining* (1982), 186.

61 DRO, CC/A10/1/6–8, standing joint committee minute books. The quarter ended 31 Nov. 1925 saw 494 convictions; the quarter ended 31 Nov. 1924 saw 592.

had been greatly upset owing to the coal strike, had worried a great deal, and been unable to sleep, although he had never threatened to take his life. He had two children.[62] The number of suicides further increased to ninety-two in 1927, a figure that might also have been related to the trouble in the coal industry, as some men were not restarted, pits were closed, and long-term unemployment beckoned, without the promise of some kind of future conclusion, which at least the strike had provided.[63] 'Anyhow, the funny thing is that I never felt tired till the Strike got settled,' commented one old miner.[64]

Again, however, not every striker faced such hardship. During the course of the lockout, several Durham pits announced that they would not be reopening upon the resumption of work, raising a question over the eligibility of their workforce for unemployment benefit. Initially such relief was refused, but in November the decision was referred to the Court of Referees in London and the judgement reversed. Over 1,000 miners received several weeks' worth of backdated benefit as a result.[65] For others lucky enough to find alternative employment, the effects of the lockout might become insignificant. J. Halliday was born in 1908 and although he had been brought up a miner's son and had worked in the pit himself, he had been made unemployed early in the 1920s and by 1926 was working as a farmhand. His memoirs consequently mentioned the strike only in passing and instead concentrated on his footballing escapades and his involvement with his church.[66]

More often, if the experience of the strike fades into insignificance in the memory of those who lived through it, it is because of the distressed years that surrounded it. Not only is 1926 frequently confused with 1921 in oral history interviews, but for many it is absorbed into wider memories of the depression years, and a discussion of the strike leads onto a conversation about the means test, the difficulty of finding employment, and the poverty of the interwar years as a whole.[67] Rather

[62] *BC*, 21 Aug. 1926.

[63] DRO, CC/H14, annual reports of the Medical Officer of Health, 1925–7.

[64] GCLOT, ii (Mr H.). In Scotland, Jennie Lee remembered a similar sense of anticlimax: 'In 1926 we had shouted and protested, we had had our meetings and our marches. In 1927 we had time to notice how shabby and disheartened everyone looked. Essential food and clothing got on credit during the lockout had now to be paid for. So had arrears in rent; bit by bit, week by week, in installments that left everyone bare to the bone.' J. Lee, *This Great Journey: A Volume of Autobiography, 1904–45* (1963), 79.

[65] *DCA*, 5 Nov. 1926.

[66] J. Halliday, *Just Ordinary, But . . . An Autobiography* (Waltham Abbey, 1959).

[67] For example, GCLOT, ii (J.S.).

than a coherent account of a heroic, tragic struggle, the strike therefore becomes only part of a story spanning years of deprivation, trauma, and heartache. This was the impression of one ex-miner who had worked at Ushaw Moor in the 1920s. For him, the lockout marked only one episode in a decade punctuated by periods in and out of employment: his bitter memories of local employment committees and his attempts to prove that he was genuinely seeking work loom far larger in his autobiography than do memories of the strike.[68] Even Bessie Johnson, who saw the general strike through the excited eyes of a teenager and remembered the nine days in May as something special, would in later years forget that the next few months had had anything to distinguish them from longer periods of depression: 'And it lasted for nine days. But, the results of it lasted for at least two year. Terrible.'[69]

## III

A study of the way in which the lockout is remembered thus has to reconcile two contradictory notions. One is the existence of a coherent narrative, developed after the strike, of a heroic, tragic struggle. The other is the fractured and diverse nature of individual memories, dependent on the personal circumstances of each participant. In order to understand the way in which both could coexist it is necessary to consider the role that a consciousness of the past played during the strike itself.

In 1926, the miners did not have to look too far back to find examples of their militant past. The coalfield was not quiescent in the early 1920s: in 1925, fifty-eight working days had been lost to stoppages across the coalfield; in 1924, there had been fifty-two; in 1923, there had been 112.[70] However, it was the experience of the three-month stoppage of 1921 in particular that meant that, five years later, tried and tested coping mechanisms could swing readily into action. Soup kitchens were re-established; school canteens were started up again. Those who had been too young to participate fully in 1921 now took part in their own right. George Hitchin was 9 years old in 1921 and recalled: 'We had no cycle and I was too young to give much help to my father when he had to carry the bag of coal two miles home. I assisted in gathering the coal,

[68] F. Proctor, *I Was There: An Autobiography*, ed. W. N. Tindall and P. Proctor (Gibsons, British Columbia, 1999).

[69] BLSA, C900/05554C1.

[70] DRO, C5/18, souvenir gala programme, 1950.

but years later during the long strike of 1926, I too carried my burden with the rest of them.'[71]

However, in the 1920s, a much more enduring sense of history cut deep into the consciousness of the miners and their families. The hereditary nature of the coal industry has already been discussed,[72] and the handing down of common experiences from father to son was obviously facilitated by such occupational continuity. Children grew up surrounded by stories of the pit, which they could readily relate to and then pass on when they started work themselves. In this environment, an oral history tradition flourished. Sid Chaplin was born in 1918 and explained:

> When I was a boy . . . I was the little pitcher with big ears sitting in the corner of any kitchen that made me welcome. There were old men who signed their mark but had the gift of memory. I remember hearing the story of the great strike of 1832 as told to an old man by his grandfather who took part in it.[73]

This was a child who would grow up to become a novelist and was perhaps more likely than most to soak up the story-telling of others, but those not so destined also absorbed the tales of their elders. One old man interviewed in 1976 could easily reel off all the horrifying stories about his ancestors that had been passed down the generations. According to family lore, his great grandfather had been killed in the pit in the 1870s on the very day that his wife gave birth to a son. That baby, his grandfather's brother, was eventually killed underground in his turn. A deputation from the pit called on his wife, but found no one at home and so left the body lying in front of the fire.[74] Writing in more optimistic times about the outlook of the British miner, Zweig expressed surprise that such grievances still had relevance to the younger generation, but 'the past has been transferred to him with his mother's milk, and he knows all about it, and resents it even more than does the old miner'.[75]

It must not be forgotten that in the 1920s some colliery villages were relatively new, and many families had only recently moved to the region. However, they joined a culture in which the telling of local history extended beyond the oral tradition. This was evident in the local press, most clearly in obituary and anniversary columns, which often emphasized the difficulties and hardships of their subjects' mining

[71] Hitchin, *Pit-Yacker*, 40.
[72] See Chap. 1, Sect. I.
[73] S. Chaplin, *A Tree with Rosy Apples* (Newcastle, 1972), 14.
[74] GCLOT, iii (H.M.).
[75] Zweig, *Men in the Pits* (1948), 19.

pasts. In June 1926, for example, the *Durham County Advertiser* drew attention to two couples celebrating their golden wedding anniversaries. One husband had experienced as a child an eighteen-week strike, when miners with their belongings had been turned out of their houses into severe weather; the other, it was reported, had begun work in Eppleton Colliery at the age of 8, and continued to work there, aged 69.[76] A few months later another golden wedding was recorded in the local press: the husband in this case had been one of the last men to have been rescued from Seaham Colliery after the explosion of 1880, in which 164 miners had been killed, the youngest of them 14 years old.[77] In 1926, four old women still received 2*s*. a week from the Seaham Colliery Relief Fund, the last surviving widows of the disaster.[78]

Moreover, those new to the area arrived in a landscape in which coal had been mined for centuries. The Great Northern coalfield was the oldest in Britain. It had had a significant coal trade since the early thirteenth century, and had been the most productive coalfield until the First World War.[79] The tragedies and hardships of the past could therefore sometimes make a more forceful reappearance in the lives of the coalfield communities than in the muted columns of local newspapers. In July, a group of miners digging into the hillside in a search for coal near Elemore Colliery unearthed the bones of ponies killed in an explosion of 1886. One miner told a local newspaper that he intended to make a door knocker from one of the bones.[80] The story echoes an anecdote reported by Bert Coombes in South Wales in the early 1920s, when an even more grisly discovery was made upon the reopening of some old pit workings. They had been closed since an explosion generations before, and inside they found several skeletons 'that could only have been the remains of children of the ages of six or seven'.[81]

But, if the miner of 1926 was aware of the struggles of an earlier generation, at times his own plight might seem similarly harsh. He did much the same work as his grandfather and great-grandfather had done before him, in the same dirty, dark, difficult conditions. This was particularly the case in Durham, where mechanization had had little effect on the underground pattern of work: only 15.6 per cent of the total output of the coalfield was cut by machine in 1924, a percentage

[76] *DCA*, 11 June 1926.
[77] *DC*, 18 Sept. 1926.
[78] DRO, D/Lo/F703 (17), accounts for Seaham Colliery Relief Fund, 1926.
[79] J. Hatcher, *The History of the British Coal Industry*, i, *Before 1700: Towards the Age of Coal* (Oxford, 1993), 70–2.
[80] *DC*, 31 July 1926.
[81] Coombes, *These Poor Hands*, 82–3.

lower than any other British coalfield except South Yorkshire (10.9 per cent) and South Wales and Monmouth (5.1 per cent).[82] Meanwhile, the disasters and tragedies that Sid Chaplin heard whispered around the fireside resonated all the more powerfully because their dangers continued to threaten. One of the benefits of a seven-month lockout was that pit accidents ceased (although men might still die from accidents suffered in earlier months or years), but, for some, grief remained fresh. With a poignancy that echoes the deaths of men killed in the trenches in the late autumn of 1918, ten men died in Durham pits in April 1926.[83] One was Thomas Dawson, a married man with six children, killed by a fall of stone at Sacriston; he was buried a week before the lockout began. His funeral, as for any man killed underground, would have been a huge ceremonial affair, with crowds lining the streets as the colliery band accompanied the coffin through the village, its banner draped in black. It was a ritual that powerfully reminded audience and participants alike of the dangers of pit work, of the support of their union, of—as with the fallen soldier—the heroic, tragic role of the miner. In Dawson's case the imagery merged. He was an ex-serviceman and his coffin was covered with the Union Flag.[84] A military analogy was repeated elsewhere in the coalfield. One of the very few lodge banners to depict an underground scene was that of West Stanley, where an explosion in 1909 had killed 168 men and boys, the worst ever disaster of the Durham coalfield. Its banner portrayed a man trapped underground, captioned 'the unknown miner'.

The memory of a history laden with grievances added to the determination of the miners to fight. With regard to the miners of the Ruhr, S. H. F. Hickey has commented that the economic importance of mining, its dangers, and the sense that it was financially and socially undervalued, increased many miners' discontent with their material lot: 'It was a moral dimension which few other trades could match.'[85] A similar resentment was present in Durham. Ron Rooney was a child in 1926, but spoke of the 1930s when he described how:

> when I started in the pit . . . when we were sitting getting our baits, at lunchtime, breaktime, the old miners would tell stories about their young days and what

[82] Report of the Royal Commission on the Coal Industry (1925), iii, Appendices (1926), 151–60. The coalfield with the highest percentage of coal cut by machines was Scotland (47%).

[83] DMA, *Fatal Accidents Book, 1920–50* (Durham, 1995). This recorded union members only.

[84] *Chester-le-Street Chronicle* (henceforth *CC*), 30 April 1926.

[85] S. H. F. Hickey, *Workers in Imperial Germany: Miners of the Ruhr* (Oxford, 1985), 168.

used to happen . . . People living now don't realise what the mines were. They don't understand why the miners are so bitter now. They're bitter because they remember what it used to be like in days gone by.[86]

The potential power of such emotion was appreciated by the union leaders, who actively encouraged their members to connect with their mining past. Events such as the opening of aged miners' homes, for example, provided a link to the past that echoed that of the newspaper columns. When the foundation stones of a new set of homes were laid at Kibblesworth in April 1926, one of the invited representatives was a Mrs Pratt. At the ceremony it was pointed out that her husband had worked at Kibblesworth Colliery between the ages of 8 and 73, and had had only a year and a half away from the colliery during that time.[87]

During the strike itself, by drawing on the grievances that were already familiar to many, the miners' leaders could reinforce the message that theirs was a just fight. In November 1926, Peter Lee gave a talk at Shildon. He told of how he had started work at the tender age of 7: 'There were no cages in those days, but a rope [with a loop at the bottom] . . . His first shift lasted thirteen hours . . . There had been great changes at the collieries and old miners would scarcely know them again. So far as the houses were concerned, however, they would be able to say, "We recognize these alright" '.[88]

A sense of outrage, sharpened by an acute historical awareness, was particularly evident with regard to the Coal Mines Act, which represented a reversal in a gradual shortening of hours since the nineteenth century. This was particularly resented in Durham, owing to the traditionally short hours of the region's hewers. In September, Jack Lawson rose in the House of Commons:

Let me tell the House what I heard in my own home the other day. A young man, a thrifty young man, with some amount of education, who loves his books, who has a wife and two or three children, and who loves his home, told me . . . 'Sixty years ago our fathers won, not an eight-hours' day, but six and a half hours, for the Durham miner at the coal face.' . . . He said that, over sixty years ago, the miners in this part of the country not only fought a great battle but were thrown out of their homes in winter time . . . This is the kind of historical spirit which animates the men of this country, for good or for ill, and he said to me: 'We have not got there yet, and what would we be worth if,

86 R. Rooney, 'Changing Times', in Austrin et al., *But the World Goes on the Same*, 37.
87 *CC*, 16 Apr. 1926.
88 *DC*, 6 Nov. 1926.

with all our intelligence, in the year 1926 we were to yield and lose what our fathers gained sixty years ago?'[89]

His words were echoed in the coalfield itself, and when Joe Batey addressed a crowd of miners in July he imagined long-dead Durham hewers returning from the grave to be told that their descendants were now being asked to work eight hours. 'They would ask, "Is this progress?" and if told it was, would reply, "Let me get back to the grave as soon as possible" '.[90] In fact, this was very much about the construction of a past to suit the purposes of the DMA, for Durham had been exceptional in the past only in regard to the short hours worked by its hewers. Even after the legislation of 1908 and 1920, surface workers in Durham usually worked two shifts to cover the hewers' three, entailing longer hours as a result.[91]

Just as an awareness of the shared struggles of the past could create solidarity through a sense of collective grievance, such solidarity could also be cemented through a complementary awareness of progress. To be most effective, the miners' history had to be heroic as well as tragic; for the past to provide inspiration it had also to provide hope. In 1926, the Samuel Report suggested one reason for the continued discontent in the coal industry:

the present grievances of the mining population are frequently viewed by them in association with the grievances of the past . . . It is of course recognized that the progress in these matters has been great. But, the men know that in many, perhaps in most cases it has been won by their own efforts, often in the face of strong opposition. The progress is frequently regarded less as a cause for gratitude, than as a reason for believing that the hardships that still exist, and are represented as unavoidable, may be as unnecessary and as open to remedy as those that have in fact been abolished.[92]

The DMA was obviously eager to promote a history which cast itself as the means through which progress could be achieved: miners were more likely to remain loyal to a union that had proved itself in the past. This history had to be built with more determination, however, for prosperity was less memorable than distress. Even before the wheels of the pit shafts ground to a halt in May 1926, therefore, the union had been keen to emphasize its historical presence. In 1919, a pamphlet published to

[89] *HPD*(C), 199, c. 348.

[90] *DC*, 31 July 1926.

[91] See S. Webb, *The Story of the Durham Miners, 1662–1921* (1921), 64–9.

[92] *Parl. Papers, 1926*, xiv (1), Royal Commission on the Coal Industry (1925) i, Report, 109.

celebrate the fiftieth anniversary of the DMA reminded its readers of its many successes:

In casting our eyes over the fifty years that have passed since the Durham Miners' Association was formed we see ample evidence of the great and fundamental changes which have been brought about by the power of Trades Unionism. Wages have been vastly increased, working hours materially shortened, and working conditions greatly improved.[93]

By far the most powerful propaganda tool available to the union was the annual miners' gala. This was the most important and effective ritual of the coalfield, and in the 1920s about a hundred banners were carried into Durham City each year. They provided a potent symbolism as the striking colours of each flew high amidst crowds of hundreds of thousands of miners and their families. The images on the banners tended to be celebratory and triumphant, and extolled the achievements of the union. The banner of Monkwearmouth lodge, for example, commissioned in 1921, depicted the abandonment of the yearly bond in 1869, around which success the DMA had been formed.

Many carried portraits of local or national leaders. In 1923, the *Durham County Advertiser* noted that several of the banners portrayed William Gladstone. 'Those lodges are scarcely up to date,' commented a reporter.[94] But, this was missing the point: the banners were not intended to reflect the immediate interests of the current generation, but carried a sense of history that was almost tangible. The gala stressed the historical continuity of a mining community stretching back into a bygone era. In 1910, forty years after the gala had been inaugurated, John Wilson had commented on the banners that celebrated 'the lives of men of the early days . . . borne aloft by men who were not at the first demonstration because they were not born'.[95] A personal continuity was also provided by the participants and spectators, many of whom continued to attend throughout their lives. Jack Lawson claimed that it was 'the boast of many that their bairns have been to the Big Meeting since the year of their birth. It is a kind of social baptism'.[96]

93 DRO, C5/17, souvenir to commemorate DMA jubilee, 1919.

94 *DCA*, 3 Aug. 1923.

95 W. Moyes, *The Banner Book: A Study of the Banners of the Lodges of the DMA* (Newcastle-upon-Tyne, 1974), 69.

96 J. Lawson, *Peter Lee* (1936), 239–40.

7. Monkwearmouth Lodge Banner, 1986
Reproduced courtesy of Derrik Scott.

It was a concern with the past that would continue throughout the century. 'Here was a walking museum,' enthused David Douglass in the early 1980s, 'A history book so vividly illustrated that you were left with a perfect chronology of struggle and conciliation.'[97] In the years after the Second World War, many of the official gala programmes chronicled and thus reinforced such a history. Frequently, the final dozen or so pages were simply taken up with long lists going back to the nineteenth century: of the names of all the miners' agents; of gala dates and invited speakers; of strikes and stoppages; of accidents, explosions, and numbers killed; of cases dealt with by the union; of changing costs of living; of wages.[98] By the 1980s, another list had been added: of pits closed since nationalization.[99] In time, the memory of 1926 would also take its part

[97] Douglass, 'Worms of the Earth', 62.
[98] See, for example, gala programmes for 1948, 1950, 1960, 1970, 1975, 1978–80.
[99] Gala programme, 1987.

in the pantheon and, in 1986, Monkwearmouth lodge commissioned a new banner on which A. J. Cook sat side by side with Arthur Scargill and Tommy Hepburn, the man who had led the Durham miners in their first attempt at building a union in the 1830s. Underneath, the motto read, 'As in the past we fight for the future' (see Illustration 7). Men and women absorbed the heroic image even if they could not grasp the political or social significance, like the old man who came to Durham on gala day to get his 'reets' [rights]: 'We dinnat knaa what they are, but we've come to get them.'[100]

In 1926 itself, the financial circumstances of the DMA meant that the gala had to be cancelled. Local lodges levied their members throughout the year to create a gala fund and by the end of June 1926 some lodges had already dipped into their savings; others wanted to cancel the event and disperse the windfall. Instead, an unofficial gala was organized by Burnhope lodge. Fifty lodges and about 35,000 people attended. Maurice Ridley remembered:

> I walked there with other young locked out miners and our parents, and it was my first real experience of the leadership that was really in charge of our struggle. I should say that the speech of Arthur Cook . . . affected me for the rest of my life . . . Even though it wasn't held in Durham [City], it was recognized that this meeting taking place at Burnhope was to be addressed by the leadership of the miners and the politicians within the labour movement who were supporting us. For any ordinary trade unionist it would have been a crime not to be able to go. I mean you just automatically had to go because you were in the midst of the struggle.[101]

## IV

Maurice Halbwachs, the sociologist who pioneered the study of collective memory, was frequently criticized for his dismissal of individual psychology as an independent influence upon memory. With reference to this, Fentress and Wickham explained that they preferred to use the term 'social memory' to indicate a conception of memory which, 'while doing full justice to the collective side of one's conscious life, does not render the individual as some sort of automaton, passively obeying the internalized collective will'.[102] In fact, Halbwachs would also have

100 Cited in Moyes, *Banner Book*, 69.

101 M. Ridley, 'Making a Contribution', 60.

102 Fentress and Wickham, *Social Memory*, p. ix.

rejected any conception of the individual as automaton. Although he argued that it was 'always necessary to revert to a combination of influences that are social in nature,' he suggested two different types of memory: individual and collective (or 'autobiographical' and 'historical', as he later went on to describe them):

> In other words, the individual participates in two different types of memory, but adopts a quite different, even contrary, attitude as he participates in the one or the other. On the one hand, he places his own remembrances within the framework of his personality, his own personal life; he considers those of his own that he holds in common with other people only in the aspect that interests him by virtue of distinguishing him from others. On the other hand, he is able to act merely as a group member, helping to evoke and maintain impersonal remembrances of interest to the group. These two memories are often intermingled.[103]

Any exploration of the memories surrounding 1926 has to be similarly layered. The collective biography of the mining community was certainly influential, but personal memories of the strike varied greatly depending on particular experiences. An individual could even interpret different strikes in different ways, depending on the point at which they took place in his life. One old miner from Horden remarked upon a huge contrast between the 1926 and 1984–5 conflicts. He remembered 1984–5 as a strike marred by trouble with the police, rifts caused by blacklegs, an egotistic national leadership, and the hardship caused by the lack of money and coal. He could remember no trouble with the police in 1926, declared that there had been no blacklegs where he lived, in Sacriston,[104] expressed no opinion of A. J. Cook, and was grateful that the union had always ensured that families had an excess of coal. 'In 1926,' he stated, 'Everybody mucked in, helped . . . [But,] it was everybody for himself, practically, in '84. There were people stabbing backs in '84 . . . In the '26 strike in Sacriston, everybody pulled together there.' But, in 1926 he was a child of 5, and his memories of the strike are perhaps half-remembered ones, or those that parents and older siblings later told him. In 1984, he was a grandfather; retired, but with two sons-in-law who worked at the pit, and it fell to him to help them out as much as he could with his savings and pension.[105]

[103] M. Halbwachs, *The Collective Memory*, introd. by Mary Douglas (New York, 1980), 48–51.

[104] A total of 196 men, not including safety men and officials, were working at the Charlaw, Sacriston, and Frankland group of collieries by the end of Nov. *DCA*, 26 Nov. 1926.

[105] Interview with R. M., Horden, County Durham, 17 Sept. 2005.

Yet, despite the myriad personal experiences, it is clear that some kind of recognizable common account emerged from the events of May to November 1926. In Durham, a constructed memory of 1926 could be easily slotted into a history that already encompassed 1920–1, 1912, 1892,[106] and further back, by a community already used to militant commemoration. At the unfurling ceremony of Heworth lodge's new banner in August 1926, for example, W. P. Richardson noted that the dispute was currently the severest in the area since the stoppage of 1844, which had lasted twenty weeks.[107] Collective memories merged with and reinforced individual ones, and thus a former Blackhall pitman, born in 1900, could explain years later that 'during the 1926 strike, people were thrown out of their houses. You got a fortnight's notice, and if you weren't out they put the bailiffs in to force you out.'[108] As one who had been a young man, not a child, in 1926, he might have been expected to know that in fact no families were evicted from colliery houses in 1926—or in 1921. It may be that he was confusing the national strike with a local dispute during which evictions may have taken place (although I have no knowledge of any such happening), but eviction as a strategy of the owners against strikers had not been used in any widespread way since the late nineteenth century, before this man was born. Evictions were, however, part of the folklore of the coalfield, even memorialized in song.[109] And, in this interview at least, it is what he 'remembered' most vividly about the lockout.

Jon Lawrence has examined the mythologized history of the Labour Party and described myths that were varied, overlapping, and constantly shifting.[110] A similar process was ongoing amongst the mining communities and, in 1926, as later, some dissenting voices challenged a single homogenized conception of the community's past. In September 1926, for example, a letter published in the *Stanley News* employed the militant legacy of the Durham miners not to stress a romantic history but to repeat the familiar complaint of the 'thrifty miner', who resented

106 The Durham and Northumberland miners did not participate in the otherwise national strike of 1893, but had fought a three-month strike of their own the previous year.

107 *DC*, 28 Aug. 1926.

108 J. Stephenson, *Being Human* (Durham, 1986), no page.

109 For example, 'Oakey's Strike,' and 'South Medomsley Strike', reproduced inT. Armstrong, *Tommy Armstrong Sings*, with intro. by T. Gilfellon (Newcastle-upon-Tyne, 1971), 40–1, 50–1.

110 J. Lawrence, 'Labour—The Myths it has Lived By', in D. Tanner, P. Thane, and N. Tiratsoo (eds), *Labour's First Century* (Cambridge, 2000), 341–66.

the payment of outdoor relief to the families of strikers, relief that he himself did not benefit from (and may have contributed to):

Remembering as I do the coal stoppages of 1879 and 1892 in the County of Durham, and having passed through all the stoppages since those dates, I wonder what has happened to the morals and self-respect of the Durham miner? . . . in 1879 and 1892 not five per cent of the miners sought relief from the Guardians for themselves or their families.[111]

Indeed, the specifics of 1926 themselves could sometimes be forgotten if circumstances demanded it, as the ringing endorsement of Ramsay MacDonald's candidature at Seaham in 1931 showed. However, this was a society in which the image of miners being obsessed by history had become in turn part of the miners' collective memory. Sid Chaplin declared: 'Well, the one thing about being born into a mining community is that "ye knaa whe [who] ye are". You know where you spring from, you know who you belong to, your roots are firmly embedded. In fact at times you feel imprisoned in a past that isn't entirely yours, a past that belongs to the community.'[112]

This was partly facilitated by the determination of the union to build a 'usable' past, aware that to commemorate collective achievements in the past helped to ensure them in the future. But, it was also due to the exceptional history of the miners, which attracted and encouraged commemoration in a way unlike many other working-class occupational groups. All history is subject to political interpretation, and the 'history of the history of the miners', as Bill Williamson has noted, was 'no different in this respect from that of engineers, steelmen, seamen or agricultural labourers, except, of course, in that, at least in the North East, there are many more histories of miners than of any other group'.[113] By 1926, the Durham miners had already had the story of their union told four times in the previous half century.[114] No other working-class group had endured such long strikes as the miners; no other workplace

[111] *Stanley News and Consett Chronicle*, 9 Sept. 1926.

[112] S. Chaplin, 'Durham Mining Villages', in M. Bulmer (ed.), *Mining and Social Change* (1978), 60.

[113] B. Williamson, 'Living the Past Differently: Historical Memory in the North East', in R. Colls and B. Lancaster (eds), *Geordies: Roots of Regionalism* (Newcastle-upon-Tyne, 2005 edn.; first pub. 1992), 162.

[114] R. Fynes, *The Miners of Northumberland and Durham: A History of their Social and Political Progress* (Sunderland, 1923 edn.; first pub. 1873); J. Wilson, *A History of the Durham Miners' Association, 1870–1904* (Durham, 1907); S. Webb, *The Story of the Durham Miners (1661–1921)* (1921); E. Welbourne, *The Miners' Unions of Northumberland and Durham* (Cambridge, 1923).

accident could catch the headlines as dramatically as a big pit disaster. Walter Citrine later attempted to defend the record of the TUC in 1926 with reference to this, explaining that 'the miners were accustomed to strikes: many other workers were not. The miners scarcely felt that they were really out until some months had elapsed. In . . . railways and road transport, there was a totally different psychology.'[115]

When, in the 1970s, a woman donated a bundle of papers to Durham County Record Office relating to a soup kitchen set up at Springwell during 1926 she commented in her cover letter, 'It is rather ironic, don't you think, that their names will be recorded for their labour in a soup kitchen, and not for their work and skill as miners.'[116] But, the work of the miner was already conceptualized in a romantic form and so such exceptional events further enhanced a militant, heroic historical past.

'These are epic days,' asserted A. J. Cook, writing in *The Miner* in October 1926, 'We of this generation shall be remembered in spite of ourselves.'[117] And so, despite the variety of experiences, 1926 remains a powerful reference point. Like the woman whose words began this chapter, for one old man, '1926' held a meaning comparable even to the events of the Second World War:

> These hard days, you know, they taught us to stick things out, and in later years when conditions were very bleak, and when for five years some of us stood out against the filth, the hunger, the insults, the degradation and humiliation which was meted out daily in prisoner of war camps . . . We had experienced it all before so it didn't affect us a great deal, certainly not as much as it did those lads who'd been more fortunate than us in the past.[118]

Another made a direct comparison: 'Following the collapse of the General Strike, there seemed to be born a feeling of isolation among the mining community; a back-to-the-wall attitude that I never experienced again until the summer of 1940, when it became apparent on a national scale.'[119] It is striking that he chose to compare the mining lockout with the collapse of France and the retreat from Dunkirk, when individual experiences and social memory would again both contradict and reinforce each other, and myth and reality merge and overlap.

115 W. Citrine, *Men and Work: An Autobiography* (1964), 130.
116 DRO, D/X411/173, C. Wardle to DRO, Mar. 1973.
117 *Miner*, 23 Oct. 1926.
118 GCLOT, iii (T.L.).
119 Hitchin, *Pit-Yacker*, 81.

# Conclusion

Despite the much more imaginative approach to coalfield history that has developed in recent years, a romantic tradition still pervades historical studies of 1926. In a recent collection of essays (and the most dedicated study of the lockout to date), one author celebrated the solidarity of the South Wales coalfield. In addition to the low rate of blacklegging, she suggested that during the lockout 'there was a strong moral code in which stealing from wealthy outsiders was sanctioned, whereas stealing from members of the mining community would have meant that the whole family would be ostracized or even driven out'.[1] It seems unlikely that this was true in South Wales; it was certainly not the case in Durham, where the image of a united community sharing a homogeneous set of beliefs, aspirations, and moral codes was not one that matched reality. Differences arising from gender, age, religion, respectability, and individual hopes and aspirations conditioned the outlook of the various inhabitants of the mining settlements, and much of the evidence recounted in this study has demonstrated the importance of such divergent influences. Meanwhile, a society in which men might steal from each other in times of hardship resonated strongly enough to be echoed in contemporary fiction.[2]

However, in 1926, these alternative identities and concerns did not prevent a solidaristic response amongst the vast majority of men and women as part of a common struggle. DMA records show that, by the end of November, 129,281 men remained eligible to vote in the coalfield ballot concerning the owners' proposals. It is an accurate minimum

[1] S. Bruley, 'Women', in J. McIlroy, A. Campbell, and K. Gildart (eds), *Industrial Politics and the 1926 Mining Lockout* (Cardiff, 2004), 238.

[2] A. J. Cronin, *The Stars Look Down* (1935), 21–2. For examples of the stealing that went on in the Durham coalfield, see Chap. 2, Sect. VI. Other commentators have also expressed surprise that, in the worst of times, even miners might steal from one another. See, for example, D. Geary, 'Unemployment and Working-Class Solidarity: The German Experience, 1929–33', in R. J. Evans and D. Geary (eds), *The German Unemployed: Experiences and Consequences of Mass Unemployment from the Weimar Republic to the Third Reich* (1987), 274.

number of those still on strike, as blacklegs were disenfranchised.[3] In Durham, despite misgivings, stealing, and suspicions, and despite a catalogue of other influences that threatened to fracture any cohesive sense of a 'mining community', the strike remained solid. Explained in the earlier accounts of labour historians as the natural result of the exemplary class consciousness of the miner, in the light of these divisions, such solidarity becomes a much more difficult phenomenon to understand.

## I

In the 1920s, life in County Durham was fundamentally entwined with the coal that lay beneath it. From their youngest days, those who grew up in the mining villages were aware of the local colliery and its affairs as an integral part of daily life. Sid Chaplin remembered that as a child in the 1920s 'coal soaked into your being, if only from the interminable coal talk, good and bad cavils, coal wet, coal dry, coal you could push down with the toeplate of your boot . . . .'[4] For Chaplin and his peers, the colliery was not only the place of employment for fathers, elder brothers, neighbours, and—not least—for their future selves, but its presence dominated the village. The pit made its physical mark on the men and, although the blackness that covered them when they emerged from their shift at the end of the day could be washed off, the blue scars worn by a pitman would identify him as such until he died. The wider physical landscape was darkened by the stark forms of the hundreds of pitheads and slag heaps that were scattered across the county, while the noise they generated became the accompaniment to daily life.

The sheer number of men employed in or around the pits, often living in purpose-built colliery houses, erected in villages which owed their existence to coal, meant that their concerns dominated the political and social life of the region. Even for those not formally connected with the coal industry, fortunes could turn on its success or failure, from the small shopkeepers dependent on the custom of miners' wives, to those other members of the working classes whose own employment was put

[3] Durham County Record Office (henceforth DRO), D/DMA109, result of ballot on owners' proposals, 29 Nov. 1926. The actual number would have been higher as half-members of the DMA (those under 18 years of age) were not permitted to vote in union ballots, though they were expected to come out on strike.

[4] S. Chaplin, 'Durham Mining Villages', in M. Bulmer (ed.), *Mining and Social Change* (1978), 68.

at risk when the collieries went into decline. The miners' numerical superiority, combined with a near-total union membership, also allowed the culture of the miners' union to pervade the region, both in industrial and pastoral matters. 'Chopwell miners' lodge was Chopwell village,' recalled one old miner fifty years later, admittedly describing a particularly militant lodge.[5] It was a dominance that was frequently recalled with great pride. Another old miner remembered with enthusiasm: 'in those days, the miners' lodge was a law which people regarded much more seriously than any laws of the land. The decision of the miners' lodge was binding on everybody, not only the miners but on the miners' wives and on everybody in the village.'[6]

This is not to suggest that alternative definitions of community did not exist in Durham in the 1920s. For the relative few who belonged to the middle class, for example, 'Durham' tended to mean Durham City rather than County Durham and, for a significant proportion of these, the most meaningful community was that which revolved around the university and its colleges. Their concerns were often very different from those of the pit villages, and when municipal elections were held for six vacancies on the city council in November 1926, two candidates standing for the Ratepayers' Association were amongst those elected. Their success was attributed at least in part to their avowed hostility to plans for the development of a new bus stand in the city centre, a project that promised to increase local rates and had attracted numerous complaints for weeks.[7] Indeed, on gala day, the one day each year in which the miners extended their numerical superiority into Durham City itself, the symbolism was noted by the local press, which reported each summer on the miners' 'invasion'.[8]

However, such different conceptions of community were not equally relevant to the miners, nor were all equally acceptable. In the pit villages, the nature of community was determined by the dominant group, not necessarily by a social elite; it therefore remained something that was defined by the strikers and was heavily shaped by the culture of the DMA. Nor did such a 'community' necessarily engender the same emotional attachment amongst all of a pit village's inhabitants. In his dictionary of 'keywords', Raymond Williams suggested that what is perhaps most important about the term 'community' is 'that unlike all other terms

[5] Gateshead Central Library, Oral History Transcripts, iv (B.M.).

[6] Ibid., iv (Mr S.). [7] *Durham County Advertiser* (henceforth *DCA*), 5 Nov. 1926.

[8] See, for example, ibid., 3 Aug. 1923.

of social organization . . . it seems never to be used unfavourably, and never to be given any positive opposite or distinguishing term'.[9] Yet, as Calhoun has argued, it is important not to forget 'the restraining which real community requires, the sacrifices which it demands, and the fears which enforce them'.[10]

Amongst the Durham pit villages, means of escape were available to a few, whether through education, sport, religion, or politics. Many more could not benefit from such opportunities. The Durham mining communities were not egalitarian and, for women in particular, social and cultural opportunities were severely limited. Men might also suffer, and in *The Back-to-Backs*, the novelist J. C. Grant painted a lonely portrait of Tom Shieldykes, a young miner whose sensitive nature was out of place in the fictional Durham village of Hagger-le-Hell:

> He kept a sort of illustrated diary of events in the outer world, cutting pictures from the newspapers and little items of news, and pasting them in old copy-books and ledgers . . . He yearned for the far-away world beyond the swamps, although he knew that Hagger would hold him fast to the end. Yet, he could travel there by way of his scrap-books any time he wished . . . He could get away from Hagger, though Hagger owned him still.[11]

Grant's novel was furiously denounced by Labour leaders who objected to his portrayal of pit villages that were unremitting in their horror.[12] But, even those who hated pitwork, those who resented the accident of birth that had placed them in a Durham mining village, still felt tied to the unwritten rules of the mining community. If the number of men who blacklegged in 1926 remained tiny, many more must have considered it in the smallest hours of the night; if the numbers who fulminated against the policy of the MFGB or even the DMA in the letters pages of local newspapers were relatively few, many more must have refrained from voicing their doubts, or confined them to private conversations with friends. Yet, despite misgivings, these men, too, remained on strike for seven months.

In such an environment, rather than conflicting identities damaging any sense of communal solidarity, they instead tended to complement

[9] R. Williams, *Keywords: A Vocabulary of Culture and Society* (1988 edn.; first pub. 1976), 76.

[10] C. J. Calhoun, 'History, Anthropology and the Study of Communities: Some Problems in Macfarlane's Proposal', *Social History*, 3 (1978), 369.

[11] J. C. Grant, *The Back-to-Backs* (1930), 25–6.

[12] For a discussion of *The Back-to-Backs* and its reception, see A. Croft, *Red Letter Days: British Fiction in the 1930s* (1990), 75–80.

a dominant occupational culture. The numerical superiority of miners throughout the county as a whole, let alone in individual pit villages, meant that to some extent alternative centres of influence had to adapt to suit their mining constituents. Within close geographical boundaries, it was through the culture of the union and the more humdrum, day-to-day culture of the miners that other local identities were filtered. A religious faith, for example, did not therefore constitute an independent, external value system that sought to impose a coherent belief-system in opposition to one of occupation or class. Rather, it was a set of values that was forced to adapt to a region in which the specific occupational interests of miners could never be ignored. This is not to suggest that it lost its distinctiveness, or that it ceased to remain an important point of reference for many miners and their families. But, as E. P. Thompson commented with regard to the Methodist labour leaders of the early nineteenth century, the 'Methodist political rebel' simply 'carried through into his radical or revolutionary activity a profound moral earnestness, a sense of righteousness and of "calling", a "Methodist" capacity for sustained organizational dedication and (at its best) a high degree of personal responsibility'.[13]

A similar pragmatic accommodation was also made within the elementary schools, which, rather than providing a separate source of identity, were often mobilized to support the needs of the miners. Meanwhile, one of the most important stratifications was that of gender. With regard to South Wales, Chris Williams has suggested that this was 'the most permanent and critical fracture' in its coalfield communities, one which 'conditioned and perhaps constrained the lives of virtually every individual'.[14] In Durham, gender divisions were also important. Men and women had very different roles and expectations in 1926, retaining a distinctive sense of their separate identities. However, such divisions occurred within an already constructed solidarity, and always remained subsidiary to the fact that both groups were fundamentally supportive of the strike. On the whole, women seem to have supported their men, often in the conviction that the cause for which the union was fighting was their cause too. Finally, a basic loyalty to the local lodge was able to shroud even the most unlikely political differences and allow apparently contradictory ideological positions to coexist. Both the lodge

[13] E. P. Thompson, *The Making of the English Working Class* (1991 edn; first pub. 1963), 431.

[14] C. Williams, *Capitalism, Community and Conflict: The South Wales Coalfield, 1898–1947* (Cardiff, 1998), 76.

official who volunteered as a special constable and the Conservative-voting striker, for example, continued to adhere to the basic principle of union solidarity—at least to the DMA, if not always to the MFGB.

Blacklegs were the exception to this, and the numerical inferiority of such men was disproportionate to their impact upon the mining village as a whole. Their decision to return to work created the bitterest divisions of the dispute, splitting families and tarnishing reputations for decades. But, by returning to work, such men forfeited their place within the mining community. Perhaps the best example of this was the fact that in 1927 the workmen of Adventure Colliery and their families planned their own separate outing to South Shields to take place on gala day. During the lockout, the high rate of blacklegging at Adventure Colliery eventually led to its split from the DMA. Its non-attendance at the gala, the ultimate expression of a united mining community, now became a physical symbol of its isolation.[15]

## II

Solidarity was not a feature of mining communities everywhere, and in other coalfield settlements, whether in Britain or abroad, it could falter when differing identities proved irreconcilable. In particular, Durham lacked the ethnic and racial conflicts which could lead to critical fractures elsewhere, particularly in the mining townships of South Africa or the southern American states, where such divisions could rarely be overcome.[16] In Imperial and Weimar Germany, race, religion, and political beliefs acted in concert to break down a cohesive union identity, as in the coal-mining areas of the Ruhr. There, the three main miners' unions reproduced the divisions of a civil society that was 'rigidly segmented by ethnicity, religion and politics'.[17] Ethnic and religious differences could also hinder unity within British coalfields.

[15] *DCA*, 13 Aug. 1927.

[16] See, for example, P. Alexander, 'A Moral Economy, an Isolated Mass and Paternalised Migrants: Transvaal Colliery Strikes, 1925–49', in S. Berger, A. Croll, and N. LaPorte (eds), *Towards a Comparative History of Coalfield Societies* (Aldershot, 2005), 239. Jeremy Krikler has explored the 1922 rising amongst gold miners in the Rand, whose famous slogan, 'Workers of the world unite and fight for a white South Africa', was only the most obvious evidence for his observation that 'Looming over what I found, suffusing all its phenomena like an inescapable fog, was race.' See J. Krikler, *White Rising: The 1922 Insurrection and Racial Killing in South Africa* (Manchester, 2005), p. x.

[17] L. James, *The Politics of Identity and Civil Society in Britain and Germany: Miners in the Ruhr and South Wales, 1890–1926* (Manchester, 2008), 31.

Alan Campbell has argued that the presence of both Sinn Fein activists and Orange lodges in many of Lanarkshire's mining communities meant that 'even if its more violent manifestations fluctuated over time and eventually subsided, sectarian pageantry was regularly visible'.[18]

Elsewhere in Britain, other factors impinged upon the development of a cohesive occupational community. It is perhaps no coincidence that the primary area of study for both Patrick Joyce and Trevor Griffiths, each of whom played down (although ultimately did not deny) the importance of class, was the North West of England.[19] Here, Peter Clarke has described a working class 'characterised by Conservative politics and aggressive Churchmanship' in the years before the First World War.[20] Even if the politics of Protestantism were fading by the interwar years, the greater number of women in employment continued to create a different gender dynamic; a significant Irish presence disrupted ethnic and religious homogeneity; and paternalistic involvement in leisure and cultural activities reduced the ability of cotton and mining unions to create a unitary discourse.

The comparative study of different coalfields is fraught with difficulty. Writing in the late 1940s, Ferdynand Zweig remarked that, 'nowhere can we find a greater range of variation than in the work and life of the coal miner. From coalfield to coalfield, from county to county, from colliery to colliery, from village to village you find astonishing and perplexing differences.'[21] Fifty years after Zweig, and at a conference gathered for the specific purpose of comparing mining communities across the world, another scholar suggested that in the light of increasingly detailed regional and local studies, perhaps the only generalization that continued to remain valid 'is that mining is always a hard and dangerous occupation'.[22]

Nevertheless, certain key themes can be identified. The importance of a union identity in the shaping of a specifically mining community is often presented as fundamental to overcoming divisions, particularly

[18] A. Campbell, *The Scottish Miners, 1874–1939*, i, *Industry, Work and Community* (Aldershot, 2000), 39.

[19] P. Joyce, *Visions of the People: Industrial England and the Question of Class, 1848–1914* (Cambridge, 1991); T. Griffiths, *The Lancashire Working Classes, c.1880–1930* (Oxford, 2001).

[20] P. Clarke, *Lancashire and the New Liberalism* (Cambridge, 1971), 74.

[21] F. Zweig, *Men in the Pits* (1948), 39.

[22] A. Shubert, 'A Divided Community: The Social Development of the Austrian Coalfields to 1934', in K. Tenfelde (ed.), *Towards a Social History of Mining in the Nineteenth and Twentieth Centuries: Papers Presented to the International Mining History Congress, Bochum, 1989* (Munich, 1992), 284.

with regard to the integration of 'outsiders'. Roger Fagge has suggested that the acceptance of the influx of Spanish immigrants into the South Wales coalfield after 1907 was hastened by the fact that they were active within the SWMF, while hostility towards foreign-born miners in West Virginia decreased due to their prominent role in strike activity and their support for the United Mine Workers of America.[23] Similar conclusions were drawn by John Laslett, who has argued that before the First World War multiethnic unity in Illinois was 'given a significant boost' by the constructive role that new European immigrant miners played in the American coal strikes of 1894 and 1897.[24] Also crucial to the power of the union was its degree of control over civic space. If the union was prevented from influencing such space, its ability to shape a collectivist discourse could be severely hindered, as historians have documented in areas as diverse as West Virginia, southern Australia, and the British Midlands.[25]

In Durham, of course, the union was able to exert a considerable influence over life outside the workplace, and was well-established within the industry. In this respect perhaps the most valid region of comparison is with South Wales, which has immediate similarities with its Northern counterpart. Like Durham, it was a large exporting coalfield that suffered more than most during the interwar years. Mining and miners dominated its social and cultural ethos; the union and Labour dominated its political life. Analyses of the meanings of community in Durham can therefore find parallels in David Gilbert's comparative work on the coalfields of Nottinghamshire and South Wales. He observed that, in 1926, the Nottinghamshire boards of guardians faced considerably fewer requests for relief than their counterparts in South Wales, but suggested that this was not because of a lesser degree of poverty in the area, nor because the Nottinghamshire miners possessed more self-respect (although the clerk to the Mansfield Board of Guardians congratulated mining families on keeping themselves 'clear of all taint of pauperism'). Rather, he argued that such differences were the result of the contrasting

[23] R. Fagge, *Power, Culture and Conflict in the Coalfields: West Virginia and South Wales, 1900–1922* (Manchester, 1996), 79, 68.

[24] J. H. M. Laslett, *Colliers across the Sea: A Comparative Study of Class Formation in Scotland and the American Midwest, 1830–1924* (Urbana, Ill., 2000), 138–9.

[25] Fagge, *Power, Culture and Conflict*, 39–52; M. Fletcher, '"Slaves of the Lamp?": Independence and Control in Two State Coal Mining Communities in Victoria, Australia', in Berger et al., *Towards a Comparative History*, 99–112; R. Waller, *The Dukeries Transformed: The Social and Political Development of a Twentieth-Century Coalfield* (Oxford, 1983), esp. 102–3.

social environments of the two coalfields. In South Wales, lodge officials provided advice to those receiving relief and coordinated applications to the boards of guardians, which were often dominated by Labour men anyway; this went some way towards creating a collective response to poverty and reducing its social stigma. In contrast, Gilbert suggested that the reactions of Nottinghamshire mining families were more 'privatized': 'without strong community leadership, in a more mixed social environment with different local mores, the Nottinghamshire miners were more likely to resort to individualistic solutions, such as drawing upon savings, or relying on a second wage (often female), or commercial outcropping.'[26]

Gilbert's work does not entirely escape another of the pitfalls of comparative history—namely, the tendency to homogenize individual regions. Some parts of the Nottinghamshire coalfield did possess a strong union presence and a solidaristic mining culture. Nor were 'privatized' coping strategies a phenomenon restricted to Nottinghamshire: examples of such could be cited from the North East. Nevertheless, his analysis has important implications for any examination of Durham, where, as in South Wales, a similarly dominant miners' union was able to confer legitimacy on a range of wider social and political activities.

However, in what remains a significant and provocative history of the SWMF, Hywel Francis and Dai Smith suggested in the strongest of terms that the culture of the South Wales coalfield was significantly different from the rest of British society; that during the lockouts of 1921 and 1926 'what emerged was in the manner of an alternative culture with its own moral code and political tradition: it was a society within a society'.[27] They claimed that South Wales was exceptional in this regard:

> The nine days of the general strike and more especially the seven month lockout revealed an alternative cultural pattern which had no comparable equivalent in the other British coalfields. The totality of commitment to the miners' cause was a form of class consciousness which translated itself into a community consciousness, so overwhelming were the miners in numbers and influence.[28]

[26] D. Gilbert, *Class, Community and Collective Action: Social Change in Two British Coalfields, 1850–1926* (Oxford, 1992), 197.

[27] H. Francis and D. Smith, *The Fed: A History of the South Wales Miners in the Twentieth Century* (Cardiff, 1980), 66.

[28] Ibid., 54–5. Both authors reiterated such views in their individual works. See D. Smith, *Aneurin Bevan and the World of South Wales* (Cardiff, 1993); H. Francis, *Miners Against Fascism: Wales and the Spanish Civil War* (1984).

In fact, in 1926, the Durham coalfield possessed many of the same attributes that Francis and Smith claimed were representative of an 'alternative culture'. They cited collectivist endeavours such as the setting up of communal kitchens and boot centres, working outcrops for coal, agitation towards poor law guardians, and demonstrations against blacklegs;[29] the present study has documented all of the above in Durham, where miners were also 'overwhelming' in numbers and influence. Indeed, it could be argued that the collectivist culture seen in Durham in 1926 exceeded that of South Wales, as it maintained a lower proportion of blacklegs throughout (see Table 2.1).

However, in 1926, Durham may have matched or even exceeded South Wales in its support of the strike, but the Welsh coalfield continued to attract a more lasting reputation for militancy. In 1912, the Unofficial Reform Committee of the SWMF produced the famous syndicalist pamphlet, *The Miners' Next Step*;[30] in 1917, a call for the abolition of capitalism was added to the union's rulebook;[31] in the early 1920s, the SWMF was alone amongst the miners' unions in attempting to persuade the MFGB to affiliate to the Red International of Labour Unions.[32] In the interwar years as a whole, the miners of South Wales were the most strike-prone of any British coalfield, and considerably more so than the miners of Durham.[33] Although not a native-born Welshman, A. J. Cook had spent his adult life in the Rhondda, and for those who feared revolution in 1926, his became the face of radicalism. Many of his district-level colleagues could boast a similar reputation for militancy (and for some it was more deserved), with Noah Ablett, S. O. Davies, Arthur Horner, and Will Mainwaring amongst them.

This is not to paint an image of the Welsh miners as irredeemable militants, for South Wales, like Durham, had a strong Lib–Lab heritage. Even after the First World War, the coalfield of Arthur Horner was also

29 Francis and Smith, *The Fed*, 55–9.

30 D. Egan suggests that 'the fame, notoriety and indeed eloquence of this tract was to play a significant part in earning a reputation for the strength of syndicalist views among the South Wales miners'. Egan, '"A Cult of their Own": Syndicalism and *The Miners' Next Step*', in A. Campbell, N. Fishman, and D. Howell (eds), *Miners, Unions and Politics, 1910–47* (Aldershot, 1996), 17.

31 D. Geary, 'The Myth of the Radical Miner', in Berger et al., *Towards a Comparative History*, 45.

32 D. Howell, *MacDonald's Party: Labour Identities and Crisis, 1922–1931* (Oxford, 2002), 106.

33 See K. G. J. C. Knowles, *Strikes: A Study in Industrial Conflict* (Oxford, 1952), 185–97.

that of Arthur Jenkins and Frank Hodges,[34] and active Communists continued to work comfortably alongside orthodox Labour Party members within the SWMF. As in Durham, any radical ideology also continued to be reflected first and foremost through the lens of the union. Will Paynter was a young man in 1926 but was later to rise to prominence within both the SWMF and the National Union of Mineworkers:

> It has often been said of me that I was a miner and trades unionist first and a communist second . . . I have to admit that it has a good deal of truth in it . . . The Fed was a lot more than a trade union; it was a social institution . . . It is not surprising, therefore, that this kind of background produces a loyalty to the union so strong and primary that the union is regarded as a substitute for a political organisation.[35]

The nature of the left in South Wales therefore has to be carefully assessed, and indeed more recent historians have provided a more nuanced picture.[36] However, it remains the case that the politics of Durham were different. Possible explanations for the differences between the two coalfields have been explored elsewhere, notably by M. J. Daunton, who has considered both the conditions of underground work and the impact of housing policy in shaping the culture of the community before the First World War.[37] In a stimulating essay, Stefan Berger and Neil Evans have also argued that the conclusions suggested by comparative studies might be affected by the different historiographical traditions of the areas compared: 'South Wales has been portrayed as a left-wing, militant coalfield: but to what extent is this because its history has been written by left-wing historians?'[38] Their argument has implications for studies of the British coalfields, where the proliferation of studies on South Wales, particularly in comparison to Durham, perhaps contributes to an impression that South Wales was exceptional.

[34] Although Frank Hodges left the SWMF in 1919 to become secretary of the MFGB. He then left this post in 1923 when he was elected MP for Lichfield, Staffordshire.

[35] Cited in Egan, 'A Cult of their Own', 30–1.

[36] See, for example, C. Williams, *Democratic Rhondda: Politics and Society, 1885–1951* (Cardiff, 1996); M. Lieven, 'A Fractured Working-Class Consciousness? The Case of the Lady Windsor Colliery Lodge, 1921', *Welsh History Review*, 21 (2003), 729–56.

[37] M. J. Daunton, 'Down the Pit: Work in the Great Northern and South Wales Coalfields, 1870–1914', *Economic History Review*, 34 (1981), 578–97; and 'Miners' Homes: South Wales and the Great Northern Coalfield, 1880–1914', *International Review of Social History*, 25 (1980), 143–75.

[38] S. Berger and N. Evans, 'Two Faces of King Coal: The Impact of the Historiographical Traditions on Comparative History in the Ruhr and South Wales', in Berger et al., *Towards a Comparative History*, 39.

Another part of the answer might be found within union and political life. In conjunction with the NUR, the SWMF provided most of the funding and many of the students for the CLC, at least until 1926, when its financial position became more precarious.[39] Several of its leaders had taken part in the Ruskin College strike that led to the CLC's foundation; others were heavily involved in the Plebs League. Many more were inducted into radical politics through alternative organizations: there were thirty-eight Communist Party branches in South Wales by 1927, with 2,300 members.[40]

Most Durham leaders lacked a grounding in such political education, and there were few outlets for the teaching of Marxist theory amongst the rank and file. Unfortunately, records are scarce and any picture is reliant on scraps of biographical material. Jack Lawson attended Ruskin and Will Lawther attended the CLC, but it was certainly not the experience of many, even amongst the miners' leaders. It may be that the more moderate Durham union—although not without its own internal divisions—encouraged a greater cohesion than its more radical Welsh counterpart. Militancy is not the same as solidarity, and the Welsh coalfield was plagued by factionalism after the lockout, faced by a strong challenge from the breakaway non-political union. It is notable that the Scottish coalfield, which also acquired a militant reputation in the interwar years, was also a divided one, although there the main challenge came from the left, with the formation of the United Mineworkers of Scotland in 1929 (in fact significant Communist activity was largely restricted to Lanarkshire and Fife). Durham escaped such fractures. While the Welsh and Scottish unions also had considerable success in building a collectivist response to the lockout, therefore, their dominance over the hearts and minds of their constituents was neither greater nor less than that forged by the DMA.

## III

Over a decade ago, Alan Campbell, Nina Fishman, and David Howell observed that 'the recent emphasis within mining historiography on the diversity of local and regional experiences has been co-opted by

[39] For the links between the SWMF and the CLC see S. Macintyre, *A Proletarian Science: Marxism in Britain, 1917–1933* (Cambridge, 1980), 81–5.

[40] Williams, *Capitalism, Community and Conflict*, 58.

some "revisionist" historians to highlight the internal divisions and inherent fragmentation within the working class'.[41] This study is not a comparative one, but nor does it seek to continue a trend towards an ever more fragmented and isolated history of coalfield communities. Durham was not representative of every coalfield, but the actions of its men and women in 1926 demonstrated that a collectivist response could be achieved, and that ideologies could come together, however tenuously. Furthermore, it was precisely the local and regional experience that was fundamental in enabling such solidarity. The occupational and union-dominated identity that remained paramount in Durham was itself heavily mediated by the local social world in which miners found themselves; the product of an intensely local experience of poverty and hardship. The well-read, Ruskin-educated Jack Lawson, MP could therefore declare that he 'preached no abstract economic theory, not even Marx. I knew the problem better than any theorist, and had plenty of material at hand from day-to-day experience to point the moral.'[42]

Nine days at the beginning of May had provided a flash of wider national and class solidarities, and in historical accounts of the general strike the remaining six and a half months of the miners' lockout are frequently seen as an epilogue. For many Durham miners, however, the general strike was the prologue to their lockout: welcome, but essentially subordinate to their own local battle with the coal owners over wages and hours. Nearly twenty years later, Mark Benney would describe the effect of the strike in his fictionalized account, *Charity Main*:

> During the first fortnight the village was bewildered: this was not like any other strike they had known . . . Then came the news that the General Council of the TUC had called off the general strike. It was a bitter blow, but in a way a relief: the village could settle down again to the kind of strike it was used to.[43]

Even the Durham leaders were cautious of the more radical wing of the MFGB, and, like the majority of their membership, were reluctant to view the stoppage as unusual. During 1926, both the DMA and the DCOA therefore conformed to the normal procedures of an industrial dispute, distancing themselves from the strike committees and OMS respectively. Instead, both sides continued with their familiar routines, negotiating compensation payments, opening aged miners' homes, and participating in joint social events.

[41] A. Campbell, N. Fishman, and D. Howell, 'Introduction', in Campbell et al., *Miners, Unions and Politics*, 6–7.
[42] J. Lawson, *A Man's Life* (rev. 2nd edn., 1944), 117.
[43] M. Benney, *Charity Main: A Coalfield Chronicle* (1946), 119–20.

As the lockout progressed, a regional emphasis was accentuated rather than broken down, as mechanisms for survival came from local institutions, and both local leaders and the rank-and-file turned inwards. Even the wider concerns implied by a national struggle continued to be mediated through the DMA and never lost their relevance to local issues. The remarkable receptions accorded to A. J. Cook, for example, were not simply an indication of his personal magnetism, but reflected the fact that his message had a particular significance in a coalfield that would suffer more than most if the threatened increase in hours and reduction in wages came about. Even the social memory of the strike, and in 1926 the social memory of an earlier mining past, was formulated as a narrative specific to Durham.

The mining settlements in which men and women lived and worked meant that an occupational consciousness was constructed within strict geographical boundaries, which tended to militate against the development of a wider working-class identity. For most, if not all, of their lives they would carry out their social interactions within a predominantly mining culture. Even when geographic and social boundaries became more fluid, later in the century, the Durham miners would retain a sense of their exclusivity. As late as 1979, the Durham County Association of Trades Councils wrote to the DMA requesting that the annual gala be extended to include other trade unionists in the county, referring to the benefit that this would bring in counteracting the recent decline in gala attendance, and speaking of the many ex-mining people who, though no longer working in the mines, would like the chance to march again. The reply from the DMA was uncompromising: 'I need not remind you that the Big Meeting is uniquely a Miners' Gala . . . As you will gather, we are jealous of our traditions and our Big Meeting . . . If you understand our traditions you will have half expected a "no".'[44]

This focus on local identity and a localized, occupational community does not mean that class ceased to be important. Eric Hobsbawm considered the career of W. P. Richardson, who lived and worked at Usworth all his life, pointing out that 'this man, who was as rooted in his village as any Herefordshire milkmaid, helped to found the local ILP branch, joined the board of the *Daily Herald*, championed the nationalization of all mines and was to become the national treasurer of the Miners'

[44] DRO, D/X935/2/120, correspondence between Nickie Nicholson and Tom Callan, Aug.–Sept. 1979.

Federation'.[45] Rather, to many who belonged to the Durham pit villages, the miners *were* the working class. This is emphasized in Lawson's biography of Herbert Smith. Like Richardson, Smith had long been involved in the higher echelons of the national union, though he remained deeply attached to his Yorkshire background. Lawson explained:

> though he had throughout life a contempt for theorizing, the fact remains he was emphatically class conscious. To have said he was 'Miner-conscious' would have pleased him most. He would have understood that better. Miners are real: classes are abstract—and Herbert Smith had no room for abstractions.[46]

## IV

Too often the concepts of 'community' and 'class' have been seen as contradictory: one narrow and inward-looking, the other broader and outward-looking. Howard Newby, for example, in his study of agricultural workers, whilst acknowledging the complexity of concepts of community, suggested that its localist character permitted manipulation by a social elite: 'The traditional English landowning class placed an ideological gloss on their monopoly of power within the locality through the concept of "community" . . . [involving] notions of stability, harmony and social order.'[47] On the other side of the world, in his study of the Australian steel town of Port Kembla, Erik Eklund preferred the term 'locality' to that of 'community', defining this as an 'ideology that elevates local interests above all others and has the effect of creating alliances or coalitions of classes that obscure class interests and mediate class conflict'. He argued that localist and class politics existed in opposition to each other, and that politics at Port Kembla was characterized by the shifting relationship between the two.[48]

However, the relationship between 'class' and 'community' is more subtle than many commentators have assumed. In Durham, it was precisely through an experience of the local that the conceptualization of wider class or occupational identity was made possible. With regard to the political dominance of the Labour Party in the coalfields of the

45 E. Hobsbawm, *Worlds of Labour: Further Studies in the History of Labour* (1984), 199.

46 J. Lawson, *The Man in the Cap: The Life of Herbert Smith* (1941), 33.

47 H. Newby, *The Deferential Worker: A Study of Farm Workers in East Anglia* (1977), 52.

48 E. Eklund, 'The "Place" of Politics: Class and Localist Politics at Port Kembla, 1900–30', *Labour History*, 78 (2000), esp. 94–5.

interwar years, Duncan Tanner reflected upon their occupational and cultural unity, arguing that such factors made it easier 'to manufacture a language of class which was locally meaningful'.[49] A similar process affected the spread of union and collectivist ideology. In a study of the 1984–5 miners' strike, Andrew Richards suggested:

the overriding fact that power in the union was essentially locally based is key to understanding both its strength and weaknesses as an institution. Thus while locally based power often proved to be a source of resilience and intense loyalty in times of crisis, it also limited the institutionalization of a wider sense of identity, and constrained the power of national leaders.[50]

In the 1920s, the picture was more subtle still. Then, the power of the union was also locally based and this did indeed prove a strength in times of crisis. It also proved a simultaneous handicap to the national union, as miners identified first and foremost with their local lodge, and in turn rewarded their regional union with a much deeper loyalty than that given to the MFGB. However, as much as it was a handicap, the strength of the local lodge was also critical to the existence of wider solidarities and remained fundamental both to the construction of the DMA, and then to the MFGB beyond.

It was an attachment to the local lodge that provided the basic building block for imagining such wider loyalties. Sid Chaplin reflected upon the solidarity of a 'mining community':

two lodges raised their banners together in times of trouble and the men were bonded together just as certainly as the two pits were linked underground. And just as certainly as we linked up with Byers Green, and Westerton, and Willington, we all linked together with the DMA at Red Hills, Red Hills facing Coal Trade Hall across the Tyne.[51]

He could have presumably have extrapolated further, looking North to Scotland, South to Yorkshire and ultimately to the London headquarters of the MFGB at Russell Square. His words illustrate the way in which the lodge banner, which represented the community of the individual pit or village, helped its members to slot themselves into a wider jigsaw of lodges and loyalties. At the annual gala, at which hundreds of lodge banners processed around Durham City, accompanied by hundreds of

[49] D. Tanner, 'The Labour Party and Electoral Politics in the Coalfields', in Campbell et al., *Miners, Unions and Politics*, 77–8.

[50] A. Richards, *Miners on Strike: Class Solidarity and Divisions in Britain* (Oxford, 1996), 80.

[51] Chaplin, 'Durham Mining Villages', 70.

thousands of miners and their families, such a message must have been impossible to miss. Identities could thus be built one on top of the other. In the summer of 1926, various local newspapers ran anonymous adverts, urging miners to return to work. In August, one took up half a page in the *Durham Chronicle*: 'For your family's sake—for your own sake—for your fellow Trade Unionists' sake—for your Country's sake—GO BACK TO WORK.'[52] One could easily envisage an almost identical advert promoted by the union. Only two words would have to be added to the final sentence: DO NOT GO BACK TO WORK.

Even so, such a process was not automatic, and the construction of the MFGB, and to a lesser extent the DMA, remained a fragile one. Their building blocks had to be continually reinforced, and it was here that the social memory of the mining communities became most important. The construction of a usable past has frequently been represented as union-driven, and in 1926 the union certainly attempted to influence the way in which the past was remembered, at both a national and regional level. Its role should not be underestimated. As Patrick Joyce has commented, 'Union discourse was more than the reflection of the workers' world. It actively shaped it.'[53] But, the imagined community of miners was so powerful, both to insiders and outsiders, because a memory of the past was also being formulated at a much more basic and unconscious level. The hereditary nature of the industry was important in allowing a strong oral tradition to develop, while during the strike itself echoes of the past rumbled constantly in the background. Lawson believed that such a history was fundamental to any analysis of 1926:

> It is questionable if any other section of workers could, under the circumstances, have kept faith and maintained their organization intact. Only miners have such a compact, warm communal life. Only miners have such a history of struggles to remind them in the darkest night that daybreak will surely come. A century of discipline kept them steady when stern facts would have broken their ranks.[54]

## V

As Michael Savage has observed, it should not be assumed that solidarity flows automatically and inevitably from homogeneity.[55] Indeed, Robert

[52] *Durham Chronicle*, 28 Aug. 1926.

[53] Joyce, *Visions of the People*, 137.

[54] Lawson, *Man in the Cap*, 221–2.

[55] M. Savage, *The Dynamics of Working-Class Politics: The Labour Movement in Preston, 1880–1940* (Cambridge, 1987), 232.

Waller suggested that the isolation of the Dukeries pits and the monolithic domination enjoyed by the colliery companies was a major factor in the moderation of their employees and their adherence to the non-political union, in contrast to more heterogeneous colliery settlements elsewhere in the Nottinghamshire coalfield.[56] However, despite the problems associated with the theory of miners as an isolated mass, it appears that in Durham the social and physical separation of miners was an important factor in the development of a collective consciousness. Indeed, frequently such isolation was not purely a social and geographical phenomenon (many miners were not 'isolated' in this way at all), but it also embodied an emotional attitude. In the way in which Durham miners conceptualized their occupation, their history—even, in fact, their strikes—they continued to think of themselves as different. Recent scholarship has suggested that, despite this, other identities such as gender, religion, respectability, and (in places other than Durham) ethnicity were destructive to an overall sense of a mining community, breaking up its cohesion. However, in the light of the conclusions of this study, it is worth rethinking the concept of community itself.

Rather than the ideal type of mining community being one in which a homogeneous occupational identity existed to the exclusion of all others, it seems that the essence of community lay in its ability to subsume and integrate other categories of identity. Multiple identities still existed in the Durham mining villages of the 1920s, but they complemented rather than rivalled each other, and men and women rarely found themselves forced to choose between them. 'Durham produced both the "red village" and the dutiful miner,' Patrick Joyce has commented: a useful reminder that the Durham miner was not simply the militant class warrior who has tended to dominate the attention of historians.[57] Certainly it did, but whether miners were dutiful, red, or any number of positions in between, they were no less a part of the community.

However, while the strength of the 'mining community' in this new sense is fundamental to an explanation of local solidarity and the potency of the response demonstrated in 1926, it fails to explain wider loyalties to the MFGB, or even the DMA. For this, it is important to see the local community as simply the first building block for the imagination of wider solidarities, grounded in a historical metanarrative that had to be

[56] Waller, *Dukeries Transformed*, 130.

[57] P. Joyce, *Work, Society and Politics: The Culture of the Factory in Later Victorian England* (Brighton, 1980), 228.

continually reinforced. Rather than a homogeneous entity, the mining communities of Durham therefore consisted of interlocking layers of identity, placed one on top of the other. These layers remained fragile and were never absolute. Nor was their integration always smooth. Any conception of a wider mining community should not be seen as static, therefore, but rather as a social variable, in which an individual's identity and its expression varied according to time, place, and social context.

Perhaps the final words should be left to Sid Chaplin. The Shildon-born writer has been cited frequently throughout this study. Though he remained associated with the coal industry all his life, he was not the typical underground miner: he began work as an apprentice blacksmith; after the Second World War he worked as a reporter for the National Coal Board magazine. His novels and writings became famous for their portrayal of Northern working-class life, although Chaplin always disliked the tag he was inevitably awarded of 'regional writer'. His words sum up the image of a community in which multiple identities varied and overlapped, yet never lost their hold:

> In every industrial community the secret spring of life is identification. You identify with the family . . . your immediate mates, with your pit or your works, with your industry, union, your chapel and your street. It is a creative identification, but it has its dangers. One for me was that I went through life looking for something else to identify with—and I suspect this is true of most of my generation . . . I look at myself and find a chronic identifier. It is a terrible compulsion.[58]

[58] S. Chaplin, *A Tree with Rosy Apples* (Newcastle, 1972), 101.

# *Bibliography*

## MANUSCRIPT AND ARCHIVAL SOURCES


**Beamish Museum, County Durham**

Oral History Recordings: 1976/125; 1991/82; 1993/5; 1997/19; 1998/4; 1999/2.

**Bodleian Library, Oxford**

Sankey Papers: MSS. Eng. hist. c.504–11; e.273; e.280.

**British Library Sound Archive, London**

Oral History Recordings: C707/155/1–3C1; C900/11043C1; C900/05554C1.

**Clayport Library, Durham City**

Unpublished pamphlets and memoirs, series B/L.

**Dean and Chapter Library, Durham Cathedral**

Diary of Herbert Hensley Henson.

**Durham County Record Office, Durham City**

*Church Records*

All Saints' Church, Penshaw: EP/Pen.
Chester Moor Wesleyan Methodist Church: M/CS 56.
Perkinsville Wesleyan Methodist Church: M/CS 195.
St Andrew's Church, Blackhall: EP/Bla.
St Andrew's Church, Beamish: EP/Bea.
St Mary's Church, Trimdon: EP/Tr.
St Nicholas' Church, Hetton-le-Hole: M42/93.

*Durham County Association of Trade Councils Records*

D/X935/2/120.

*Local Government Records*

Blaydon District Education Sub-Committee: E/N/A7.
Chester-le-Street District Education Sub-Committee: E/NC/A6.
Chester-le-Street Petty Sessional Divisional Records: CS/PS.
Chester-le-Street Urban District Council: UD/CS.
Durham County Council Health Department Files: CC/H14.
Durham County Council Standing Joint Committee Minutes: CC/A10.
Durham County Council, Education Department: CC/Ed283; CC/Ed398; CC/E143/6.

Easington District Education Sub-Committee: E/NE/A5.
Stanley Urban District Council: UD/Sta.

*Marriage Registers (Parishes in Easington Deanery)*
Cassop cum Quarrington: M42/562.
Cornforth: M42/706.
Coxhoe: M42/736.
Deaf Hill: M42/745.
Easington: M42/475.
Ferryhill: M42/964.
Haswell: M42/764.
Hawthorn: M42/765.
Kelloe: M42/839.
Shadforth: M42/919–20.
Shotton: M42/589.
South Hetton: M42/432
Thornley: M42/792.
Trimdon: M42/471.
Trimdon Grange: M42/873.
Wheatley Hill: M42/798.
Wingate: M42/907.

*National Coal Board Records*
NCB1 Co86 (895); NCB/1/X124.

*Personal Papers*
Cuthbert Headlam Diaries: D/He 22.
Londonderry Papers: D/Lo/F613.
Robert Shotton Papers: D/Sho
Sam Watson Papers: D/DMA (Sam Watson).
Sidney Webb Papers: D/X1268/32–9.

*Political Organizations*
Durham County Teachers' Labour Group: D/X1099/4.
Durham Labour Women: D/X1048.
Durham Municipal and County Federation: D/MCF25; D/MCF175.

*Poor Law Union Records*
Chester-le-Street Poor Law Union: U/CS.
Easington Poor Law Union: U/Ea.

*School Records*
Bearpark Council School: E/C68; E/C/G6.
Blackhall Colliery Council School: E/E41.

Brandon Council Infants' School: E/C8.
Browney Council School: E/C/G155.
East Stanley Council School: E/St/G20.
Houghton-le-Spring Council School (Boys' Dept.): EP/Ho685.
New Seaham Council Girls' School: E/NE55.
Seaham Harbour Church of England Infants' School: E/NE81.
Seaham Harbour Church of England School: E/NE60; E/NE63; E/NE/F6.
Seaham Harbour Council Girls' School: E/NE36.
Seaham Harbour National School (Girls' Dept.): E/E10.
Seaham Harbour Roman Catholic School: E/NE31.
Seaham Harbour Upper Standard Council School (Boys' Dept.): E/NE69.
Sherburn Hill Council Infants' School: E/C62.
Shotton Colliery Council Boys' School: E/E43.
Wheatley Hill Council Boys' School: E/E90.
Wheatley Hill Council Infants' School: E/E82.
Wheatley Hill Council Senior Girls' School: E/E24.

*Springwell Men's Soup Kitchen*
D/X411/173; D/X411/172.

*Typescript Memoirs and Published Pamphlets*
I. A. Embleton, 'When I was Growing Up in a Pit Village'.
E. Farbridge, 'Recollections of Stanley'.
G. Purdon, *It was Grand Toffee: Horner's of Chester-le-Street Remembered*.
L. Turnbull, *Chopwell's Story*.
Souvenir gala programme, 1950.
Souvenir to commemorate DMA Jubilee, 1919.
L. Wild, *The Years Between*.
Durham Mining Records Project, miscellaneous: D/MRP.

*Union Records*
DMA Ascertainments, Balance Sheets, Notices, Posters, etc: D/DMA 108; D/DMA 109.
DMA Executive Committee files: D/DMA (Acc: 2157 (D)), 208 (box).
DMA minutes of arbitration committee: Library, C47–8; D/DMA 331/2.
Eden Lodge minutes: D/DMA 334/5.
Langley Park Lodge minutes: D/DMA (Acc: 2157 (D)), 187(vol).
MFGB Files: D/DMA (Acc. 2157(D)), 194 (box).

**Durham University Library**
Auckland Castle Visitation Returns, 1924: AUC/4/13.
Auckland Castle Visitation Returns, 1928: AUC/4/14.
Jack Lawson Papers

Data gathered by Prof. Robert Moore for his book, *Pitmen, Preachers and Politics: The Effects of Methodism in a Durham Mining Community* (Cambridge, 1974): Misc. accession 1996/7:* 4; Qualidata reference QD35/Moore.

**Gateshead Library**
General Strike Oral History Transcripts (4 vols.)

**John Rylands University Library, Manchester**
Ramsay MacDonald Papers: RMD/1/4/81; RMD/1/5/14.

**London School of Economics**
Beveridge Coal Commission Papers: MF Archives.
Independent Labour Party Files: ILP 6/16/5.
Sidney and Beatrice Webb Papers: Passfield 2/3/1, 2/4/H, 4/15, 4/5; Stoatley Rough School 8/4.

**Modern Records Centre, University of Warwick**
Central Labour College Papers: MSS.127; MSS.241.
Mining Records: MSS.88; MSS.292.

**The National Archives: Public Record Office, London**
Admiralty Records: ADM 116/2439.
Annual Reports of Inspectors of Mines: POWE 7.
Board of Education Papers: ED 18, 19, 21, 35, 50, 53, 54/23, 88, 99, 120.
Cabinet Office Papers: CAB 21/296.
Friendly Society Returns: FS 11/411, 12/10.
Home Office Papers: HO 144/6902; 144/6894.
James Ramsay MacDonald Papers: PRO 30/69.
Ministry of Health Papers: MH 57; 132.
Ministry of Labour Papers: LAB 27.
Rail Company Papers: RAIL 1057/2788.
Treasury Papers: T 172/1558.

**Nuffield College, Oxford**
G. D. H. Cole papers

**Trades Union Congress Library Collections, London Metropolitan University**
Ministry of Health Files: HV4546.
Workers' Educational Association Files: WEA/Central 4/2/1/1; WEA/CJAC 4/2.

**Tyne and Wear Archives Service**
Records of the Petty Sessional Division of Houghton-le-Spring: MG.HS/91/1.
'Reminiscences of a Walbottle Colliery Miner': DX 201/2.

## NEWSPAPERS AND PERIODICALS

*Blaydon Courier*
*Catholic Directory*
*Chester-le-Street Chronicle*
*Church Times*
*Colliery Guardian*
*Co-operative News*
*Crockford's Clerical Directory*
*Daily Telegraph*
*Durham Chronicle*
*Durham County Advertiser*
*Evening Chronicle*
*Guardian*
*Labour Magazine*
*Labour Woman*
*Miner*
*Morning Post*
*Newcastle Chronicle*
*North Mail*
*Seaham Weekly News*
*Stanley News and Consett Chronicle*
*Tablet*
*Wearside Catholic News*
*Workers' Weekly*

## OFFICIAL PUBLICATIONS

*Census of England and Wales, 1921*, ii, *County of Durham* (HMSO, 1923).
*Census of England and Wales, 1921, General Report with Appendices* (HMSO, 1927).
*Hansard Parliamentary Debates* (Commons), fifth series.
*Hansard Parliamentary Debates* (Lords), fifth series.
*Parl. Papers 1883*, lxxx (1), Census of England and Wales, 1881, iii (ages, condition as to marriage, occupations, and birthplaces).
*Parl. Papers 1893–4*, cvi (1), Census of England and Wales, 1891, iii (ages, condition as to marriage, occupation, birthplaces, and infirmities).
*Parl. Papers 1902*, cxviii (673), Census, County of Durham, 1901.
*Parl. Papers 1913*, lxxviii (1), Census of England and Wales, 1911, ix (birthplaces).
*Parl. Papers 1914–16*, xxviii (307), Second General Report of the Departmental Committee Appointed to Inquire into the Conditions Prevailing in the Coal Mining Industry due to the War, 1916.

*Parl. Papers 1919*, xi (373), Coal Industry Commission (1919), i, Reports and Minutes from the First Stage of the Enquiry, Mar. 1919.

*Parl. Papers 1919*, xii (1), Coal Industry Commission (1919), ii, Reports and Minutes from the Second Stage of the Enquiry, Apr. 1919.

*Parl. Papers 1920*, xxvi (1), Report of the Departmental Committee on Workmen's Compensation, 1920.

*Parl. Papers 1920*, xxvi (87), Report of the Departmental Committee on Workmen's Compensation, Minutes of Evidence, i.

*Parl. Papers 1920*, xxvi (605), Report of the Departmental Committee on Workmen's Compensation, Minutes of Evidence, ii.

*Parl. Papers 1926*, xiv (1), Report of the Royal Commission on the Coal Industry, 1925, i, Report.

*Parl. Papers 1927*, xi (1123), Chester-le-Street Union. Report of the Board of Guardians on the Administration for the Period 30 Aug. 1926–31 Dec. 1926.

*Parl. Papers 1928*, ix (39), Report of the Board of Education and the Statistics of Public Education for the Year 1926–7.

*Parl. Papers 1928*, ix (781), Return showing details of Membership, Income, Expenditure and Funds of Registered Trade Unions with 10,000 or more Members in the Years 1925 and 1926.

*Parl. Papers 1931–2*, xiv (879), Report by the Government Actuary on the Third Valuation of the Assets and Liabilities of Approved Societies.

*Parl. Papers 1933–4*, xiii (313), Reports of Investigations into the Industrial Conditions in Certain Depressed Areas, Part II: Durham and Tyneside, 1934.

Report of the Royal Commission on the Coal Industry, 1925, ii, Minutes of Evidence (HMSO, 1926).

Report of the Royal Commission on the Coal Industry, 1925, iii, Appendices (HMSO, 1926).

## PUBLISHED CONFERENCE REPORTS

Labour Party Annual Conferences, 1925–6.

Co-operative Congress Annual Conferences, 1926–7.

MFGB Proceedings, 1926.

TUC Proceedings, 1926.

## MEMOIRS, DIARIES, AND CONTEMPORARY WRITING

All-Russian Council of Trade Unions, *Red Money: A Statement of the Facts Relating to the Money Raised in Russia during the General Strike and Mining Lockout in Britain* (1926).

Alsop, G., 'A Kind of Socialism', in T. Austrin, et al. (eds), *But the World Goes on the Same: Changing Times in Durham Pit Villages* (Whitley Bay, 1979), 20–33.

Armstrong, T., *Tommy Armstrong Sings*, with introd. by T. Gilfellon (Newcastle-upon-Tyne, 1971).

Arnot, R. Page, '1926 Remembered and Revealed', *Llafur*, 2 (1977), 9–17.

Ashby, Henry, 'Send Your Sons into the Mines . . .' in K. Armstrong and H. Beynon (eds), *Hello, Are You Working? Memories of the 'Thirties in the North-East of England* (Whitley Bay, 1977), 33–7.

Ashby, Hilda, 'Wait Till the Banner Comes Home!' in K. Armstrong and H. Beynon (eds), *Hello, Are You Working? Memories of the 'Thirties in the North-East of England* (Whitley Bay, 1977), 37–40.

Baldwin, S., *On England* (1927 edn.; first pub. 1926).

Beavis, D., 'Never Again', in K. Armstrong and H. Beynon (eds), *Hello, Are You Working? Memories of the 'Thirties in the North-East of England* (Whitley Bay, 1977), 17–23.

Bell, F., *At the Works: A Study of a Manufacturing Town* (1907).

—— 'Women at the Works—and Elsewhere', in *Landmarks: A Reprint of Some Essays and Other Pieces Published between the Years 1894 and 1922* (1929), 53–73.

Benney, M., *Charity Main: A Coalfield Chronicle* (1946).

Bestford, E., 'My Father and Brothers are Miners!', in K. Armstrong and H. Beynon (eds), *Hello, Are you Working? Memories of the 'Thirties in the North-East of England* (Whitley Bay, 1977), 85–7.

Bulman, H. F. and R. A. S. Redmayne, *Colliery Working and Management* (rev. 4th edn., 1925).

Carr, B., 'Memories of the General Strike', *North East Group for the Study of Labour History Bulletin*, 10 (1976), 14–21.

Chaplin, S., *A Tree with Rosy Apples* (Newcastle-upon-Tyne, 1972).

Citrine, W., *Men and Work: An Autobiography* (1964).

Claxton, W. J., *Journeys in Industrial England* (1914).

Cole, G. D. H., *Labour in the Coal Mining Industry, 1914–21* (Oxford, 1923).

Cook, A. J., *The Nine Days: The Story of the General Strike Told by the Miners' Secretary* (1926).

Coombes, B. L., *These Poor Hands: The Autobiography of a Miner Working in South Wales* (1939).

Craddock, M., *North Country Maid* (Maidstone, 1995 edn.; first pub. 1960).

—— *Return to Rainton* (1963).

Cronin, A. J., *The Stars Look Down* (1935).

Davies, I., *The Angry Summer: A Poem of 1926* (1943).

—— *The Complete Poems of Idris Davies*, ed. D. Johnston (Cardiff, 1994).

Farbridge, E., *Recollections of Stanley* (Durham County Record Office typescript, 1973).

Fenton, J. H., *Going Back A Bit* (Southport, 1994).

Fynes, R., *The Miners of Northumberland and Durham: A History of their Social and Political Progress* (Sunderland, 1923 edn.; first pub. 1873).

Gibson, F. A., *The Coal Mining Industry: Supplement with Statistics for the Year 1925* (Cardiff, 1927).

Grant, J. C., *The Back-to-Backs* (1930).
Greening, E., '1926 in Aberdare', *Llafur*, 2 (1977), 31–8.
Hall, E. (ed.), *Durham Miners' Association: Fatal Accidents Book, 1920–50* (Durham, 1995).
Hall, W. S., *A Historical Survey of the Durham Colliery Mechanics' Association, 1879–1929* (Durham, 1929).
Halliday, J., *Just Ordinary, But . . . An Autobiography* (Waltham Abbey, 1959).
Hamilton, W., *Blood on the Walls: Memoirs of an Anti-Royalist* (1992).
Headlam, C., *Parliament and Politics in the Age of Baldwin and MacDonald: The Headlam Diaries, 1923–35*, ed. S. Ball (1992).
Healy, W., *Between the Wars: Childhood Memories of Horden* (Durham, 1996).
Henson, H. Hensley, *Quo Tendimus? The Primary Charge Delivered at his Visitation to the Clergy of his Diocese in November 1924* (1925).
——*Retrospect of an Unimportant Life*, 3 vols. (Oxford, 1942–50).
——*Letters of Herbert Hensley Henson*, ed. E. F. Braley (1950).
——*More Letters of Herbert Hensley Henson*, ed. E. F. Braley (1954).
Henty, G. A., *Facing Death, A Tale of the Coal Mines* (1908 edn.; first published 1883).
Hitchin, G., *Pit-Yacker* (1962).
Kirkup, J., *The Only Child: An Autobiography of Infancy* (1957).
Lawrence, D. H., *Sons and Lovers* (1913).
——'Return to Bestwood', in *Phoenix II: Uncollected, Unpublished and other Prose Works*, ed. W. Roberts and H. Moore (1968), 257–66.
——'The Miner at Home', in K. Sagar (ed.) *The Mortal Coil and Other Stories* (Harmondsworth, 1971), 109–14.
Lawson, J., *Peter Lee* (1936).
—— *The Man in the Cap: The Life of Herbert Smith* (1941).
——*A Man's Life* (rev. 2nd edn., 1944).
—— *Who Goes Home? Broadcasts and Sketches* (1945).
Lawther, W., 'Transcripts from an Interview with Will Lawther', *North East Group for the Study of Labour History Bulletin*, 2 (1968), 10–12.
Lee, J., *This Great Journey: A Volume of Autobiography, 1904–45* (1963).
Llewellyn, R., *How Green Was My Valley* (1939).
Manley, G., 'The Durham Meteorological Record, 1847–1940', *Quarterly Journal of the Royal Meteorological Society*, 67 (1941), 363–80.
Mess, H. A., *Industrial Tyneside: A Social Survey Made for the Bureau of Social Research for Tyneside* (1928).
Millar, J. P. M., *Bias in the Schools* (1936).
Moran, L., *Seven Decades of Destiny* (Durham, 1996).
Muckle, W., *No Regrets* (Newcastle-upon-Tyne, 1981).
Newsom, J., *Out of the Pit: A Challenge to the Comfortable* (Oxford, 1936).
Ottey, R., *The Strike: An Insider's Story* (1985).
Parkin, Miss, 'The Three R's', in T. Austrin, et al. (eds), *But the World Goes on the Same: Changing Times in Durham Pit Villages* (Whitley Bay, 1979), 50–1.
Parry-Jones, T. J., *The Other Story of Coal* (1925).

Phillips, M., *Women and the Miners' Lockout: The Story of the Women's Committee for the Relief of Miners' Wives and Children* (1927).
Pilgrim Trust, *Men Without Work: A Report made to the Pilgrim Trust*, with introd. by the Archbishop of York and preface by Lord Macmillan (Cambridge, 1938).
Postgate, R. W., E. Wilkinson, and J. F. Horrabin (eds), *A Worker's History of the Great Strike* (1927).
Priestley, J. B., *English Journey* (1935 edn.; first published 1934).
Proctor, F., *I Was There: An Autobiography*, ed. W. N. Tindall and P. Proctor (Gibsons, British Columbia, 1999).
Quiller-Couch, M., *A Little Book on Thrift* (Oxford, 1912).
—— *A Little Book on Temperance* (1914).
Reid, A., *Reid's Handy Colliery Guide and Directory for the Counties of Northumberland and Durham* (Newcastle-upon-Tyne, 1923).
Ridley, M., 'Making a Contribution', in T. Austrin, et al. (eds), *But the World Goes on the Same: Changing Times in Durham Pit Villages* (Whitley Bay, 1979), 57–70.
Rooney, R., 'Changing Times', in T. Austrin, et al. (eds), *But the World Goes on the Same: Changing Times in Durham Pit Villages* (Whitley Bay, 1979), 36–41.
Seed, J., *The Jimmy Seed Story: Forty-Three Years in First-Class Football as Player and Manager* (1957).
Smillie, R., *My Life for Labour* (1924).
Squires, E., *Pit Pony Heroes* (Newton Abbot, 1974).
Stephenson, J., 'A Comment by James Stephenson of Winlaton', *North East Group for the Study of Labour History Bulletin*, 4 (1970), 25–32.
Stephenson, J. J., *Being Human* (Durham, 1986).
Swan, J. E., *The Mad Miner: A Saga of the North* (1933).
TUC and Labour Party, *Mines and Men: The Position in the Mining Industry* (1924).
Wade, M., *To the Miner Born* (Stocksfield, 1984).
Webb, B., *Beatrice Webb's Diaries, 1912–24*, ed. M. Cole, 2 vols. (1952–6).
—— *The Diary of Beatrice Webb*, ed. N. MacKenzie and J. MacKenzie, 4 vols. (1982–5).
Webb, S., *The Story of the Durham Miners (1661–1921)* (1921).
Welbourne, E., *The Miners' Unions of Northumberland and Durham* (Cambridge, 1923).
Welldon, J. E. C., *The English Church* (1926).
—— *Forty Years On: Light and Shadows (A Bishop's Reflections on Life)* (1935).
Wheatley, J., *Miners, Mines and Misery* (Glasgow, 1920).
Whitfield, F., 'And Of Course, I've Got Some Dust', in K. Armstrong and H. Beynon (eds), *Hello, Are You Working? Memories of the 'Thirties in the North-East of England* (Whitley Bay, 1977), 57–63.

Wilkinson, E., *Clash*, with a new introd. by B. Vernon (1989 edn.; first pub. 1929).

Wilson, J., *A History of the Durham Miners' Association, 1870–1904* (Durham, 1907).

——*Memories of a Labour Leader: The Autobiography of John Wilson JP, MP* (Firle, 1980 edn.; first pub. 1910).

## ORAL INTERVIEWS

Telephone conversation with Mrs E. Stoves, 13 June 2005.

Interview with Mr R. M., 17 Sept. 2005, Horden.

## PRINTED SECONDARY WORKS

Adeney, M. and J. Lloyd, *The Miners' Strike 1984–5: Loss Without Limit* (1986).

Alexander, P., 'A Moral Economy, an Isolated Mass and Paternalised Migrants: Transvaal Colliery Strikes, 1925–49', in S. Berger, A. Croll, and N. LaPorte (eds), *Towards a Comparative History of Coalfield Societies* (Aldershot, 2005), 238–52.

Anderson, B., *Imagined Communities: Reflections on the Origin and Spread of Nationalism* (1983).

Anderson, S. A. and S. Butler, 'Language and Power in the Classroom: An Interview with Harold Rosen', *English Journal*, 71 (1982), 24–8.

Armstrong, K. (ed.), *Horden Miners: The Lives of Two Horden Miners in their Own Words* (Peterlee, 1984).

——and H. Beynon (eds), *Hello, Are You Working? Memories of the 'Thirties in the North-East of England* (Whitley Bay, 1977).

Arnot, R. Page, *The Miners: A History of the Miners' Federation of Great Britain, 1889–1910* (1949).

——*The Miners; Years of Struggle: A History of the Miners' Federation of Great Britain (from 1910 onwards)* (1953).

Austrin T. et al. (eds), *But the World Goes on the Same: Changing Times in Durham Pit Villages* (Whitley Bay, 1979).

Baggs, C., 'The Miners' Institute Libraries', *Planet*, 118 (1996), 45–52.

Ball, S., A. Thorpe, and M. Worley, 'Elections, Leaflets and Whist Drives: Constituency Party Members in Britain between the Wars', in M. Worley (ed.), *Labour's Grass Roots: Essays on the Activities of Local Labour Parties and their Members, 1918–45* (Aldershot, 2005), 7–32.

Barron, H., '"'Tis very Embarrassing, Say What You Like, To Be a Good Vicar in a Valley on Strike": The Church of England and its Relationship with the Durham Miners at the time of the 1926 Lockout', *Twentieth Century British History*, 17 (2006), 350–72.

——'Women of the Durham Coalfield and their Reactions to the 1926 Miners' Lockout', *Historical Studies in Industrial Relations*, 21 (2006), 53–83.

Beddoe, D., *Out of the Shadows: A History of Women in Twentieth-Century Wales* (Cardiff, 2000).

Bell, C. and H. Newby, *Community Studies: An Introduction to the Sociology of the Local Community* (1971).

Bellamy, J. M., et al. (eds), *Dictionary of Labour Biography*, 12 vols. (1972–2005).

Berger, S., 'Working-Class Culture and the Labour Movement in the South Wales and the Ruhr Coalfields, 1850–2000: A Comparison', *Llafur*, 8 (2001), 5–40.

____ A. Croll, and N. LaPorte, *Towards a Comparative History of Coalfield Societies* (Aldershot, 2005).

____ 'Introduction', in S. Berger, A. Croll, and N. LaPorte (eds), *Towards a Comparative History of Coalfield Societies* (Aldershot, 2005), 1–11.

____ and N. Evans, 'Two Faces of King Coal: The Impact of the Historiographical Traditions on Comparative History in the Ruhr and South Wales', in S. Berger, A. Croll, and N. LaPorte (eds), *Towards a Comparative History of Coalfield Societies* (Aldershot, 2005), 29–42.

Beynon, H. (ed.), *Digging Deeper: Issues in the Miners' Strike* (1985).

____ and T. Austrin, *Masters and Servants: Class and Patronage in the Making of a Labour Organisation* (1994).

Blauner, R., 'Work Satisfaction and Industrial Trends in Modern Society', in W. Galenson and S. M. Lipset (eds), *Labor and Trade Unionism: An Interdisciplinary Reader* (New York, 1960), 339–60.

Bourke, J., *Working-Class Cultures in Britain, 1890–1960: Gender, Class and Ethnicity* (1994).

Brogden, M., 'Interviews at Armthorpe, December 1984', in R. Samuel, B. Bloomfield, and G. Boanas (eds), *The Enemy Within: Pit Villages and the Miners' Strike of 1984–5* (1986), 166–202.

Bruley, S., 'Women', in J. McIlroy, A. Campbell, and K. Gildart (eds), *Industrial Politics and the 1926 Mining Lockout: The Struggle for Dignity* (Cardiff, 2004), 229–48.

____ 'The Politics of Food: Gender, Family, Community and Collective Feeding in South Wales in the General Strike and Miners' Lockout of 1926', *Twentieth Century British History*, 18 (2007), 54–77.

Bulmer, M., (ed.), *Working-Class Images of Society* (1975).

____ 'Sociological Models of the Mining Community', *Sociological Review*, 23 (1975), 61–92.

Burge, A., 'In Search of Harry Blount: Scabbing Between the Wars in One South Wales Community', *Llafur*, 6 (1994), 58–69.

Burnett, J., D. Vincent, and D. Mayall, *The Autobiography of the Working Class: An Annotated, Critical Biography*, ii, *1900–1945* (Brighton, 1987).

Byrne, D., 'Immigrants and the Formation of the North Eastern Industrial Working Class', *North East Labour History Bulletin*, 30 (1996), 29–36.

Bythell, D., 'Lawson, John James', *Oxford Dictionary of National Biography* (Oxford, 2004), xxxii, 899–901.

Calhoun, C. J., 'History, Anthropology and the Study of Communities: Some Problems in Macfarlane's Proposal', *Social History*, 3 (1978), 363–73.

Callcott, M., 'Labour Women in North East England', *North East Labour History Bulletin*, 17 (1983), 35–9.

—— '1926: Women Support the Miners', *North East Labour History Bulletin*, 19 (1985), 40–2.

Campbell, A., *The Scottish Miners, 1874–1939*, 2 vols. (Aldershot, 2000).

—— N. Fishman, and D. Howell (eds), *Miners, Unions and Politics, 1910–47* (Aldershot, 1996).

—— —— —— 'Introduction', in *Miners, Unions and Politics, 1910–47* (Aldershot, 1996), 1–8.

Carr, G., *Pit Women: Coal Communities in Northern England in the Early Twentieth Century* (2001).

Catterall, P., 'Morality and Politics: The Free Churches and the Labour Party between the Wars', *Historical Journal*, 36 (1993), 667–85.

Catterall, S., 'Police', in J. McIlroy, A. Campbell, and K. Gildart (eds), *Industrial Politics and the 1926 Mining Lockout: The Struggle for Dignity* (Cardiff, 2004), 249–68.

Chadwick, O., *Hensley Henson and the Durham Miners, 1920–1939* (Durham, 1983).

—— *Hensley Henson: A Study in the Friction between Church and State* (Oxford, 1983).

Chaplin, S., *A Tree with Rosy Apples* (Newcastle, 1972).

—— 'Durham Mining Villages', in M. Bulmer (ed.), *Mining and Social Change* (1978), 59–82.

Chase, M., *Chartism: A New History* (Manchester, 2007).

Childs, M., 'Labour Grows Up: The Electoral System, Political Generations, and British Politics, 1890–1929', *Twentieth Century British History*, 6 (1995), 123–44.

Church, R., *The History of the British Coal Industry*, iii, *1830–1913: Victorian Pre-eminence* (Oxford, 1986).

—— and Q. Outram, *Strikes and Solidarity: Coalfield Conflict in Britain, 1889–1966* (Cambridge, 1998).

Clarke, P., *Lancashire and the New Liberalism* (Cambridge, 1971).

Cohen, A., *The Symbolic Construction of Community* (1985).

Cohn, S., *When Strikes Make Sense—and Why: Lessons from Third Republic French Coal Miners* (New York, 1993).

Coleman, D., 'Population and Family', in A. H. Halsey (ed.), *Twentieth-Century British Social Trends* (3rd edn., 2000), 27–93.

Colls, R., *The Collier's Rant: Song and Culture in the Industrial Village* (1977).

Colls, R., *The Pitmen of the Northern Coalfield: Work, Culture and Protest, 1790–1850* (Manchester, 1987).

____ '"Oh Happy English Children!": Coal, Class and Education in the North-East', *Past and Present*, 73 (1976), 75–99.

____ 'Coal, Class and Education in the North-East: A Rejoinder', *Past and Present*, 90 (1981), 152–65.

____ Review of Alan Metcalfe, *Leisure and Recreation in a Victorian Mining Community: The Social Economy of Leisure in North-East England, 1820–1914* (2006), *Social History*, 32 (2007), 88–90.

____ 'When We Lived in Communities: Working-Class Culture and its Critics', in R. Colls and R. Rodger (eds), *Cities of Ideas: Civil Society and Urban Governance in Britain, 1800–2000* (Aldershot, 2004), 283–307.

Corbin, D. A., *Life, Work and Rebellion in the Coal Fields: The Southern West Virginia Miners, 1880–1922* (Urbana, Ill., and London, 1981).

Craig, F. W. S., *British Parliamentary Election Results, 1918–49* (Glasgow, 1969).

Crew, D. F., 'Rapport/Bericht', in K. Tenfelde (ed.), *Towards a Social History of Mining in the Nineteenth and Twentieth Centuries: Papers Presented to the International Mining History Congress, Bochum, 1989* (Munich, 1992), 53–8.

Croft, A., *Red Letter Days: British Fiction in the 1930s* (1990).

Cronin, J. E., *Industrial Conflict in Modern Britain* (1979).

Currie, R., A. Gilbert, and L. Horsley, *Churches and Churchgoers: Patterns of Church Growth in the British Isles since 1700* (Oxford, 1977).

Daunton, M. J., 'Miners' Homes: South Wales and the Great Northern Coalfield, 1880–1914', *International Review of Social History*, 25 (1980), 143–75.

____ 'Down the Pit: Work in the Great Northern and South Wales Coalfields, 1870–1914', *Economic History Review*, 34 (1981), 578–97.

Davies, P., *A. J. Cook* (Manchester, 1987).

Deacon, A., *In Search of the Scrounger: The Administration of Unemployment Insurance in Britain 1920–1931* (1976).

Dennis, N., F. Henriques, and C. Slaughter, *Coal is our Life: An Analysis of a Yorkshire Mining Community* (1979 edn.; first pub. 1956).

DeStefanis, A., 'Violence and the Colorado National Guard: Masculinity, Race, Class, and Identity in the 1913–1914 Southern Colorado Coal Strike', in J. J. Gier and L. Mercer, *Mining Women: Gender in the Development of a Global Industry, 1670–2005* (New York and Basingstoke, 2006), 195–212.

Douglass, D., 'Pit Talk in County Durham: A Glossary of Miners' Talk, Together with Memories of Wardley Colliery, Pit Songs and Piliking', *History Workshop Pamphlets*, 10 (Oxford, 1973).

____ 'The Durham Pitman', in R. Samuel (ed.), *Miners, Quarrymen and Saltworkers* (1977).

____ 'Worms of the Earth: The Miners' Own Story', in R. Samuel (ed.), *People's History and Socialist Theory* (1981), 61–7.

Egan, D., ' "A Cult of their Own": Syndicalism and *The Miners' Next Step*', in A. Campbell, N. Fishman, and D. Howell (eds), *Miners, Unions and Politics, 1910–47* (Aldershot, 1996), 13–33.

Eklund, E., *Steel Town: The Making and Breaking of Port Kembla* (Melbourne, 2002).

—— 'The "Place" of Politics: Class and Localist Politics at Port Kembla, 1900–30', *Labour History*, 78 (2000), 94–115.

Emery, N., *The Coalminers of Durham* (Stroud, 1992).

—— *Banners of the Durham Coalfield* (Stroud, 1998).

Evans, N., 'Immigrants and Minorities in Wales, 1840–1990: A Comparative Perspective', *Llafur*, 5 (1991), 5–26.

Fagge, R., *Power, Culture and Conflict in the Coalfields: West Virginia and South Wales, 1900–1922* (Manchester, 1996).

Farman, C., *The General Strike: May 1926* (1972).

Feldman, G. D. and K. Tenfelde (eds), *Workers, Owners and Politics in Coal Mining: An International Comparison of Industrial Relations* (Oxford, 1990).

Fentress, J. and C. Wickham, *Social Memory* (Oxford, 1992).

Fletcher, M., ' "Slaves of the Lamp?": Independence and Control in Two State Coal Mining Communities in Victoria, Australia', in S. Berger, A. Croll, and N. LaPorte (eds), *Towards a Comparative History of Coalfield Societies* (Aldershot, 2005), 99–112.

Foot, P., *'An Agitator of the Worst Kind': A Portrait of the Miners' Leader A. J. Cook* (1986).

Francis, H., *Miners Against Fascism: Wales and the Spanish Civil War* (1984).

—— and D. Smith, *The Fed: A History of the South Wales Miners in the Twentieth Century* (Cardiff, 1980).

—— and G. Rees, 'No Surrender in the Valleys', *Llafur*, 5 (1989), 41–71.

Garside, W. R., *The Durham Miners, 1919–1960* (1971).

Geary, D., 'Unemployment and Working-Class Solidarity: The German Experience, 1929–33', in R. J. Evans and D. Geary (eds), *The German Unemployed: Experiences and Consequences of Mass Unemployment from the Weimar Republic to the Third Reich* (1987), 261–80.

—— 'The Myth of the Radical Miner', in S. Berger, A. Croll, and N. LaPorte (eds), *Towards a Comparative History of Coalfield Societies* (Aldershot, 2005), 43–64.

Geary, R., *Policing Industrial Disputes, 1893–1985* (Cambridge, 1985).

Gibb, M., 'Memories from the Past: The First Durham Labour Women's Gala', *North East Labour History*, 17 (1983), 40–1.

Gier-Viskovatoff, J. and A. Porter, 'Women of the British Coalfields on Strike in 1926 and 1984: Documenting Lives Using Oral History and Photography', *Frontiers: A Journal of Women's Studies*, 19 (1998), 199–230.

Gilbert, D., *Class, Community, and Collective Action: Social Change in Two British Coalfields, 1850–1926* (Oxford, 1992).
—— 'Imagined Communities and Mining Communities', *Labour History Review*, 60 (1995), 47–55.
Gilbert, M., *Churchill: A Life* (1991).
Graham, B., *The History of Sunderland AFC, 1879–1995* (Houghton-le-Spring, 1995).
Graves, P., *Labour Women: Women in British Working-Class Politics, 1918–39* (Cambridge, 1994).
Gregory, A., *The Silence of Memory: Armistice Day, 1919–1946* (Oxford, 1994).
Gregory, R., *The Miners and British Politics, 1906–1914* (Oxford, 1968).
Griffin, A. R., *The Miners of Nottinghamshire, 1914–1944: A History of the Nottinghamshire Miners' Unions* (1962).
—— *Mining in the East Midlands, 1550–1947* (1971).
Griffiths, T., *The Lancashire Working Classes, c.1880–1930* (Oxford, 2001).
Grimley, M., *Citizenship, Community and the Church of England: Liberal Anglican Theories of the State between the Wars* (Oxford, 2004).
—— 'Henson, Herbert Hensley', *Oxford Dictionary of National Biography* (Oxford, 2004), xxvi, 612–15.
Halbwachs, M., *The Collective Memory*, with an introd. by Mary Douglas (New York, 1980).
Hall, V. Gordon, 'Contrasting Female Identities: Women in Coal Mining Communities in Northumberland, England, 1900–1939', *Journal of Women's History*, 13 (2001), 107–31.
Harrison, R. (ed.), *Independent Collier: The Coal Miner as Archetypal Proletarian Reconsidered* (Hassocks, 1978).
—— 'Introduction', in Harrison (ed.), *Independent Collier: The Coal Miner as Archetypal Proletarian Reconsidered* (Hassocks, 1978), 1–16.
Hatcher, J., *The History of the British Coal Industry*, i, *Before 1700: Towards the Age of Coal* (Oxford, 1993).
Heesom, A. J., 'Problems of Church Extension in a Victorian New Town: The Londonderrys and Seaham Harbour', *Northern History*, 15 (1979), 138–55.
—— and B. Duffy, 'Coal, Class and Education in the North-East', *Past and Present*, 90 (181), 136–51.
Hickey, S. H. F., *Workers in Imperial Germany: Miners of the Ruhr* (Oxford, 1985).
Hobsbawm, E., *Worlds of Labour: Further Studies in the History of Labour* (1984).
—— 'Methodism and the Threat of Revolution in Britain', in *Labouring Men: Studies in the History of Labour* (1964), 23–33.
—— and T. Ranger, *The Invention of Tradition* (Cambridge, 1983).
Howard, S., 'Leisure in the Pit Villages: Meaning and Change', *North East Labour History Bulletin*, 27 (1993), 14–23.

Howard, S., 'Miners' Autobiography: Text and Context', *Labour History Review*, 60 (1995), 89–98.

Howell, D., *British Workers and the Independent Labour Party, 1886–1906* (Manchester, 1983).

——*MacDonald's Party: Labour Identities and Crisis, 1922–1931* (Oxford, 2002).

—— 'Where's Ramsay MacKinnock? Labour Leadership and the Miners', in H. Beynon (ed.), *Digging Deeper: Issues in the Miners' Strike* (1985), 181–98.

—— ' "All or Nowt": The Politics of the MFGB', in A. Campbell, N. Fishman, and D. Howell (eds), *Miners, Unions and Politics, 1910–47* (Aldershot, 1996), 35–58.

Hudson, M., *Coming Back Brockens* (1994).

Inglis, K. S., *Churches and the Working Classes in Victorian England* (1963).

Jaffe, J. A., 'The "Chiliasm of Despair" Reconsidered: Revivalism and Working-Class Agitation in County Durham', *Journal of British Studies*, 28 (1989), 23–42.

James, L., *The Politics of Identity and Civil Society in Britain and Germany: Miners in the Ruhr and South Wales, 1890–1920* (Manchester, 2008).

—— 'Trade Union Development in the Ruhr and South Wales, 1890–1914', in S. Berger, A. Croll, and N. LaPorte (eds), *Towards a Comparative History of Coalfield Societies* (Aldershot, 2005), 253–66.

Jeremy, P., 'Life on Circular 703: The Crisis of Destitution in the South Wales Coalfield during the Lockout of 1926', *Llafur*, 2 (1977), 65–75.

John, A. V., ' "A Miner Struggle": Women's Protests in Welsh Mining History', *Llafur*, 4 (1984), 72–92.

Johnson, P., *Saving and Spending: The Working-Class Economy in Britain, 1870–1939* (Oxford, 1985).

Jones, R., 'Women, Community and Collective Action: The "Ceffyl Pren" Tradition', in A. John (ed.), *Our Mother's Land: Chapters in Welsh Women's History, 1830–1939* (Cardiff, 1991), 17–41.

Joyce, P., *Work, Society and Politics: The Culture of the Factory in Later Victorian England* (Brighton, 1980).

—— *Visions of the People: Industrial England and the Question of Class, 1848–1914* (Cambridge, 1991).

Keith-Lucas, B. and P. G. Richards, *A History of Local Government in the Twentieth Century* (1978).

Kerr, C. and A. Siegel, 'The Interindustry Propensity to Strike: An International Comparison', in A. Kornhauser, R. Dubin, and A. Ross (eds), *Industrial Conflict* (New York, 1954), 189–212.

Kertzer, D., *Ritual, Politics and Power* (New Haven, Conn. and London, 1988).

Kinnear, M., *The British Voter: An Atlas and Survey since 1885* (1968).

Kirby, M. W., *Men of Business and Politics: The Rise and Fall of the Quaker Pease Dynasty of North-East England, 1700–1943* (1984).

Knowles, K. G. J. C., *Strikes: A Study in Industrial Conflict* (Oxford, 1952).
Krikler, J., *White Rising: The 1922 Insurrection and Racial Killing in South Africa* (Manchester, 2005).
Laslett, J. H. M., *Colliers across the Sea: A Comparative Study of Class Formation in Scotland and the American Midwest, 1830–1924* (Urbana, Ill., 2000).
Lawrence, J., 'Labour—The Myths it has Lived By', in D. Tanner, P. Thane, and N. Tiratsoo (eds), *Labour's First Century* (Cambridge, 2000), 341–66.
Laybourn, K., *The General Strike of 1926* (Manchester, 1993).
Lee, R., *The Church of England and the Durham Coalfield, 1810–1926: Clergymen, Capitalists and Colliers* (Woodbridge, 2007).
____'Class, Industrialisation and the Church of England: The Case of the Durham Diocese in the Nineteenth Century', *Past and Present*, 191 (2006), 165–88.
Lieven, M., *Senghennydd: The Universal Pit Village, 1890–1930* (Llandysul, 1994).
____'A "New History" of the South Wales Coalfield?', *Llafur*, 8 (2002), 89–106.
____'A Fractured Working-Class Consciousness? The Case of the Lady Windsor Colliery Lodge, 1921', *Welsh History Review*, 21 (2003), 729–56.
Light, J., 'Manufacturing the Past: The Representation of Mining Communities in History, Literature and Heritage: "Fantasies of a World that never was?" ', *Llafur*, 8 (2000), 19–31.
Lockwood, D., 'Sources of Variation in Working-Class Images of Society', *Sociological Review*, 14 (1966), 249–67.
Lummis, T., *Occupation and Society: The East Anglican Fishermen, 1880–1914* (Cambridge, 1985).
Lynn, P., 'The Influence of Class and Gender: Female Political Organisations in County Durham during the Inter-War Years', *North East History*, 31 (1997), 43–64.
Macfarlane, A., 'History, Anthropology and the Study of Communities', *Social History*, 2 (1977), 631–52.
Machin, G. I. T., *Churches and Social Issues in Twentieth Century Britain* (Oxford, 1998).
Macintyre, S., *Little Moscows: Community and Working-Class Militancy in Interwar Britain* (1980).
____*A Proletarian Science: Marxism in Britain, 1917–1933* (Cambridge, 1980).
Martin, R., *Communism and the British Trade Unions, 1924–1933: A Study of the National Minority Movement* (Oxford, 1969).
Mason, A., *The General Strike in the North East* (Hull, 1970).
Mates, L., 'Durham and South Wales Miners and the Spanish Civil War', *Twentieth Century British History*, 17 (2006), 373–95.
Matthew, C., R. McKibbin, and J. Kay, 'The Franchise Factor in the Rise of the Labour Party', *English Historical Review*, 91 (1976), 723–52.
McCord, N., *North East England: An Economic and Social History* (1979).

McIlroy, J., A. Campbell, and K. Gildart (eds), *Industrial Politics and the 1926 Mining Lockout: The Struggle for Dignity* (Cardiff, 2004).

——'Revolutionaries', in J. McIlroy, A. Campbell, and K. Gildart (eds), *Industrial Politics and the 1926 Mining Lockout: The Struggle for Dignity* (Cardiff, 2004), 269–97.

——'Nottinghamshire', in J. McIlroy, A. Campbell, and K. Gildart (eds), *Industrial Politics and the 1926 Mining Lockout: The Struggle for Dignity* (Cardiff, 2004), 208–26.

——and A. Campbell, 'Fighting the Legions of Hell', in J. McIlroy, A. Campbell and K. Gildart (eds), *Industrial Politics and the 1926 Mining Lockout: The Struggle for Dignity* (Cardiff, 2004), 49–106.

McKibbin, R., *Classes and Cultures: England, 1918–1951* (Oxford, 1998).

McLeod, H., *Religion and the Working Class in Nineteenth-Century Britain* (1984).

Metcalfe, A., *Leisure and Recreation in a Victorian Mining Community: The Social Economy of Leisure in North East England, 1820–1914* (Abingdon, 2006).

Miller, A., *Death of a Salesman* (first performed New York, 1949).

Moore, R., *Pitmen, Preachers and Politics: The Effects of Methodism in a Durham Mining Community* (Cambridge, 1974).

Moran, L., *The History of Brandon Colliery, 1856–1960* (Houghton-le-Spring, 1988).

Morgan, J., *Conflict and Order: The Police and Labour Disputes in England and Wales, 1900–39* (Oxford, 1987).

Morgan, K. O., 'A Time for Miners to Forget History', *New Society*, 71 (1985), 283–5.

Morris, M., *The General Strike* (Harmondsworth, 1976).

Morris, R. W., 'The General Strike in the North East', *North East Group for the Study of Labour History Bulletin*, 10 (1976), 2–12.

Moyes, W., *The Banner Book: A Study of the Banners of the Lodges of the DMA* (Newcastle-upon-Tyne, 1974).

Newby, H., *The Deferential Worker: A Study of Farm Workers in East Anglia* (1977).

Nicholson, T., 'Community and Class: The Cleveland Ironstone Field, 1850–1914', *Labour History Review*, 60 (1995), 79–89.

Noel, G., *The Great Lockout of 1926* (1976).

Nordlinger, E., *The Working-Class Tories: Authority, Deference and Stable Democracy* (1967).

Outram, Q., 'The Stupidest Men in England? The Industrial Relations Strategy of the Coalowners between the Lockouts, 1923–4', *Historical Studies in Industrial Relations*, 4 (1997), 65–95.

——'Class Warriors: The Coalowners', in J. McIlroy, A. Campbell, and K. Gildart (eds), *Industrial Politics and the 1926 Mining Lockout: The Struggle for Dignity* (Cardiff, 2004), 107–35.

Parkin, F., 'Working-Class Conservatives: A Theory of Political Deviance', *British Journal of Sociology*, 18 (1967), 278–90.

Passerini, L., *Fascism in Popular Memory: The Cultural Experience of the Turin Working Class*, trans. R. Lumley and J. Bloomfield (Cambridge, 1987; first pub. 1984).

Perkins, A., *A Very British Strike: 3–12 May 1926* (2006).

Phillips, G., *The General Strike: The Politics of Industrial Conflict* (1976).

Pitt, M., *The World on Our Backs: The Kent Miners and the 1972 Miners' Strike* (1979).

Pope, R., *Building Jerusalem: Nonconformity, Labour and the Social Question in Wales, 1906–1939* (Cardiff, 1998).

Pugh, M., 'The Rise of Labour and the Political Culture of Conservatism, 1890–1945', *History*, 87 (2002), 514–37.

Renshaw, P., *The General Strike* (1975).

Richards, A., *Miners on Strike: Class Solidarity and Divisions in Britain* (Oxford, 1996).

Rimlinger, G. V., 'International Differences in the Strike-Propensity of Coal Miners: Experience in Four Countries', *Industrial and Labour Relations Review*, 12 (1959), 389–405.

Roberts, E., *A Woman's Place: An Oral History of Working-Class Women, 1890–1940* (Oxford, 1984).

Robinson, J., *Tommy Turnbull: A Miner's Life* (Newcastle-upon-Tyne, 1996).

Rowe, D. J., 'The North-East', in F. M. L. Thompson (ed.), *The Cambridge Social History of Britain, 1750–1950*, i, *Regions and Communities* (Cambridge, 1990), 415–70.

Royle, E., *Radicals, Secularists and Republicans: Popular Freethought in Britain, 1866–1915* (Manchester, 1980).

Runciman, W. G., *Relative Deprivation and Social Justice: A Study of Attitudes to Social Inequality in Twentieth-Century England* (1966).

Russell, A., 'Local Elites and the Working-Class Response in the North-West, 1870–95: Paternalism and Deference Reconsidered', *Northern History*, 23 (1987), 153–73.

Ryan, P., 'The Poor Law in 1926', in M. Morris (ed.), *The General Strike* (Harmondsworth, 1976), 358–78.

Samuel, R., 'Preface', in R. Samuel, B. Bloomfield, and G. Boanas (eds), *The Enemy Within: Pit Villages and the Miners' Strike of 1984–5* (1986), pp. ix–xviii.

____ 'Introduction', in R. Samuel, B. Bloomfield, and G. Boanas (eds), *The Enemy Within: Pit Villages and the Miners' Strike of 1984–5* (1986), 1–39.

____ Bloomfield, B., and G. Boanas (eds), *The Enemy Within: Pit Villages and the Miners' Strike of 1984–5* (1986).

____ and Thompson, P. (eds), *The Myths We Live By* (1990).

Savage, M., *The Dynamics of Working-Class Politics: The Labour Movement in Preston, 1880–1940* (Cambridge, 1987).

Scates, B., 'Mobilizing Manhood: Gender and the Great Strike in Australia and Aotearoa/New Zealand', *Gender and History*, 9 (1997), 285–309.

Shubert, A., 'A Divided Community: The Social Development of the Austrian Coalfields to 1934', in K. Tenfelde (ed.), *Towards a Social History of Mining in the Nineteenth and Twentieth Centuries: Papers Presented to the International Mining History Congress, Bochum, 1989* (Munich, 1992), 284–93.

Skelley, J. (ed.), *The General Strike, 1926* (1976).

Smith, D., *Aneurin Bevan and the World of South Wales* (Cardiff, 1993).

—— 'A Place in the South of Wales', in *Wales! Wales?* (1984), 55–97.

—— 'Tonypandy 1910: Definitions of Community', *Past and Present*, 87 (1980), 158–84.

Stacey, M., 'The Myth of Community Studies', *British Journal of Sociology*, 20 (1969), 134–47.

Stead, J., *Never the Same Again: Women and the Miners' Strike, 1984–5* (1987).

Stedman Jones, G., *Outcast London: A Study in the Relationship between Classes in Victorian Society* (Harmondsworth, 1984 edn.; first pub. 1971).

Storm-Clark, C, 'The Miners, 1870–1970: A Test-case for Oral History', *Victorian Studies*, 15 (1971), 49–74.

Studdert-Kennedy, G., *Dog-Collar Democracy: The Industrial Christian Fellowship, 1919–29* (1982).

Supple, B., *The History of the British Coal Industry*, iv, *1913–46: The Political Economy of Decline* (Oxford, 1987).

Szreter, S., *Fertility, Class and Gender in Britain, 1860–1940* (Cambridge, 1996).

Tanner, D., *Political Change and the Labour Party, 1900–1918* (Cambridge, 1990).

—— 'The Labour Party and Electoral Politics in the Coalfields', in A. Campbell, N. Fishman, and D. Howell (eds), *Miners, Unions and Politics, 1910–47* (Aldershot, 1996), 59–92.

Taylor, Andrew, 'So Many Cases but So Little Comparison: Problems of Comparing Mineworkers', in S. Berger, A. Croll, and N. LaPorte (eds), *Towards a Comparative History of Coalfield Societies* (Aldershot, 2005), 12–28.

Taylor, Avram, ' "You Never Told Your Mam": Women's Management of Household Finances and Credit during the Interwar Period', *North East Labour History Bulletin*, 30 (1996), 51–65.

Thompson, E. P., *The Making of the English Working Class* (1991 edn.; first pub. 1963).

—— 'Rough Music', in *Customs in Common* (1991), 467–538.

—— 'Patrician Society, Plebian Culture', *Journal of Social History*, 7 (1974), 382–405.

—— 'On History, Sociology and Historical Relevance', *British Journal of Sociology*, 27 (1976), 387–402.

Thompson, P., *The Voice of the Past: Oral History* (Oxford, 2000 edn.; first pub. 1978).

Thompson, S., ' "That Beautiful Summer of Severe Austerity": Health, Diet and the Working-Class Domestic Economy in South Wales in 1926', *Welsh History Review*, 21 (2003), 552–74.

Thorpe, A., 'The Membership of the Communist Party of Great Britain, 1920–45', *Historical Journal*, 43 (2000), 777–800.

Tönnies, F., *Community and Civil Society*, ed. J. Harris, trans. J. Harris and M. Hollis (Cambridge, 2001).

Waller, R., *The Dukeries Transformed: The Social and Political Development of a Twentieth-Century Coalfield* (Oxford, 1983).

Wearmouth, R. F., *The Social and Political Influence of Methodism in the Twentieth Century* (1957).

Webb, S. and B. Webb, *English Poor Law History; Part II: The Last Hundred Years*, with introd. by W. A. Robson (1963 edn.; first pub. 1929).

Webster, C., 'Healthy or Hungry Thirties?', *History Workshop Journal*, 13 (1982), 110–29.

Welshman, J., 'School Meals and Milk in England and Wales, 1906–45', *Medical History*, 41 (1997), 6–29.

Williams, C., *Democratic Rhondda: Politics and Society, 1885–1951* (Cardiff, 1996).

—— *Capitalism, Community and Conflict: The South Wales Coalfield, 1898–1947* (Cardiff, 1998).

Williams, J., 'Churches, Sport and Identities in the North, 1900–39', in J. Hill and J. Williams (eds), *Sport and Identity in the North of England* (Keele, 1996), 113–36.

Williams, R., *Keywords: A Vocabulary of Culture and Society* (1988 edn.; first pub. 1976).

Williamson, B., *Class, Culture and Community: A Biographical Study of Social Change in Mining* (1982).

—— 'Living the Past Differently: Historical Memory in the North East', in R. Colls and B. Lancaster (eds), *Geordies: Roots of Regionalism* (Newcastle-upon-Tyne, 2005 edn.; first pub. 1992), 149–68.

Wilson, A. and H. Levy, *Workmen's Compensation*, 2 vols. (1939–41).

Wilson, K., 'Chartism and the North East Miners: A Reappraisal', in R. W. Sturgess (ed.), *Pitmen, Viewers and Coalmasters: Essays on North East Coalmining in the Nineteenth Century* (Newcastle-upon-Tyne, 1986), 81–104.

Winter, J. M., *The Great War and the British People* (1986).

Worley, M., *Class against Class: The Communist Party between the Wars* (2002).

—— *Labour inside the Gate: A History of the British Labour Party between the Wars* (London and New York, 2005).

Worthen, J., *D. H. Lawrence: The Early Years, 1885–1912* (Cambridge, 1991).

Zweig, F., *Men in the Pits* (1948).

## UNPUBLISHED THESES

Barron, H., 'Meanings of Community in the Durham Coalfield during the 1926 Miners' Lockout' (Oxford University DPhil thesis, 2006).

Emery, N., 'Pease and Partners in the Deerness Valley: Aspects of the Social and Economic History of Waterhouses, Esh Winning and Ushaw Moor' (Durham University MA thesis, 1984).

Wilson, K., 'Political Radicalism in the North East of England 1830–60: Issues in Historical Sociology' (Durham University PhD thesis, 1987).

# *Index*


Ablett, Noah 263
adult education, *see* education
Adventure Colliery 130, 131fn, 155, 184, 230–1, 259
aged miners' homes 116–7, 121–2, 245, 266
agricultural workers 268, *see also* farmers
Alexander, A. V. 222
All-Russian Trade Union Council 59, 107
allotments 236, 238
Alma Colliery 119
Alnwick Castle 47
Alsop, George 135
Anderson, Benedict 5
Anglicanism, *see* Church of England
Anglo-Russian Miners' Committee 107
Annfield Plain, County Durham 40, 126
anti-clericalism, *see* secularism
Archbishop of Canterbury 172
Armistice Day 63–5, 165
army 72, 105
Arnot, Robin Page 60, 96
Asquith, Herbert Henry 184
Associated Society of Locomotive Engineers and Firemen 32
Astor, Lady Nancy 135
Auckland Castle 45, 48, 170
Australia 16, 51, 261, 268
Austrin, Terry 12

*Back-to-Backs, The* 257
Baggs, Chris 223
Baldwin, Stanley 1, 7, 63, 75, 88, 89, 91, 104, 105, 123, 198
banners of the DMA 64
  of Chopwell miners' lodge 151
  of Elemore miners' lodge 64fn
  emphasis on education 221
  of Heworth miners' lodge 251
  of Monkwearmouth miners' lodge 247, 248–9
  of Morrison miners' lodge 110
  of Murton miners' lodge 64fn
  portrayal of A. J. Cook 88
  portrayal of religious imagery 196
  portrayal of underground scenes 244
  portrayal of war memorials 64
  of Tursdale miners' lodge 64fn
  of West Stanley miners' lodge 244
  symbolism of 247, 269–70
  *see also* gala
Barnard Castle, County Durham:
  and Methodism 168
  parliamentary constituency 75fn, 80, 81, 87, 91, 96fn, 123
  Unionist Association 124
Barnsley Football Club 16
Barrington, Shute 173
Batey, Joe 50fn, 82, 117, 186, 237, 246
Beamish parish church 169
Bearpark, County Durham 53, 200, 212
Belgian miners 61
Bell, Colin 4
Bell, Lady Florence 121fn
Bell, Molly 121fn
Bell, Sir Hugh 117, 121, 135
Belmont, vicar of 179
Benfieldside parish church 182
Benney, Mark, see *Charity Main*
Berger, Stefan 124, 264
Bevan, Aneurin 222
Bewicke Main Colliery 116
Beynon, Huw 12
Birtley, County Durham 17, 39, 44, 53, 89fn, 145, 148, 152, 214
  anti-blackleg agitation 154
  Armistice Day commemorations 64
  Conservative Party support in 124
  distribution of relief 84, 100
  disturbances at 74
  Labour women's organization 158
  Methodist chapels in 188–9
Bishop Auckland, County Durham 81, 98, 218
Bishops' memorandum 88, 172–3
'Black Friday' (1921) 31
Blackhall, County Durham 53, 169, 180, 208, 251
Blackhill, County Durham 63, 64, 220

blacklegs 7, 17, 27, 41, 65, 84, 106, 113, 124, 127–32, 145, 187, 232, 235, 250, 255, 257
- at Adventure Colliery 230–1, 259
- courage of 129
- divisions caused by 229–30, 259
- historiographical treatment of 9fn, 81
- masculinity, supposed lack of 128, 151, 160, 162
- memories of 149, 230–1
- middle- and upper-class volunteers 36
- mistreatment of 129, 154–7, 159–64, 176, 199, 223–4, 229–30
- motives of 111, 131, 134, 138–9, 142, 149–51
- in Nottinghamshire 230
- numbers of men returning to work 78–9, 255
- profile of 36–7, 129–31
- and religious beliefs 131, 184, 187
- repercussions for wives and children 150, 163–4, 211, 230
- son of Chopwell vicar 194
- in South Wales 230, 254, 263
- stereotypes of 127–29

Blackwell, Joseph 140
Blaydon, County Durham 53, 61, 75, 81, 98, 112, 124
Board of Education 207, 211, 218
Boards of Guardians 207, 252
- of Chester-le-Street 100–3, 127, 155, 158, 209, 238
- criticism of 39, 102–3, 109–10, 126–7, 155
- of Durham 54
- of Easington 33, 102
- of Houghton 103
- Labour sympathies of 70, 83, 100
- of Lanchester 103, 110
- membership of teachers 209
- of Sedgefield 54
- of South Wales and Nottinghamshire 261–2, 263
- support given by 70, 100–3, 237–8
- variation in relief rates 54, 238
- *see also* relief

Board of Guardians (Default) Act 1926: 101
Boldon, County Durham 119
Bolton, Harry 75, 113
Bondfield, Margaret 123
boots, distribution of, *see* fundraising and charity
Bourke, Joanna 6
Bourne, Francis Cardinal 173
Bowburn football club 77
Boyne, Viscount 117
Bradlaugh, Charles 192
Brancepeth, County Durham 34, 60, 117
Brandon, County Durham 38, 53, 64, 94, 192
Brass, Captain 121
Bristol 41, 49
British Empire 51, 60–1, 202–3, 223
- Exhibition (1924) 202–3
- *see also* Empire Day

British Legion 63, 64, 65
Browney, County Durham 94, 117, 179, 212, 213
Bruley, Sue 141–2, 162
Bulmer, Martin 5, 9, 225
Burge, Alan 129
Burnhope, County Durham 83, 104
- gala organized during lockout 65, 199, 249

Burnopfield, County Durham 38, 74
Burt, Tommy 192
by-election, Wallsend (1926) 123
Byers Green, County Durham 54, 269
Byrne, David 48

Calhoun, C. J. 257
Callaghan, James 15
Cambrian Combine dispute 226
Cambridge University 36, 174, 177, 213
Campbell, Alan:
- on Scottish coalfield 11, 26, 43fn, 70–1, 119, 159fn, 260, 265
- on W. P. Richardson 136

Canada, emigration to 51
Cape, Thomas 153
Carr, Bill 128, 130
Carr, Griselda 159
Carter, Jimmy 15
Castle Eden, County Durham 132, 165
Castletown, County Durham 53
Catholicism:
- Diocese of Hexham and Newcastle 168

political allegiances of adherents 131, 151, 186
presence in the Durham coalfield 49, 167–9, 186, 218
response of Church hierarchy to general strike 173
rivalry with Anglicans 169
sense of fellowship engendered 169, 182–3
source material for study of 169
Catterall, Peter 168, 189–90
cavils, system of 12, 205
Central Labour College (CLC) 87, 93, 223, 265
Chamberlain, Neville 101, 127
Chaplin, Sid 52, 54, 242, 244, 252, 255, 269, 272
charity, *see* fundraising and charity
*Charity Main* 17, 54, 119, 266
Charlaw and Sacriston Colliery Company 117
Charlton Athletic Football Club 220
Chartism 111
Chester-le-Street, County Durham 32, 33, 68, 93, 164, 190, 208
Amateur Swimming Club 56
Board of Guardians 100–3, 127, 155, 158, 209, 238
Deanery 174, 178, 181
district council 98–99, 101, 126
local elections (1925) 98–9
parliamentary constituency 14, 81
Ratepayers' Association 38
school attendance in 206
Chester Moor, County Durham 194
children 17, 46, 70, 85, 101, 102, 158, 162, 172, 175, 179, 184, 238, 239, 243
attitude to blacklegs 128, 199, 223–4
awareness of social differences 27, 45, 57–8
help given to parents during lockout 34, 148, 199–200, 205, 234, 238
immersion in colliery life 13, 200, 242, 255
local rivalries 54–5
memories of the lockout 36, 57, 136, 233–4, 250
movement between schools 52
and religious bodies 165, 172, 175, 176, 179, 182, 190, 193
repercussions if father blacklegged 150, 211, 229–30
used in propaganda 139, 140, 199
as victims 34, 36, 82, 100, 108, 140–1, 147, 175, 179
*see also* elementary schools
children's feeding centres:
conduct of children in 46, 57
donations to 38, 118, 125, 183, 191
in elementary schools 70, 99–101, 142, 206–11, 233, 241
tensions at 162
theft from 133
volunteers in 159, 179
*see also* soup kitchens
Chilton, County Durham 219
Chopwell, County Durham 12, 75, 135, 144, 154, 160, 193, 208, 219, 256
anti-blackleg activity in 157, 158, 194
banner 112, 151
colliery 54, 55
deprivation in 54
radicalism of 104, 111–14
Christian Socialism 176, 191–2
Church of England:
as apolitical body 180–1, 183, 196
association with coal owners 167, 169–70, 196
association with Conservatism 169–72
available source material 168
concern with social questions 172–3
estimated number of adherents 167
financial consequences of strike upon 179–80
involvement in fundraising 178–9, 183
in Lancashire 260
neglect of by mining historians 166, 168
presence in colliery villages 165–7, 196–7
receipt of royalties 170, 192
sense of fellowship engendered 182–3, 193
*see also* clergymen; Henson; Welldon
Church, Roy 10
Churchill, Winston 6, 18, 104, 226, 229, 237–8
Churchyard, Revd H., *see* clergymen

Circular 703: 238
Citrine, Walter 31, 104, 253
Clarke, Peter 260
class consciousness 3–4, 42, 171
  assumed of miners 5, 81, 255
  of Durham miners 76, 266–8, 271
  fragmentary nature of 135
  impact of religion on 166, 182–9, 197–8, 258
  intensified during lockout 45
  interplay with other identities 23, 151
  in Lancashire 10, 260
  local nature of 266–70
  recent historiography on 10, 266
  relationship with 'community' 268–70
  in South Wales 262
  of women 43, 151, 161, 163
Claxton, W. J. 203
clergymen:
  attitudes to strike and Labour Party 175–7
  changing social backgrounds of 177
  Churchyard, Revd H. 202
  Davison, Revd Alfred 180
  Duncan, Revd James 170, 177
  Fenton Fyffe, Revd Thomas 181
  Hayward, Revd Harry 179
  Hodgson, Revd William 177–8
  Knight, Revd Percival 177
  Law, Revd W. 184
  Maish, Revd Edward 179
  opinion of miners 174–5, 177–9
  participation in colliery life 165, 177, 183
  Peck, Revd H. J. 175
  Rust, Revd Edward 175–6, 178, 194
  separation from parishioners 174, 177
  Watts, Revd Harry 176–8, 184
coal industry, *see* Durham coalfield; mining industry
Coal Industry Commission (1919), *see* Sankey Commission
*Coal is Our Life* 8–9, 74fn, 225
coal miners, *see* Durham miners; miners
Coal Mines Act (1926) 28fn, 67
  Durham opposition to 91, 245
  Labour opposition to 28, 63, 92, 93, 123, 153
coal owners 9, 32, 81, 106, 131fn, 135, 180, 192, 196, 201, 223, 251
  cooperation with union 116, 120–2, 128–9
  demands of in 1926: 1, 3, 28, 67–8, 88–90, 135, 198
  differentiation in colliery hierarchy 46, 119
  hostility towards 46–7, 82, 111, 120–2
  involvement in miners' welfare 116–19, 122
  support received from government 28, 123, 197–8
  *see also* colliery managers; Londonderry; paternalism
Cohen, Anthony 4
Cohn, Samuel 59
Coinsborough, Yorkshire 9
Cole, G. D. H. 21–2, 93
collective memory:
  of Durham mining communities 225–8, 241–9, 252–3, 270
  theories of 249–50
  *see also* memories of lockout
colliery deputies 9, 13, 90, 189
colliery managers 46, 111, 115–7, 119, 120–2, 123, *see also* coal owners
Colls, Robert 15–16, 223
Communism 95, 104–5, 110, 146
  in Chopwell 111–14
  hostility to 110–11, 136–7, 151
  in Scotland 265
  in South Wales 11, 113, 264–5
  support for 109–14
  *see also* Soviet Union; Young Communist League
Communist Party of Great Britain (CPGB) 87, 109–10, 208, 265
community:
  definitions of 4–7, 256–7, 268–72
  relationship with 'class' 268–70
  *see also* mining communities
compensation committee 116, 121–2
conditions of work, *see* mining industry
Conference on Christian Politics, Economics and Citizenship (COPEC) 172–3

Conservative Government
(1924–29) 3, 32, 72, 85, 104
attitude to strike and strikers 61, 82, 89, 105, 123, 197–8
and Coal Mines Act 28, 123
conception of community 7
criticism of 28, 63, 123, 140, 230, 237
miners' support for 106–7
and non-renewal of subsidy 1–2, 62
and policing of strike 72–3
propaganda 28
Conservative Party 135
female attitudes towards 143
hostility towards supporters of 81, 84, 123, 127
in Lancashire 260
in South Wales 9
support for in Durham coalfield 81–2, 122–27, 259
*see also* local Conservative associations; Conservative Government (1924–29)
Consett, County Durham 40, 64, 73, 81, 172
Consett Iron Company 33, 111
Cook, A. J. 45, 66, 71, 86, 140, 177, 194, 250
biography 7fn, 87, 92, 263
on blacklegs 65, 129, 150
at Burnhope gala 65, 249
on causes of industry's decline 93
on 'community' 7
comparison to Scargill 230, 248–9
criticism of 71
on Durham miners 65
during First World War 64
on general strike 28
on Hensley Henson 170
on lockout 253
radical reputation of 85, 87, 108, 263
recent scholarship on 87, 87–8fn
on Royal Commissions 94
support for 72, 85–8, 96, 152, 161, 249, 267
on TUC leadership 32
Coombes, Bert 15, 216, 232, 243
Cooper, Robert 116
Co-operative Society 34, 57, 111, 130, 145–6, 185, 191, 193
affected by lockout 38, 239
support given to miners 37–8
tensions with smaller tradesmen 38–9
Cornforth, County Durham 181
Cornsay, County Durham 134
corruption, amongst union officials 132, 134
cotton unions 260
Councils of Action 104
Coxhoe, County Durham 39, 42, 54
Coxon, William 210fn
Craddock, Mary 136–7, 159–60
Cramlington, Northumberland 47
Crawcrook parish church 182
Crawford, William 97, 185
cricket 56, 147, 182–3, 208
Cronin, A. J. see *Stars Look Down, The*
Crook, County Durham 214
Cumberland 116fn
Currie, Robert 195

Daunton, M. J. 18, 264
Davidson, Randall, *see* Archbishop of Canterbury
Davies, S. O. 263
Davison, Revd Alfred, *see* clergymen
Dawdon, County Durham 39fn, 98, 118, 132, 170, 209
strike (1929) 148
Dawes plan 93
Dawson, A. J. 203–4, 213
Deaf Hill, County Durham 223
Dean and Chapter miners' lodge 134
*Death of a Salesman* 51
Deerness Valley, County Durham 115, 139, 166, 167–8, 185, 187, 217
Divisional Labour Party 210–11
*see also* Moore, Robert
deference 81, 114, 120, 122, 196, *see also* paternalism
Dennis, N., see *Coal is Our Life*
Dillon, Malcom 108, 202
Dipton, County Durham 39fn, 96
disasters 253
Seaham (1880) 226, 243
Usworth (1885) 92
Stanley (1909) 227, 244
Easington (1951) 227
district settlements, fight against 3, 66–7, 83
doctors 33, 43, 45, 103, 216
domestic servants, *see* female employment
Douglass, Dave 229, 248

drunkenness 239
Duncan, Revd James, *see* clergymen
Dunkirk 253
Dunn, Jack 214
Dunnico, Herbert 40, 172
Durham Cathedral 196
Durham City 36, 247, 256, 269
Durham coalfield:
accidents in 92, 122, 221, 233, 242, 244, 248, *see also* disasters
bracketed with Northumberland 12
comparison with South Wales 13, 261–5
early attempts at unionization 65, 247
early history of 243
early stronghold of Liberalism 97
geological conditions of 18, 92
map of 2
mechanization 243
migration to and from 48–52
movement within 52–5
neglect by historians 12–13, 264
paternalism in 114–22
profitability of 67
rivalry with other districts 68–9
strikes and stoppages in 148, 156, 173, 225, 241, 242, 251, 252, *see also* lockout (1926); strikes and stoppages, (1921); strikes and stoppages (1984–5)
typicality of 11–13, 266
variation within 11–12, 53–4, 119–20, 236–7
wider relevance of 13
*see also* Durham miners
Durham Coalowners' Association (DCOA) 52, 106, 115–16, 121, 128, 131fn, 266
Durham Constabulary, *see* police
Durham, County:
dominance of miners and mining industry 13, 23–5, 76–7, 255–6, 258, 263
middle-class presence 42, 256
mobility of population 48–52
number of radio licences 76
Durham County Association of Trades Councils 267
Durham County Colliery Enginemen, Boiler Minders and Firemen's Association 110
Durham County Council, *see* local government
Durham County Labour Women's Advisory Council 144, 146, 163
Durham County Record Office 253
Durham County Teachers' Labour Group 210
Durham Football Association 55, 112
Durham miners:
attitudes to union 21, 65–6, 69–72, 78, 80
attitudes to work 15–17
class consciousness of 76, 266–9
dialect 27
histories of 252
individualism of 132–4
intermarriage within communities 26–7
international perspective of 59–60, 93
national loyalties of 58–65
occupational hierarchy within the industry 14, 18, 46–7, 55, 90, 246
occupational variety among 14
oral history tradition 242–5, 270
rank-and-file militancy 82–3
recruitment of during First World War 62
and respectability 56–8, 236
sense of regional identity 48–58, 60, 65–72, 75–6, 265–72
solidarity of 3, 65–6, 78–9, 254–6, 265–72
support for A. J. Cook 86–8
unromantic view of 133–4
*see also* miners; Durham Miners' Association
Durham Miners' Association (DMA) 12, 13–14, 41, 50, 52, 70, 81, 92, 95, 96, 122, 131, 191, 196, 230–1, 254, 259
Approved Society 78
attitude to education 221
attitude to female involvement during lockout 159
attitude to general strike 106, 266
construction of the past 246–9, 252, 270
cooperation with colliery hierarchy 115–16
criticism of 84–5, 135

divisions between leaders and rank and file 82–85, 96
duties of officials 78
earlier history of 65, 97, 185
importance of 135, 256, 267
influence on local government 34, 83–4, 97–9
links with Labour Party 80, 98–9
links with Methodism 185–6
miners' loyalty to 78, 136, 258–9, 265, 269–71
opposition to Communism 110
payment of lockout benefit 70, 84–5, 111, 106–7, 132, 238
relationship with MFGB 65–72, 97, 125, 259, 266, 269
successes of 69, 247
support for Spanish Republicans 59
*see also* banners of the DMA; gala

Durham Municipal and County Federation 210
Durham parliamentary constituency 81
Durham University 169, 174, 177, 256
  Joint Tutorial Committee 216

Easington, County Durham 64, 126fn, 194, 208, 236
  Deanery 26, 73, 174, 178, 181
  local Conservative club 125
  pit disaster (1951) 227
  Poor Law Union 33, 100fn, 102
  school attendance in 206
  vicar of 177
  women's Labour section 176
Eastwood, Nottinghamshire 156
Eden, County Durham 53, 84, 107, 125, 130, 132
Edinburgh University 222
education:
  access to university 213, 215, 217, 220
  adult 215–17, 219–20, 223
  alienation as a result of 216–7, 222
  financial obstacles to 214
  parental attitudes to 214–5, 217–8
  regard for 215, 219–20, 221
  religious 165–6, 202
  technical 218–20
  values promoted 201–4, 212–13, 222–3
  *see also* elementary schools; secondary schools; social mobility

Education (Provision of Meals) Act (1906) 206
Education ('Butler') Act (1944) 200
Egan, David 263fn
Eklund, Eric 268
elementary schools 52, 184, 258
  curriculum 203, 212–13
  distribution of boots 206, 224
  division between home and school 200
  during lockout 205–11, 223–4, 258
  influenced by coal industry 204–5, 207–8, 222–3
  involvement of colliery hierarchy 201–2
  libraries 203–4
  maintained by a religious institution 165
  numbers attending 200, 205–7
  provision of meals 70, 99–101, 126, 142, 206–11, 233, 236–7, 241
  teachers 98, 196, 208–12, 219, 220, 222, 223
  use of premises 208
  *see also* education
Elemore Colliery 116, 243
Emergency Powers' Act, *see* Emergency Regulations
Emergency Regulations 69, 105, 107, 108–9, 128, 155
Empire, *see* British Empire
Empire Day 202–3
engineering industry, affected by lockout 33
enginemen 13, 14, 66, 110
Eppleton, County Durham 64, 116, 243
Escomb, vicar of 177
Esh, vicar of 180, 195
ethnicity 10, 49, 56, 200fn, 259–61, 271
Europe 59, 61–2, 212, 261, *see also* Belgian miners; French miners; Germany; Holland
Evans, Neil 264

Fagge, Roger 115, 261
fancy dress competitions 101, 116, 128, 154fn
Fanny pit 129–30
farmers 12, 38–40
Fatfield, County Durham 53
Federation of British Industries 185

Felling, County Durham 53
female employment 13, 142, 213, 260
  domestic servants 43–4, 142, 148, 213
  increase due to lockout 146
feminism 162
Fenton Fyffe, Revd Thomas, *see* clergymen
Fentress, James 225, 249
Ferryhill Ratepayers' Association 126
First World War 76, 95–6, 151, 158, 203, 244
  commemoration of during lockout 62–5
  imprisonment of A. J. Cook during 64
  recruitment of miners during 62
  strikers compared to Germans 58, 171
  strikers dependent on war pensions 238
Fishman, Nina 265
Fletcher, Meredith 16
Flint Hill, County Durham 65
*Flying Scotsman, The*, derailing of 47, 105
football 16, 55–6, 77, 97, 105, 112, 134, 147, 153, 217, 222, 235–6
  church-based teams 182
  as means of escape 220–1
Fox, Colonel Lane 140fn
Francis, Hywel 22–3, 59, 262–3
Frankland Colliery 130
French miners 59, 61
Fry and Sons Ltd 41
fundraising and charity 234, 239
  attitude of other occupational groups 36, 38, 41
  boot funds 36, 64, 125, 154fn, 180, 183, 191, 208–9, 210
  boots, distribution of 57–8, 84, 206, 224, 263
  competition for funds 34
  distributed by MFGB 70
  donation of Prince of Wales 140
  international 59
  involvement of religious bodies 179–80, 183, 191
  involvement of teachers 210
  national scope of 66
  provision of meals, *see* children's feeding centres; soup kitchens
  from Soviet Union 59, 107–8, 175
  stigma attached to 57–8
  supported by colliery hierarchy 116–19
  supported by local Conservative groups 125
  and the Webbs 119
  women's involvement in 153–4, 161
  *see also* soup kitchens
Fynes, Richard 65

gala 196, 247–9, 256, 269
  (1925) 171–2
  (1926) unofficial gala at Burnhope 65, 83, 199, 249
  (1927) 259
  (post-1945) 248–9, 267
  women's 144
  *see also* banners of the DMA
Gallacher, Hugh 235
Garside, William 12
Gateshead 12, 36, 53, 104
Gateshead Central Library, oral history project 12, 35–6, 77, 89, 149–51, 158, 162
Geary, Dick 114
Geddes Axe 213
gender:
  relationships in mining villages 74, 141–2, 145–6, 157–8
  roles 138, 141–3, 147–8, 152–3, 155–64, 258, 260
  *see also* female employment; masculinity; women
general elections:
  (1910) 143
  (1918) 143
  (1922) 143
  (1924) 81, 97, 143–4
  (1929) 123, 143
  (1931) 82, 252
general strike 7, 21, 35, 58, 60, 75, 108, 235, 241, 253
  apathy of miners 94
  blacklegging during 36–7, 106–7
  chronology of 2
  distinct from lockout 105, 106–7, 266
  historiography of 3fn
  nostalgia about 235
  opposition to amongst miners 106–7
  policing during 72–4

reaction of DMA 106
reaction of government 7, 105
reactions of religious leaders 3, 171–3, 176, 180
response of working class 28, 31–3
revolutionary implications of 103–4, 107–8
in South Wales 262
tension caused by 31–2
victimization following 31
violence during 47, 104–5, 155
youth response to 96
generational differences:
in attitudes to education 219–20, 221
in attitudes to migration 51
in attitudes to respectability 57
in memories of lockout 233–5, 239
in political attitudes 70–1, 95–7, 143
Germany 58, 93, 151, 171
mining communities of 259
rivalry with 61
support from 59
*see also* Ruhr coalfield
Gier-Viskovatoff, Jaclyn 141
Gilbert, Alan 195
Gilbert, David 10, 22–3, 75, 261–2
Gilliland, James 89, 113, 186
Gladstone, William 113, 247
gold standard, return to 1, 3, 67, 93
Gordon Hall, Valerie 141
government, *see* Conservative Government (1924–29); local government
Graham, Duncan 61, 152
Grange Villa, Labour women's section 158
Grangetown, County Durham 118
Grant, J. C., see *Back-to-Backs, The*
Great Lumley, County Durham 98–9
Greece 93
Greenside, County Durham 94, 165, 219
Greenwood, Arthur 140fn
Gregory, Adrian 64
Gregory, Roy 8
Grenfell, David 62
Griffiths, Trevor 10, 49, 260
Griffin, Alan 142, 156fn
Grimley, Matthew 173
Halbwachs, Maurice 249–50
Halévy, Elie 166
Hall, G. H. 62
Hamsteels, vicar of 175
Hamsterley, County Durham 53, 124
Handley W. H. 101
Hardie, George 121, 140fn
Hardie, Keir 112, 151
Harrison, Royden 9–10
Harvest Festival 183, 195
Haswell, County Durham 191
Hayward, Revd Harry, *see* clergymen
Hazard Colliery 52
Headlam, Cuthbert 87, 91, 95, 104, 109, 121, 123, 134, 143, 210
health, *see* lockout, effect on health
Hebburn, County Durham 34, 53
Hedley Hill School 210
Henriques, F., see *Coal is Our Life*
Henson, Herbert Hensley 45, 48, 94–5, 175, 176, 196
criticised by miners 170–2, 194, 196
on miners' housing 47, 212
political views 3–4, 108, 170–3, 176–7
Henty, G. A. 203–4
Hepburn, Tommy 248–9
Hetton, County Durham 33, 52, 110, 126fn
Hetton-le-Hole parish church 183
hewers 55, 126
as archetypal miners 14–15, 25
as elite of mining workforce 14, 46
shorter hours of 67, 91, 245–6
wages 30, 90
Heworth, County Durham 53, 251
Hickey, S. H. F. 18, 244
Hill, Dr Eustace 207, 237
Hitchin, George 27, 39, 134, 216–7, 219, 238, 241
Hobsbawm, Eric 166, 228, 267
Hodges, Frank 66, 264
Hodgson, Revd William, *see* clergymen
Holland, support from 59
Horden, County Durham 36, 56, 86, 132, 227, 250
Horner, Arthur 263
Horsley, Lee 195

Houghton-le-Spring, County Durham 33, 42, 53, 133, 184, 191, 209
  Board of Guardians 103
  Catholic church 168
  Deanery of 174, 178, 181
  local Conservative club 124fn
  miners' lodge 94
  parish church 183
  parliamentary constituency 81, 105, 106fn
  school attendance in 205–6
hours 36, 89fn, 197–8, 245
  changes proposed by owners 1, 3, 88–9, 135, 267
  effect of increase on accident rate 93
  effect of increase on European workers 61
  effect of increase on miners' wives 152–3
  effect of increase on rest of working class 28
  effect of increase on unemployment 92
  increase in less preferable to wage cuts 91
  middle-class ambivalence towards increase 44, 185
  opposition to increase in 3, 28, 61, 63, 67–9, 83, 88–9, 91–3, 107, 123, 152–3, 199, 215, 221, 245–6, 267
  and relationship between DMA and MFGB 65, 71
  and Samuel Report 17fn, 91–2
  and Sankey Report 67
  shorter hours traditional in Durham 67–8, 88, 91, 245–7
  worked in various industries 29–30
  *see also* Coal Mines Act
housing 78, 255, 264
  home ownership 127, 134
  increase in rent arrears as result of lockout 33
  influence of on community 264
  overcrowding 47, 212
  provision of free coal 34, 117–18, 121
  tied 25–6, 46–7, 114–15, 117, 121, 126, 243, 245, 251
  *see also* aged miners' homes
*How Green Was My Valley* 201
Howard, Stuart 185, 226
Howell, David 192, 265
Hudson, Mark 56, 226
Hull 41, 73
Hunsdon, Lord (Herbert Gibbs) 58fn
Husemann, Friedrich 59

Incorporated Association of Retail Distributors 37
Independent Labour Party (ILP) 109, 267
individualism 132–4
intermarriage within mining communities 26–7, 73
International Federation of Miners 59
internationalism 23, 59–60
Ireland, immigration from 48–9
Irish:
  independence, attitudes towards 23, 59
  presence in Lancashire 260
  as strike-breakers 131
Italy 93

Jarrow 34, 53
Jenkins, Arthur 264
Jeremy, Paul 88
Jesmond 43
John, Angela 145
Johnson, Bessie 152, 161, 235, 241
Johnson, Paul 57
Jones, Jack 28
Joyce, Patrick 260, 270, 271

Kerr, Clark 21–3
Kelloe, County Durham 45, 70
Kent 62, 214
Kibblesworth, County Durham 245
Kimblesworth Colliery 121
Kinnear, Michael 167–8
Kirkup, James 27
Kirkwood, David 60
Knight, Revd Percival, *see* clergymen

Labour Party 31, 40, 181, 185, 230, 268
  association with miners 8, 21
  attitudes of parish priests towards 175–7
  constituency activities 154
  cracks in support for 81–2, 97, 102–3, 109, 127, 202

lack of engagement with young people 96
links with DMA 98, 135
and local government 70, 80, 84, 98–103, 202, 237
mythologized history of 251
participation in parliamentary debates 28, 60–3, 67, 73–4, 89, 91, 105, 106, 108, 121, 122, 123, 135, 140fn, 152–3, 215, 221, 222, 237, 245
and religious belief 176, 181, 185–6, 187, 189, 190
rise of 144
in South Wales 9, 78, 261, 264
support for in County Durham 80, 97, 101, 110, 123
support for by teachers 209–11
support given to strikers 34, 84, 97, 99–101, 116, 208, 237
women's organizations 143–6, 153–4, 158, 161, 163, 176, 181
Lambton Colliery 53
Lancashire 10, 48, 49, 63, 260
Lanchester, County Durham 87, 100fn, 103
landlords 13, 33
Langley Moor, County Durham 66
Langley Park, County Durham 56, 107, 179fn
Lansbury, George 194
Larkhall, Lanarkshire 27
Laslett, John 261
Law, Revd W., *see* clergymen
Lawrence, D. H. 156, 201
Lawrence, Jon 251
Lawson, Jack 69–70, 82, 200, 221, 245, 266
biography 14fn, 48, 50fn, 119, 265
on Coal Mines Act 123
on colliery managers 119, 122
criticism of 85
on education 216–7, 221–2
on gala 247
on Herbert Smith 16, 222, 268
and Methodism 185, 186, 189, 193, 216–17
on miners' union 78
on nineteenth-century coalfield 48, 69–70
on policing 74
portrayal of the miner 14–17, 25–6, 78, 122, 270
on school feeding 45, 100
on women of the coalfield 148, 153fn
Lawther, Andy 144
Lawther, Emmy 144, 146
Lawther, Will 75, 86, 88, 106, 112, 113, 143, 265
Lawton, Tommy 42, 234
League of Nations Union 60
Lee, Jennie 222, 240fn
Lee, Peter 66fn, 106, 119, 142, 245
biography 42–3fn, 48, 50–1, 185, 186, 245
chairman of Durham County Council 43fn, 100, 142
on miners' solidarity 42, 88
moderation of 82–3, 97, 110
Lee, Robert 48fn, 174fn, 177
Lenin, Vladimir Ilyich 104, 111–13, 149, 151
Lib-Labs 97, 143, 185, 263
Liberalism:
local organizations 97–8
party candidates 123
in South Wales 9
support for in County Durham 81fn, 82, 97–8, 123, 125, 187, 189, 192, 210
libraries 27, 203–4, 216–17, 223
Lieven, Mike 9
Lincolnshire 73
Littleburn, County Durham 94
Llanelli, Monmouthshire 226
Llewellyn, Richard 201
Lloyd George, David 48fn, 214
LNER rail company 236
local Conservative associations:
social clubs 124–5
support given to strikers by 126
women's groups 97, 126, 181
*see also* ratepayers' associations
Local Conservative League 125
local elections:
(1925) 98–9, 101, 175
(1928) 146, 177
local government:
criticism from Left 102–3, 109–10, 155
criticism from Right 126–7, 210
dominated by Labour Party 70, 80, 83, 99–102, 203, 237

local government: (*cont.*)
Durham County Council 43fn, 70, 80, 99–101, 110, 118, 126, 175, 177, 201–3, 210, 213
influence of miners' union 34, 83, 98–100
involvement of religious figures 175, 177, 184
and Peter Lee 43fn, 100, 142
provision of poor relief, *see* Boards of Guardians
provision of school meals during lockout 70, 99–101, 126, 142, 206–7, 236–7
taxation, *see* rates
lockout (1926):
apathy towards amongst miners 94–5
attitudes of wider working-class community 33–7, 60, 76–7, 117–18
awareness of wider context of 93
ballot to end lockout (November 1926) 83, 95, 119–20, 231, 254–5
chronology of events 1–3
effect on concepts of respectability 56–8
effect on health 54, 207, 236–40
effect on religious belief 195–6
effect on shopkeepers 37–42
effect on sport 55–6, 147
effect on wider community 33–4
as 'festival of the oppressed' 234–6
government reaction to 61, 82, 89, 105, 123, 197–8
historiography of 3fn, 254
increase in leisure time as consequence of 56, 147, 215–7, 234, 236
international support for 59–60
memories of the First World War during 62–5, 244
motivation of miners during 88–94
numbers returning to work 78–9, 254–5
payment of lockout benefit 70, 84–5, 106–7, 111, 132, 238
policing during 72–5
in South Wales 261–3
terminology of 1fn, 89
unromantic view of 19
variety of experience during 232–41
violence during 74–5, 103, 104–5, 108, 129, 134, 155–7, 159–62
weather during 94, 147, 205–6, 231–2, 234
*see also* blacklegs; fundraising and charity; general strike; memories of lockout
Lockwood, David 81
London 21, 47, 50
Londonderry, Lord (Charles Vane-Tempest-Stewart) 40, 47, 54, 98, 108, 150, 170, 177, 202
and miners' welfare 117–9
Lumphinnans, Fife 57
Lyons, vicar of 184

McCord, Norman 126
MacDonald, James Ramsay 82, 104, 113, 140, 150, 176, 209, 210fn, 252
Macfarlane, Alan 5
McIlroy, John 67, 136, 156fn
Macintyre, Stuart 57, 111
Macmillan, Harold 63
Mainsforth Colliery 155
Mainwaring, Will 263
Mardy Jones, T. I. 106
Marley Hill, County Durham 74
Marx, Karl 111–13, 151, 191–2, 265
masculinity, supposed lack of in blacklegs 128, 151, 160, 162
Mason, Tony 105
Mates, Lewis 59
Maughan, Annie 155, 161
mechanics 13, 14, 87
mechanization 243–4
Medical Officer of Health for County Durham, *see* Hill, Dr Eustace
memories of lockout:
development of a coherent narrative of 18, 226–32, 241, 248–9, 251–2, 266
merge into memories of interwar years 240–1
varied 232–41, 250–2
*see also* collective memory
Merthyr Tydfil judgement (1900) 102
Metcalfe, Alan 15
Methodism:
and blacklegging 131, 184, 187
and class consciousness 166, 182, 184–9, 197, 258

difference between Primitive and Wesleyan sects 166, 188–9
divisions between chapel hierarchy and rank and file 189–90, 197
divisions between Methodists and non-Methodists 193, 217
estimated number of adherents 167–8, 195
fundraising and provision of relief 38, 187–8, 190–1
lack of source material 168–9
links with Liberalism 185, 187, 189
links with trade union leaders 16, 166, 185–6
presence in colliery villages 166, 177, 190
sense of fellowship engendered 182
social mobility 187, 193, 217
Meyer, Sir Anthony 8
middle classes 42–6, 126, 256, *see also* doctors; farmers; landlords; publicans; shopkeepers
Middlesbrough 160
migration 22, 48–52
Millar, J. P. M. 222
Miller, Arthur 51
Mineowners' Association of Great Britain (MAGB) 46, 69fn, 184
*Miner, The* 66, 141, 197–8
miners:
as archetypal proletarians 9–10, 23, 81, 135
association with Labour Party 8, 81
autobiography 226
contrasting images of 15–18
fertility levels 142
as independent collier 71
occupational hierarchy within working class 16–17, 25
in popular imagination 15, 225–7
visually distinctive 27
*see also* Durham miners; hewers; putters
Miners' Federation of Great Britain (MFGB) 14, 50, 83, 85, 88, 89, 135, 140, 231, 237, 263, 267–8
and A. J. Cook 85
aims of 2–3, 66
construction of loyalties to 269–71
criticism of 70–1, 95, 97, 125, 257, 259
and First World War 62, 64
fundraising and distribution of relief 59, 70, 107, 111, 153, 238
and George Spencer 3
and Labour Party 97, 143fn
memberships of constituent districts 80
official history of 60, 229
propaganda during lockout 7, 66
relationship with DMA 65–72, 97, 125, 259, 266, 269
and women 141, 153, 159
*Miners' Next Step, The* 263
Miners' Welfare Fund 219–20
mining communities:
'Ashton' 8
cemented by women 140
comparative study of 9, 260
definitions of 5
homogeneity of 5, 254
ideal type 5, 225
importance of the union 260–1
'isolated mass', concept of 5, 9, 21–3, 271
militancy of 16
romantic view of 8, 133, 227
separation of 75, 271
solidarity of 8, 18, 21–3, 42, 259–60
strike propensity of 10
regional variations 11, 66, 260, 265
revisionist study of 9–10, 265–6
variety within 9
*see also* Durham miners; miners
mining industry:
accident rate 92,
conditions of work 17, 92, 185, 243
dangerous nature of 18, 92–3, 153, 221, 226–7, 243–4
dominance of in County Durham 13, 23–5, 76–7, 255–6, 258, 263
hereditary nature of 25–6, 213, 217–18, 242, 270
regional differences 66
*see also* hours; wages
Mining Industry Act (1926) 17fn
Ministry of Health 39, 100–102, 126, 158, 179, 207, 237–8
Minority Movement 87, 110
Moffat, Abe 57
monarchy, attitudes towards 202–3, 223

Mondism 185
Monk Hesleden, County Durham 165
Monkwearmouth, County Durham 133, 247–9
Monmouthshire 80fn, *see also* South Wales and Monmouth
Moore, Robert 46
on blacklegs 131, 139, 184, 187
on Church of England 196
estimates of religious belief 167–8
on Methodism 166, 182, 184–8, 191, 193, 196–7
on paternalism 115, 184
on social mobility 217
sources used by 168
Moorsley, County Durham 52
Morgan, Kenneth O. 229
Morpeth, Northumberland 192
Morris, Margaret 32
Mothers' Union 97
Murton, County Durham 53, 56, 64fn, 86, 128
myth, historical treatment of 228

National Blind Institution 34
National Conservative League 124
National Council of Labour Colleges 222
National Federation of Colliery Enginemen and Boilermen 66
National Government 31fn, 40fn, 94fn
National League of Liberal Trade Unionists 97–8
National Secular Society 192
national settlement, fight for 3, 66–7, 83
National Union of General and Municipal Workers 28
National Union of Mineworkers (NUM) 159, 230, 264
National Union of Railwaymen (NUR) 32, 265, *see also* railway workers
nationalization 47fn, 93–4, 185, 248, 267
New Kyo, County Durham 110
Newbottle, County Durham 179fn, 200
Newby, Howard 4, 268
Newcastle United Football Club 221, 235–6
Newcastle upon Tyne 12, 34, 36, 87fn, 105, 106, 168, 234
Newsom, John 50, 55, 134
newspapers, local 76, 242–3
Nicholson, Tony 4
non-political trade union movement 71, 78, 108, 110fn, 129, 265, 271
Nonconformists 59, 172
Baptist 167, 172
Methodist New Connexion 186, 188
Presbyterian 186fn
Primitive Methodist, *see* Methodism
Quaker 184
Salvation Army 186fn
United Methodist 186fn
Wesleyan Methodist, *see* Methodism
Norfolk 73
North Hetton welfare committee 52
Northern General Omnibus Company 36
*Northern Light, The* 154
Northumberland:
accident record of 92
and blacklegging 65, 79
bracketed with Durham coalfield 12, 50, 65, 71, 156
coalfield, early history of 243, 251fn
derailing of *The Flying Scotsman* 47, 105
and domestic service 43
and effect of lockout 92, 239
memories of residents 17, 39, 51, 162, 174, 204, 233
miners' housing in 46–7
miners' response to First World War 62
Wallsend by-election 123
women's Labour organization 144
Northumberland and Durham joint strike committee 96
Northumberland, Duke of (Alan Percy) 47
Nottinghamshire coalfield:
and blacklegging 3, 67, 142, 230
comparison with South Wales 261–2
disturbances in 156
Dukeries coalfield 114–5, 271
Mansfield 261
moderation of 11, 22, 114–5, 261–2
response to lockout 67

Nottinghamshire Miners' Association (NMA) 3, 110fn, 114

oral evidence, use of 12, 227–8
Orange lodges 260
Organization for the Maintenance of Supplies (OMS) 106, 266
Ottey, Roy 230
Ouston, County Durham 158, 214, 239
Outram, Quentin 10, 46
Owen, Robert 112, 222
Oxford University 36, 174, 213

pacifism 59, 64, *see also* League of Nations Union
Paddon, Captain 117
Palmer, Sir Alfred 104
Passerini, Luisa 227
paternalism 46, 114–22, 135, 169, 184, 223, 260, *see also* deference; housing
Paynter, Will 264
Pease, A. F. 184–5
Pease, J. A., Baron Gainford of Headlam 184–5
Pelaw, County Durham 53
Pelaw Main Collieries 131
Pelton, County Durham 127, 164
Pelton Fell, County Durham 239
Phillips, Marion 153
pit baths 17
pit ponies 14, 116
pitmatic 27
Pittington, County Durham 72, 125, 182
Plebs League 223, 265
Plymouth 105
police 105, 130, 136, 154–6, 160, 177, 187
  imported from other areas 73–4
  relationship with mining communities 72–5, 109
  *see also* special constables
Pontycymer, Mid Glamorgan 227
Pontypridd, Mid Glamorgan 106
Poor Law Unions, *see* Boards of Guardians
Pope, Robert 189
Port Kembla, NSW, Australia 268
Porter, Abigail 141
Potts, Jack 220–1
Priestley, J. B. 21, 27, 45
Prince of Wales, Edward 140
Protestantism, *see* Church of England
provision of meals, *see* children's feeding centres; soup kitchens
publicans 13
putters 14, 46, 55, 90, 135

Quebec, County Durham 174
Quiller-Couch, Mabel 203

Race, John 134
radio licences 76
railway workers 29–30, 43, 60, 77, 98, 253 *see also* National Union of Railwaymen
Rainton, County Durham 52, 101, 152, 160, 230–1, 235
Rainton Colliery Company 131fn
Ranger, Terence 228
ratepayers' associations 256
  at Annfield Plain, Stanley and Tanfield 126
  at Chester-le-Street 38
  at Ferryhill 126
  formation of during lockout 126, 175
  involvement of vicar 175
rates:
  affected miners 126–7
  complaints about 44, 126–7, 211
'Red Clydeside' 60
Red International of Labour Unions 263
Redmayne, Sir Richard 44
reference groups, theory of 22
region, concept of 11
Relief Committee for the Miners' Wives and Children 140, 153
relief (provided by state):
  effect on local taxation rates 44, 126–7
  fraudulent claims 132
  issue of payment to strikers 85, 102–3, 159, 179, 239
  seen as degrading 252, 261–2
  theft of 132
  variation in rates of benefit 54, 238, 240
  *see also* Boards of Guardians
religious belief 258
  apathy towards 193–4
  extent of 165, 167–8, 195
  hostility towards 192–5

religious belief (*cont.*)
and the mining community 196–8
*see also* Catholicism; Church of England; Methodism; secularism
Report of the Departmental Committee on Workmen's Compensation (1920) 116
respectability 56–8, 135, 193, 236, 254
Rhondda 9, 127fn, 189, 263, *see also* South Wales and Monmouth
Richards, Andrew 228, 269
Richardson, Robert 105, 108, 186
Richardson, W. P. 66, 85, 92, 101, 110, 136, 186, 251, 267–8
Ridley, Maurice 96, 249
Ritson, Joshua 40, 50fn, 68, 70, 73, 123, 153, 173, 186, 196
Roberts, Elizabeth 163
Robson, James 50, 68, 96, 122, 186
Ropner, Major (later Colonel) Leonard 82
'rough music' 156–7
Rowlands Gill, County Durham 70, 129
Royal Commission on the Coal Industry (1925), *see* Samuel Commission
Royal National Lifeboat Institution 202
Royal Society of St George 75
royalties 46–8, 116, 170, 192
royalty owners 46–7, 117
Royle, Edward 192
Ruhr coalfield 18, 93, 114–5, 244, 259
Runciman, W. G. 22
Ruskin College 144, 265, 266
strike (1909) 265
Russia, *see* Soviet Union
Ryhope, County Durham 41, 105fn, 126fn, 177
Ryton, County Durham 60

Sacriston, County Durham 46, 53, 76, 94, 182, 183, 244, 250
Saklatvala, Shapurji 110
Samuel Commission 17fn, 44, 94
on coal royalties 47
on discontent in coal industry 246
on effect of increased hours 91–2
on miners' wages 28–30, 90–1
on occupational and geographical mobility 25, 49
on pit baths 17
Samuel, Raphael 228
Sankey Commission 47, 66, 67, 92, 94, 116
Sankey, Sir John 47fn, 94fn
Savage, Michael 270
Scargill, Arthur 230, 248–9
schools, *see* elementary schools; secondary schools
Scottish coalfield 11, 50, 57, 61, 200, 221, 240fn
domestic servants 43fn
emigration from 48–9
and employer control 119
endogamous nature of 26–7
ethnic divisions in 260
Fife 57, 222, 265
Lanarkshire 260, 265
Larkhall 27
Lumphinnans 57
and soup kitchens 159fn
union divisions in 71, 265
United Mineworkers of Scotland 265
Seaham, County Durham 73, 132, 140
colliery 54
Colliery Relief Fund 243
effect of lockout in 236
Labour women's group 44, 93, 110, 144–5, 154fn, 161
parish churches 168, 182, 183, 191
parliamentary constituency 81, 82, 123, 209, 252
pit disaster (1880) 226, 243
police court 108
schools 52, 201–3, 207, 209, 218
typicality of 236
and the Webbs 44, 93, 98–9, 110, 144–5, 161, 236
Second World War 63, 225, 253
secondary schools 213–15, 217
secularism 192–4, 198
Sedgefield, County Durham 54, 80–1, 82
Seed, Jimmy 220
Shaw, Tom 61
Sheffield 128, 222
Sherburn, County Durham 19, 125, 191
Sherburn Hill, County Durham 35, 64fn, 93, 208

Shildon, County Durham 176, 183, 245, 272
Shinwell, Emanuel 194
shipping industry 33, 43, 92
Shirkie, Robert 66
shopkeepers 13, 77, 255
  affected by lockout 37–8, 41–2
  support given to miners 38–42, 60
  tensions between smaller tradesmen and Co-operative Society 38–9
Shotton, County Durham 45, 118, 208, 209, 212
Siegel, Abraham 21–3
Silksworth, County Durham 64fn, 128, 191
Sinn Fein 260
Slaughter, C., see *Coal is Our Life*
smallpox epidemic 237
Smillie, Robert 62, 66, 89fn
Smith, David (Dai) 6, 23, 59, 262–3
Smith, Herbert 66
  on Armistice Day 65
  biography 16fn, 92
  criticism of 85, 125
  intransigence of 87
  Lawson's portrayal of 16, 222, 268
  on mining women 140
  radical reputation of 108
social memory, *see* collective memory
social mobility:
  destinations of school leavers 25, 213–5
  and education 201, 213–20, 257
  obstacles to 21, 213–5, 218–9, 257
  prejudice against 217
  and religion 187, 257
  represented in literature 201
  and sport 220–1, 257
  within coal industry 218–9
solidarity:
  of Durham miners 3, 11, 42, 65–6, 78, 79, 82, 88, 197, 208, 254–5, 259, 266–72
  of mining communities (perceived) 8, 18, 21, 23, 227
  of wider working class 28, 31–3
  *see also* class consciousness
Somerset 69, 135
Somme, Battle of the (1916) 63
*Sons and Lovers* 201
soup kitchens 119, 229, 238, 241, 253, 263
  donations to 38–9, 125
  memories of 36, 57, 239
  for single men 38, 159, 239
  volunteers in 39, 141, 158–9, 179, 187–8, 239
  *see also* children's feeding centres
South Africa 51, 259
South Dene, County Durham 66
South Durham Iron and Steel Company 33
South Moor, County Durham 40, 117fn
South Shields, County Durham 27, 41, 43, 53, 98, 259
South Wales and Monmouth 216, 243, 263
  and A. J. Cook 71, 263
  accidents in 92fn, 243
  and alternative political identities 9–10, 124
  assumed irrelevance of Church of England 166
  and blacklegging 37, 106, 129, 160, 263
  and boards of guardians 261–2, 263
  centrality of union 78, 261
  colliery recruitment 25, 49
  and Communist support 113, 263–5
  comparison with Durham 13, 261–5
  dominance of Labour Party 81, 261
  existence of 'alternative culture' 262
  and fundraising 66
  internationalism of 22–3, 59
  Mardy ('Little Moscow') 113
  mechanization 244
  memories of residents 147, 227, 232
  Miners' Institute libraries 223
  and Nonconformity 189
  and owner occupation 127
  Pontycymer 227
  Pontypridd 106
  radical reputation of 11, 113
  recruitment of special constables 106
  represented as unique 262
  response to owners' proposals 88
  Rhondda Valley 9, 127fn, 189, 263
  romanticized image of 254
  soup kitchens 158, 263
  Spanish immigration into 261
  Tonypandy riots 6, 226
  and women 141–2, 156, 160, 162, 258

South Wales Miners' Federation (SWMF) 261
divisions between leaders and rank and file exaggerated 85
duties of officials 78
internationalism of 22–3, 59–60
radicalism of 262–5
Soviet Union 59, 60, 95, 104, 108, 110, 111, 175
attitudes towards 59, 61, 110–11
support from 59, 107
Spanish Republic, attitudes towards 23, 59
special constables 40, 106–7, 134, 259
Spen, County Durham 94
Spencer, George 3, 110fn, 215
Spencerism 110
Spennymoor, County Durham 81, 82fn, 116, 117
sport 73, 125, 232
church-based 182–3
effect on of lockout 55–6, 147
as means of escape 220–1, 257
*see also* football; cricket
Springwell, County Durham 38, 253
Staffordshire 42, 62, 128
Stanley, County Durham 40, 53, 56, 94, 98, 109, 119, 172, 239
library 216–7
pit disaster (1909) 227, 244
Primitive Methodist Circuit 191
Ratepayers' Association 126
West Stanley Teachers' Association 210
*Stars Look Down, The* 138, 201, 221
stealing, examples of 133–4, 254
Stella Coal Company 129
Stephenson, James 129
Straker and Love Colliery Company 118
strike-breakers, *see* blacklegs
strikes and stoppages:
(1810) 173
(1832) 242
(1844) 251
(1879) 252
(1892) 148, 156, 173, 251, 252
(1893) 226, 251fn
(1912) 226, 251
(1920) 226, 251
(1921) 37, 39, 41–2, 43, 100, 101, 118, 159, 240, 241, 251, 262
(1926), *see* lockout (1926)
(1972–4) 225, 226
(1984–5) 6, 8, 18, 22, 63, 74, 157, 159, 162, 226, 228–31, 250, 269
Storr, Revd E. B. 191
suicides 239–40
Sunday schools 190, 193, 195, 204
Sunderland 41, 53, 73
Sunderland AFC 220, 236
Swalwell, County Durham 63
Swan, Joseph E. 96, 122, 186, 194
syndicalism 11, 263fn

Tanfield, County Durham 62, 126, 131
Tanner, Duncan 269
taxation, *see* rates
Taylor, Andrew 23
teachers, *see* elementary schools, teachers
temperance 187fn, 203
Temple, William 172–3
Territorial Army 72
Thatcher, Margaret 8
theft, *see* stealing
Thomas, J. H. 31–2
Thompson, E. P. 4–5, 120, 197, 258
Thompson, Paul 228
Thornley, County Durham 33, 41, 125, 183
Thorpe, Andrew 109
tied housing, *see* housing
Tönnies, Ferdinand 4, 5
Tonypandy riots 6, 226
Towers, John 210–11
Trade Disputes and Trade Unions Act (1927) 80, 146
Trades Union Congress (TUC):
involvement in general strike 2, 28, 31–2, 103–4, 107, 222, 253, 266
relationship with miners' leaders 31–2
transport industry, affected by lockout 32
Treaty of Versailles 93
Trevelyan, C. P. 121fn
Trimdon, County Durham 45, 48, 180
Trotsky, Leon 113
Trotter, Tom 186
Tudhoe, vicar of 175
Turin, Italy 227
Tyneside 109

unemployment 33, 54, 91–2, 220
United Mineworkers of Scotland 265
United States of America 51, 59, 115, 259, 261
Unofficial Reform Committee 263
Ushaw Moor, County Durham 241
Usworth, County Durham 92, 53, 66fn, 267

vegetable shows 116, 124, 179, 236
victimization 31, 136
violence 74–5, 103, 104–5, 108, 129, 134, 155–7, 159–62
volunteers to break strike, *see* blacklegs

Wade, Mary 17, 51, 204, 233
wages 17, 32, 40, 71, 73, 122, 125, 172, 197–8, 248
  call for a 'living wage' 28, 171, 191
  and Catholic tithes 169
  of children 148, 214
  comparison of Durham to other coalfields 67–9
  comparison with poor relief 238
  comparison with royalties 47
  contemporary assessments of 35–6
  cuts preferred to increase in hours 91
  cuts proposed by owners 1, 3, 67, 88, 91, 121, 135, 267
  difficulties of ascertaining 'average' 89
  in Durham 14, 68–9, 90, 247
  effect of cuts on miners' wives 152
  effect of cuts on rest of working class 28
  effect of cuts on standard of living 91
  method used to calculate 90fn
  opposition to cuts in 3, 66, 88–9, 107, 124, 171, 191, 221, 266
  rates in various industries 29–31
  and Samuel Report 17fn, 91
  variation in coal industry 66–9, 89
  of women 148, 262
Waller, Robert 114–5, 270–1
Wallsend by-election 123
Wardley, County Durham 86
Warwickshire 69
Waterhouses, County Durham 187, 191
Watson, Ellen 159, 161
Wearmouth, Robert 166
weather during lockout 94, 147, 205–6, 231–2, 234
Webb, Beatrice 83
  on A. J. Cook 87
  on atmosphere during lockout 161, 163, 236
  on miners' relief fund 118
  on North-East politics 97, 110, 124
  and Seaham women 44, 45, 93, 110, 124, 144–5, 161
  on women and politics 143, 144, 161
Webb, Sidney 44, 118–19, 123, 144
  on attitudes of rank and file 83
  on Methodism 185
  on miners' housing 47
  nomination for Seaham 99
Weekes, Revd F. 190
Welbourne, E. 156
Welldon, J. E. C. 21, 171–2, 179, 194, 196
West Bitchburn miners' lodge 106
West Riding 73
West Stanley Teachers' Association 210
Westcott, Brooke Foss 173
Westerton, County Durham 269
Wheatley Hill 126fn, 174
Wheatley, John 221
Whickham, County Durham 39
White, Matt 187
white-shirting rituals 160
Whiteley, William 61, 67, 70–1, 73–4, 186
Whittonstall Colliery 55
Wickham, Chris 225, 249
Wilkinson, Ellen 69, 75–6, 135, 160
Williams, Chris 78, 85, 189, 258
Williams, Jack 182
Williams, Raymond 256
Williamson, Bill 161, 239, 252
Willington, County Durham 33, 38, 60, 66, 195, 269
Wilson, John 97, 143fn, 166, 185, 226, 247
Windlestone Hall 43
Wingate, County Durham 109, 181
women:
  and A. J. Cook 152, 161
  activism during strikes, history of 156–7
  and adult education 216–17, 219
  and class consciousness 45, 161, 163
  differing representations of the miner's wife 138–41

women: (*cont.*)
domestic responsibilities of 141, 142, 147, 152–3, 161
effect of owners' proposals on 152–3
feminism 162
impact of lockout upon 147, 236
limited opportunities of 55, 257
male hostility to involvement of 159–60
political allegiances of 45, 125, 143–6, 151
prosecutions of 155–6
as reason to blackleg 131, 138–9, 149–51
as recipients of relief 85, 102, 155, 158, 238
Relief Committee for the Miners' Wives and Children 140, 153
repercussions for if husband blacklegged 150, 163–4
representation as heroines 15, 139–40
representation as victims 82, 140–1, 147
support for lockout 129, 140, 142, 145, 148–9, 151–64, 194, 258
*see also* female employment; gender
Women's Institute 208
Workers' Educational Association (WEA) 93, 215–6, 219
workmen's compensation 116, 121–2
Workmen's Compensation Act (1897) 116
Worley, Michael 98, 123
Wrekenton, County Durham 53

Yallourn, Victoria, Australia 16
Yorkshire coalfield 6, 8, 9, 68–9, 225, 244
Young Communist League 96

Zweig, Ferdynand 225, 242, 260

The manufacturer's authorised representative in the EU for product safety is Oxford University Press España S.A. of el Parque Empresarial San Fernando de Henares, Avenida de Castilla, 2 – 28830 Madrid (www.oup.es/en or product.safety@oup.com). OUP España S.A. also acts as importer into Spain of products made by the manufacturer.

www.ingramcontent.com/pod-product-compliance
Lightning Source LLC
LaVergne TN
LVHW101443200926
848483LV00001B/34

* 9 7 8 0 1 9 9 5 7 5 0 4 6 *